The Widening Harvest

The Widening Harvest

The Neolithic Transition in Europe: Looking Back, Looking Forward

Edited by

ALBERT J. AMMERMAN

AND

PAOLO BIAGI

ARCHAEOLOGICAL INSTITUTE OF AMERICA

BOSTON, MASSACHUSETTS

Colloquia and Conference Papers 6

Cover image by Joel Katz

∞ The paper in this book meets the guidelines for permanence and durability of the Committee on Production Guidelines for Book Longevity of the Council on Library Resources.

Cover and text designed by Peter Holm, Sterling Hill Productions
Printed in Canada by Transcontinental Printing

06 05 04 03 5 4 3 2 1

Library of Congress Cataloging-in-Publication Data

The widening harvest : the Neolithic transition in Europe—looking back, looking forward / edited by Albert J. Ammerman and Paolo Biagi.
p. cm. – (Colloquia and Conference Papers ; 6)
Proceedings of a conference held in Venice, Italy, Oct. 29–31, 1998.
Includes bibliographical references and index.
ISBN 1–931909–05–9 (alk. paper)
1. Neolithic period—Europe—Congresses. 2. Mesolithic period—Europe—Congresses. 3. Agriculture—Origin—Congresses. 4. Human population genetics—Europe—Congresses. 5. Language and languages—Origin—Congresses. 6. Europe—Antiquities—Congresses. I. Ammerman, Albert J., 1942– II. Biagi, Paolo, 1948– III. Colloquia and conference papers ; no. 6.

GN776.2.A1W52 2003
630'.9—dc21 2003044331

CONTENTS

ILLUSTRATIONS

ESSAY CONTRIBUTORS

ALBERT J. AMMERMAN
Department of the Classics
Colgate University

PAOLO BIAGI
Scienze dell'Antichità e del Vicino Oriente
University of Venice

PETER BOGUCKI
School of Engineering
Princeton University

LUCA CAVALLI-SFORZA
Department of Genetics
Stanford University

JEAN GUILAINE
College de France

DAVID HARRIS
Institute of Archaeology
University College London

GORDON HILLMAN
Institute of Archaeology
University College London

MALGORZATA KACZANOWSKI
Institute of Archaeology
Jagiellonian University

JANUSZ K. KOZŁOWSKI
Institute of Archaeology
Jagellonian University

ANDREW M.T. MOORE
College of Liberal Arts
Rochester Institute of Technology

T. DOUGLAS PRICE
Department of Anthropology
University of Wisconsin

COLIN RENFREW
Department of Archaeology
Cambridge University

PETER ROWLEY-CONWY
Department of Archaeology
University of Durham

CURTIS RUNNELS
Department of Archaeology
Boston University

ROBIN SKEATES
Department of Archaeology
University of Durham

BRYAN SYKES
Department of Cellular Science, Institute of Molecular Medicine
Oxford University

PATTY JO WATSON
Department of Anthropology
Washington University

JOÃO ZILHÃO
Instituto Portuguès de Arqueologia

CONFERENCE PARTICIPANTS

ALBERT J. AMMERMAN
Department of the Classics
Colgate University

PAOLO BIAGI
Scienze dell'Antichità e del Vicino Oriente
University of Venice

PETER BOGUCKI
School of Engineering
Princeton University

LUCA CAVALLI-SFORZA
Department of Genetics
Stanford University

JEAN GUILAINE
College de France

DAVID HARRIS
Institute of Archaeology
University College London

GORDON HILLMAN
Institute of Archaeology
University College London

RUPERT HOUSLEY
Department of Archaeology
University of Glasgow

MALGORZATA KACZANOWSKI
Institute of Archaeology
Jagiellonian University

LAWRENCE KEELEY
Department of Anthropology
University of Illinois, Chicago

JANUSZ K. KOZŁOWSKI
Institute of Archaeology
Jagellonian University

ANDREW M.T. MOORE
College of Liberal Arts
Rochester Institute of Technology

T. DOUGLAS PRICE
Department of Anthropology
University of Wisconsin

COLIN RENFREW
Department of Archaeology
Cambridge University

PETER ROWLEY-CONWY
Department of Archaeology
University of Durham

CURTIS RUNNELS
Department of Archaeology
Boston University

ROBIN SKEATES
Department of Archaeology
University of Durham

BRYAN SYKES
Department of Cellular Science, Institute of Molecular Medicine
Oxford University

TJEERD VAN ANDEL
Godwin Institute for Quaternary Research, Department of Earth Science
Cambridge University

PATTY JO WATSON
Department of Anthropology
Washington University

JOÃO ZILHÃO
Instituto Portuguès de Arqueologia

Preface

Albert J. Ammerman

The Neolithic transition in Europe continues to be a topic of wide interest and much debate. While there has been a rapid growth in the literature over the last three decades, it is not always easy to distinguish between short-term swings in fashion and more lasting gains in our understanding of the shift from foraging to food production as a way of life. This volume presents the proceedings of "The Neolithic Transition in Europe: Looking Back, Looking Forward," a conference held in Venice, Italy on 29–31 October 1998. The year of the meeting coincided with the 25th anniversary of the publication of the "wave of advance" model for the spread of early farming in Europe. Thus, the Venice conference offered an opportunity to recognize the contribution of Luca Cavalli-Sforza in pioneering the interdisciplinary study of archaeology and human genetics. The purpose of the meeting was to do more than just review the current state of work in the field. The participants were encouraged to make a conscious effort to look back on how their own thinking on the subject and how their field experience have developed over the course of the years.

The idea for the meeting came to one of the editors in 1996 when he was working on a quite different problem—the many different attempts that have been made to reconstruct the ancient city of Rome since the time of the Renaissance. As explained in chapter 1, one of the lessons to learn from tracing the development of such a long series of reconstructions is that a reconstruction is not just a record of the archaeological evidence that happens to be available at a given time. All attempts at reconstruction have to be seen, in addition, as involving the expression of the wider cross-currents of intellectual fashion and practice in their own time. The import for the Neolithic transition in Europe, a comparatively young field of study, is the suggestion that we need to take a much longer and more historical view of the whole enterprise. Thus, why not take the initiative and make the first active step in this direction. Instead of waiting for years to pass and then trying to piece the story together after the fact (as in the case of the reconstructions of ancient Rome), why not ask those engaged in the studies themselves to step back for a moment and take stock of what they have done.

In terms of organization, the conference focused for the most part on archaeological aspects of the Neolithic transition in Europe. It was our goal to strike a balance between keeping the conference to a small size and yet including participants from a broad range of backgrounds. We knew that a small, informal meeting was more likely to encourage the open exchange of ideas that we hoped to achieve. At the same time, it was important to invite archaeologists who had conducted fieldwork for a sustained period of time in a given region, since the interplay between ideas and practice in the field was a topic of major interest. Above all, we wanted the participants—in terms of their training, their approach to the problem, and their regions of specialization within Europe—to represent a broad spectrum of experience. In order to broaden the perspective on the origins and spread of agriculture, David Harris and Patty Jo Watson, who have both worked in other parts of the world, were invited to participate in the conference. We asked Bryan Sykes and Luca Cavalli-Sforza to provide an overview of their experiences in the rapidly developing field of human genetics. There were, in addition, two other considerations that entered into the selection of the participants. English was chosen as the official language for the conference. In terms of chronology, the decision was made to place emphasis on the transition before 6,000 B.P. (uncalibrated radiocarbon years). By this time, early farming was already well established in most parts of Europe with the exception of the northern periphery, which has tended to receive wider attention in the American and British literature on the Neolithic transition. A list of the conference participants, 21 in all, is given after the list of essay contributors. The papers for the meeting were prepared ahead of time and distributed to the participants in advance; thus, time would not have to be spent in reading the papers at the meeting. The speakers in a given session first summarized the main arguments in their papers and then moved on to discussion.

The charge to the meeting participants was for each person to reflect upon how their own personal experience—their education and training, their fieldwork and collaboration on projects, their publications and the debates that they have occasioned, and the wider climate of thought in which each person worked—all contributed to shaping the positions that the individual held at different points in their career. For example, how did the training and the concepts that one had at the start of a project condition the decisions that were made about where and how to conduct fieldwork? And how did the results actually obtained in the field lead, in turn, to the reformulation of thought about the problem? In short, the participants were invited to trace the pathway of their intellectual development with reference to the problem. By looking back and comparing a range of experiences, there was the prospect of gaining a better understanding of where we stand in the field today and of how we reached this point. As one might expect, the participants chose to interpret this charge in quite different ways. We decided from the start that no attempt would be made to impose a standard form on the contributions. Since the archaeologist is not in the habit of writing autobiographically, it would have been inhibiting and counterproductive to do otherwise at this initial stage in the experiment of looking back. Diversity was thus given wide latitude both at the meeting and in the revised papers that appear here. This all constituted part of the attempt to add a new dimension to the widening harvest.

By way of introduction, it may be useful at this point to make a few brief comments about the current situation in the field. In any given place, the shift from foraging to food production can be interpreted as happening in one of three different ways: (1) the local domestication of wild plants and animals in that place, (2) the introduction of agriculture there from the outside by means of cultural diffusion, and (3) the spread of agriculture by means of demic diffusion. In the literature of the 1980s, all three were still active players in the debate over the Neolithic transition. More recently, the first line of explanation has faded from the scene in the case of Europe. The main reason for its demise was the introduction of the new AMS method of radiocarbon dating (based on accelerator mass spectrometry), which made it possible to date individual grains of cereal for the first time. Since the mid 1980s, not only have earlier claims for the local domestication of wheat and barley in Europe been called into question, but no reliable new claims have been advanced in the literature. In effect, the debate now centers on the two modes of diffusion.

It is worth recalling here that the notion of diffusion, in any form, was seen in a negative light by most prehistorians in the years between 1965 and 1990. This stemmed in part from an understandable reaction against the use and abuse of diffusion by a previous generation of scholars. One of the leading ambitions at that time was the interest in investigating the local, endogenous origins of agriculture in the place where the archaeologist chose to work. Now the tides of intellectual fashion have changed once again. Today there is an increasing awareness of the limitations of indigenism and a renewed interest in the question of the spread of agriculture, not just in Europe but elsewhere in the world as well.

As the chapters in this volume reveal, it is premature at this point in time to think in terms of a consensus for Europe as a whole. In many ways, the heart of the matter—in terms of what one can observe in the archaeological record—boils down to whether or not one can document continuity over the transition. In short, in a given place, did the final foragers and the first farmers live in much the same way, as one would expect under the cultural mode of diffusion? Or was there discontinuity between the two ways of life, as one would expect under the demic mode of diffusion? It is common in the literature of the 1970s and 1980s to encounter claims for continuity over the Neolithic transition in various regions of Europe. Many of these arguments were proposals of a general nature (part of the fashion of the time) and not claims that were well documented. With the exception of northern Europe (see, e.g., the case of Denmark, which is reviewed by Price in ch. 14), few of these claims now turn out to be well substantiated upon closer examination. The evaluation of a specific claim for continuity calls for the comparison of the final Mesolithic in a given area with the early Neolithic in the same area on the basis of settlement patterns, lithic technologies, dietary preferences, and burial practices. There are still few places in southern and Central Europe where such an evaluation can be made in a detailed and systematic manner—that is, a comparison that involves all four lines of evidence. In the debate over the question of continuity, it is important to remember, in terms of the history of scholarship on the transition, that a natural inclination exists on the part of those who engage in the study of the Mesolithic period to want to see their period as the prelude to the Neolithic, not just in a chronological sense but in a functional one as well. While such partisanship is fully understandable, it may not always serve the best interests of the balanced evaluation of the question of continuity.

It was striking to learn at the Venice meeting just how slim the evidence for final Mesolithic settlement still is in most parts of Greece, Italy, southern France, and Portugal (see the respective chapters in this volume by Runnels, Biagi, Guilaine, and Zilhão). This is the case even in those places where intensive surveys have been conducted in these countries. What emerged from the conference above all was the pronounced difference between southern Europe and northern Europe on the question of continuity. The participants also came away from the meeting with a realization that much remains to be done when it comes to sorting out the relative importance of cultural diffusion and demic diffusion in most parts of Europe. One of the challenges that still stands before us is that of building second-generation models that do a better job of explaining how the spread of early farming actually took place on the ground.

A shift in perspective occurred in the last session of the conference, which was devoted to looking forward. An attempt was made to call attention to promising new methods and research strategies and to identify leading issues that archaeologists will have to address in the next two decades. The conference participants were divided into two panels for this purpose. In turn, each of the panelists put forward his or her own thoughts on where the field was headed, and an open discussion followed. The main points that came to light are summarized in the last chapter of this volume. It will be of interest in historical terms to see how well what was proposed in looking forward holds up as the current century unfolds. As part of the final session, two resolutions were unanimously approved by the conference participants: the first recognized the fundamental importance of AMS dating in the field today (and the need to foster its wider implementation in countries that still do not have their own laboratories), and the second acknowledged the significant contribution that genetic research is now making to the study of the Neolithic transition in Europe. The full text of each resolution is given in the appendix at the end of the last chapter. The volume can be viewed in one way as a snapshot—a group portrait showing where we stand at the start of a new century. What is new is the attempt to place in the context of time our own work on the widening harvest.

Support for the conference was provided by the Wenner Gren Foundation for Anthropological Research, the University of Venice, and the Ligabue Foundation in Venice. We wish to express our appreciation to Gustavo Traversari and the staff of the Archaeological Section of the University of

Venice for their generous help in running the meeting at the Palazzo Bernardo in Venice. Pat Ryan provided valuable assistance in the preparation of the manuscript. Finally, our special thanks go to the Archaeological Institute of America for its assistance in publishing the proceedings of the meeting and, in particular, to Marni Blake Walter, the AIA publications editor, who so ably guided the volume through various stages of its production.

PART I

Introduction

— 1 —

Looking Back

Albert J. Ammerman

The spread of early farming in Europe has become a subject of broad interest to a number of disciplines in recent years. These range from archaeology and anthropology to human genetics and linguistics. Today it is well known that wheat and barley, the cereals that were cultivated by the first farmers in Europe, were originally domesticated in the Near East. This happened some 10,000 years ago. From there, early farming spread first to Greece and the Balkans and then over the rest of Europe. We have called this shift from foraging to food production, as a new way of life, the Neolithic transition (Ammerman and Cavalli-Sforza 1984, 3). Previously, the prehistorian V. Gordon Childe had used a different term, the Neolithic revolution, to identify this transformation. It is now well established that it took more than 2,500 years—that is, some 100 human generations—for early farming to spread from Greece to Scandinavia. In short, this was a long, slow process and not a revolution in the normal sense of the term. Transition is thus a more appropriate name, in the case of Europe, for the kinds of changes that were involved.

The question of how early farming began in Europe is one that has witnessed wide swings in the literature over the last 30 years. The aim of the conference held in Venice was to do more than just take stock of the current state of knowledge in the field. Each of the participants was encouraged to look back and consider how their thinking on the question and how their choices with regard to fieldwork have developed over the years. In other words, a conscious effort was made to pay greater attention to the historical context in which we, as archaeologists, attempt to solve the problem of the Neolithic transition in Europe. In the rush to obtain an answer from the archaeological evidence that is immediately at hand, there is a tendency to lose sight of the wider context in which we conduct research. In short, the historical framework in which the work itself is undertaken does not receive the full attention it should.

The paper that I wrote for the Venice conference was intended to explore this new approach and to provide an example of how one might look back on one's own experience in working on the Neolithic transition.[1] At the same time, I wanted to take a closer look at the question of indigenism. Indigenism is the position in anthropology and archaeology that wishes to treat each society as independent or autonomous and to see its development essentially in endogenous terms. It is at the heart of much of the recent debate over the Neolithic transition. As we shall see later in this chapter, indigenism also continues to constitute an issue of broad political significance in Europe today.

The idea for the Venice meeting, as mentioned in the Preface, first came to me when I was a fellow at the Center for Advanced Studies in the Visual Arts (located in the National Gallery of Art in Washington, D.C.). It may be useful, by way of introduction, to say a few more words here about the experience there that led me to this idea. In the spring of 1996, I had just spent the last year studying the many different attempts at reconstructing the ancient city of Rome (Ammerman 1996, 48–51).[2] Over the last five centuries, the reconstructions of Rome have taken many different forms: from written accounts to paintings and maps, from ice sculptures and scale models in cork, to the restoration of the monuments themselves. One of the lessons to learn from tracing such a long and rich series of reconstructions was the importance of paying closer attention to the historiography of an archaeological problem. Over time, what is observed in the case of ancient Rome is a restless dialogue between three different things—new physical remains that come to light, new interpretative keys that are brought forward to explain the past, and shifts in the wider cultural and historical context that reshape the whole endeavor. In other words, a reconstruction is not just a record of the archaeological evidence that happens to be available at a given time. All attempts at reconstructing the past have to be seen, at least in part, as the expression of the wider crosscurrents of intellectual fashion in their own time. What this implied for the Neolithic transition in Europe was the need to take a more historiographical approach to the question than we have done before. This was the insight that led to the organization of the conference in Venice, "The Neolithic Transition in Europe: Looking Back, Looking Forward."

In addition to offering an account of my own experience in the field, this opening chapter also introduces the book as a whole. With this in mind, it is worth recalling the three main lines of explanation that have been put forward for the start of early farming in Europe. One of these holds that the origins of agriculture took place in situ within Europe itself (e.g., Barker 1985). In short, there was the local, primary domestication of wild forms of wheat and barley in different regions of Europe. Hunters and gatherers, the argument goes, sometimes have a close relationship with wild plants and animals; they also possess the potential for innovation in the management and manipulation of such resources. Accordingly, local domestication becomes a working hypothesis. At the time of the renewed interest in Mesolithic studies in the 1960s and 1970s, this became a fashionable idea. In turn, various claims for local domestication were put forward in different parts of Europe. Since the early 1980s and the advent of the ^{14}C method of dating based on accelerator mass spectrometry (AMS), however, few of these early claims have stood up to rigorous evaluation. The direct dating of individual cereal grains by this new method provides chronological control that was not available before.[3] As a result, new claims for the local domestication of wheat and barley in Europe have not been made in the last decade, and there has been a shift away from this hypothesis. Now even claims for local animal domestication in Europe—a separate but related issue examined by Rowley-Conwy in chapter 6—are being called increasingly into question.

The alternative is that early farming spread to Europe from the Middle East, where the exploitation of wild forms of einkorn and emmer is well attested at early sites and where domesticated forms of these cereals also have earlier dates than they do in Europe (e.g., Moore et al. 2000; Özdogan and Basgelen 1999). In terms of biogeography and plant genetics, one is dealing with the dispersal of domesticated forms of wheat and barley in Europe (Heun et al. 1997). There are two different ways to explain a dispersal of this kind. The first involves cultural diffusion or the passage of cereals and farming techniques from one local group to the next without the geographic displacement of the groups. The second is "demic diffusion," in which the movement of the farmers themselves caused the spread of agriculture. While the two modes of diffusion need not be mutually exclusive, they warrant being clearly distinguished at the conceptual level. In the former, the diffusionary process is based on the transmission of the cereals between those who have become farmers and those who still are foragers and have not practiced farming before. In the demic mode of diffusion, the spread takes place when the farmers relocate their settlements. In the formulation of cultural diffusion proposed by Zvelebil (1986, 1996), for example, emphasis is placed on continuity between the late Mesolithic and the early Neolithic in a given area.[4] This would include the same settlement patterns and the same basic level of local population density over the transition. In addition, the change in the subsistence strategy would start with a low reliance upon domesticated plants and animals and unfold as a slow and gradual process over time. On the other hand, demic diffusion would entail both a package of domesticates and a greater reliance on food production from the start in a new area where the spread was taking place. And there would be more change—between the late Mesolithic and the early Neolithic in a given place—over the transition. In short, there are quite different expectations that are associated with the two modes of diffusion.

Working on the Problem

Cavalli-Sforza and I first published on this topic in an article titled "Measuring the Rate of Spread of Early Farming in Europe" in 1971. We then went on to develop a new formulation of the problem in "A Population Model for the Diffusion of Early Farming in Europe" (Ammerman and Cavalli-Sforza 1973). In *The Neolithic Transition in Europe and the Genetics of Populations in Europe*, we put forward a synthesis of the various studies that we had done on the subject over the years (Ammerman and Cavalli-Sforza 1984). Here I review three of the main elements in our formulation of the Neolithic transition. One of these is the concept of demic diffusion itself. As mentioned above, the spread of early farming in Europe involves a process of diffusion that can be explained in two different ways. One hypothesis is that of cultural diffusion: the transmission of the cereals and farming techniques between local groups without their displacement. The other hypothesis is that of demic diffusion, or the movement of the farmers themselves. In the case of the "wave of advance" model, the spread may be caused by the frequent relocation of early farming settlements over short distances. One of the problems that we encountered when we first began working on the question was the lack of a suitable terminology in the literature for a spread of early farming based on small-scale relocations of this kind. Commonly used terms such as colonization and migration were not really appropriate. Colonization, in its conventional meaning, refers to the intentional settlement by a group of people usually in a distant land. A familiar example of colonization would be the one practiced by the ancient Greeks in southern Italy. In prehistory, the term migration is again commonly used when a group moves over some distance and settles in a new and different place (that is, not in the same local area and not in the same social context that it lived in before moving). In short, there was a poverty of language for describing the full range of ways in which the movement of the first farmers may have taken place. Hence, we coined a new term: "demic diffusion." In effect, colonization and the wave of advance model represent the opposite ends of a spectrum of different models all belonging to the general class of the demic mode of diffusion.

The second element in our formulation was the wave of advance model. Here we drew upon the previous work of R.A. Fisher (1937) in genetics and J.G. Skellam (1951) in ecology. This now made it possible to model, in formal terms, the slow and continuous dispersal of a population in space. It can be shown mathematically that, if an increase in population numbers coincides with a modest local migratory activity (the short-distance settlement relocations mentioned above), a wave of population expansion will set in and progress outward at a steady radial rate (fig. 1.1). The full name of this model, without going into the details of the mathematical treatment

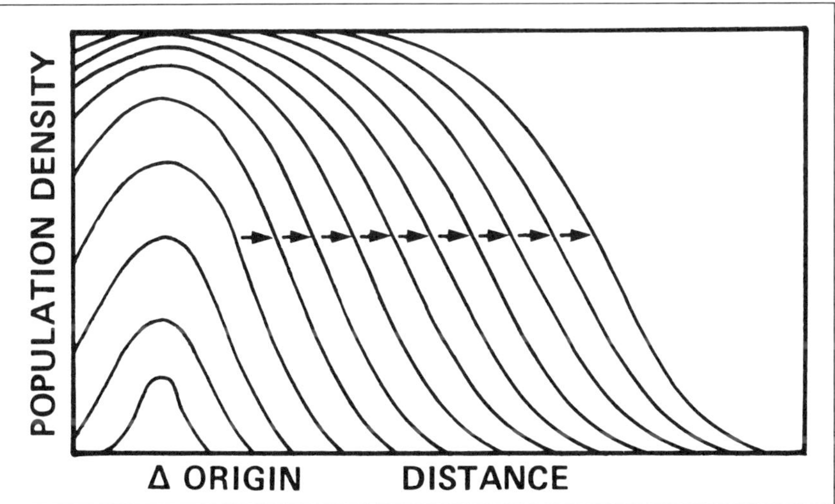

Fig. 1.1. Fisher's model of a population wave of advance. The graphic representation shows the rise in local population density expected with increasing distance from the origin as time elapses.

here, is the "diffusional population wave of advance," or more simply the "wave of advance."[5] At the time that Cavalli-Sforza and I began working together in 1970, the use of formal models already had a long tradition behind it in the fields of economics and the biological sciences. By contrast, models of this kind had seldom been used before in archaeology.[6] Thus, the proposal of such a formal model constituted something new in itself for archaeologists.

As often happens in the case of an innovation, however, there is the risk of misunderstanding. My own experience over the years indicates that the archaeologist does not always have a firm grasp of the distinction between a model, a hypothesis, and a statement that is made as a claim to historical knowledge. This has been a major source of confusion in the literature on the Neolithic transition. A model is a tool for thinking—a framework for addressing a problem. The value of a model, as the economic historian David Landes (1969, 540) notes, is accordingly heuristic; it does not tell us what happened in the past; it helps us to discover and to understand what may have happened. I would like to add here that the wave of advance model is not to be confused, as is sometimes done in the literature, with the demic hypothesis.[7] They are two quite different things. Under the demic hypothesis, how the early farmers actually moved is left open to a wide range of different possible forms or models (from the short-range relocation of sites to long-distance maritime colonization).

A third element of our formulation has to do with the interactions that populations of early farmers and late hunter-gatherers can have with one another in a frontier situation (that is, where both life styles may coexist in the same area). In 1973 and again in 1984, we outlined some of the main ways in which early Neolithic and late Mesolithic populations may have interacted with one another. These include mutualism, acculturation, competition over resources, and the transmission of disease (Ammerman and Cavalli-Sforza 1973, 353; 1984, 116–8). In addition, we realized that different scenarios may have prevailed in different regions of Europe: demic diffusion may have played the leading role in some regions, while cultural diffusion took the lead in others. The real question may well be that of evaluating the relative importance of the two modes of diffusion in different parts of Europe (Ammerman and Cavalli-Sforza 1984, 6, 134–5).

How did this formulation come about? I first met Luca Cavalli-Sforza at the conference on quantitative archaeology held at Mamaia on the Black Sea in 1970 (Hodson et al. 1971). He invited me to give a seminar at the University of Pavia, where he taught before going to Stanford. At the time, he was doing a field study of the Babinga Pygmies in Central Africa. Thus, we shared an interest in the study of hunter-gatherers. In 1967, I had moved to London to begin a course of studies in environmental archaeology. There, I had the good fortune—by accident and not design—to attend the meeting on the domestication of plants and animals held at the Institute of Archaeology at the end of my first year (Ucko and Dimbleby 1969). I became interested in the question of the origins of agriculture and went on to write my dissertation on the late hunter-gatherers of Italy. In the 1960s and 1970s, the study of the spread of agriculture was out of fashion; instead, there was a marked enthusiasm for the study of hunters and gatherers. The archaeologist with an interest in the origins of agriculture now wished to trace the independent pathway to plant and animal domestication taken by the local population of late hunter-gatherers that once lived in the area where the person happened to conduct fieldwork. When I went out to Italy, this is what I had hoped to do. I soon came to realize that this was not really in the cards for the Italian peninsula.

In Italy, I decided to look closely at the locations and the environmental settings of both Mesolithic sites and early Neolithic sites. But this only revealed the cave-bound character of Italian prehistory before the Neolithic period.[8] If one considered the evidence for sites other than caves and rock shelters, no real argument for continuity in terms of settlement could be made over the transition. In addition, a consciousness slowly began to form in my mind with regard to the limitations of studying the Neolithic transition on the basis of cave deposits alone: that is, the problem of disturbance and the mixing of cave levels (with the resulting ambiguity and confusion that often arose in the literature) and the even larger question of what the remains from cave sites (often with their special uses and their low intensities of occupation) actually represent in terms of the larger picture of human behavior in either the late Mesolithic period or the early Neolithic period. At the time that we met in Mamaia, I was still hard at work on my thesis and puzzling over such problems.

For the seminar in Pavia, I examined the pattern of the first appearance of agriculture in different parts of Europe as revealed by radiocarbon dates from early Neolithic sites. By the early 1970s, enough ^{14}C dates had just become available in Europe so that one could attempt to trace the pattern of the spread. The dates, when plotted on a map, revealed a trend running from southeast to northwest across the continent. What was new was the quantitative approach that we now took to the analysis of the problem. Soon we were able to make a first estimate of the average rate of spread of early farming across Europe—a rate of about 1 km per year. The slowness of the rate was striking. It indicated movement over a distance of only about 25 km per generation (taking 25 years to be the length of a human generation at the time). The initial analysis also revealed that regional rates for the western Mediterranean and for central Europe could run twice as high as for the average value for Europe as a whole (Ammerman and Cavalli-Sforza 1971, 683–4).[9] And the rate measurement brought us, in turn, to the question of how to explain the spread. Cavalli-Sforza recalled the model that Skellam had put forward in ecology for the dispersal of a population as a diffusionary process and illustrated with the spread of the muskrat in Central Europe (fig. 1.2; Skellam 1950, 200; Ammerman and Cavalli-Sforza 1984, 69–70). At Pavia, we began to play with the model, and this initial exploration forced us to think back through the whole problem and to formulate the hypothesis of demic diffusion.

In 1970, however, diffusion as a form of explanation in anthropology and prehistory had fallen completely out of favor. The use and abuse of diffusionism by a previous generation of scholars, including V. Gordon Childe,

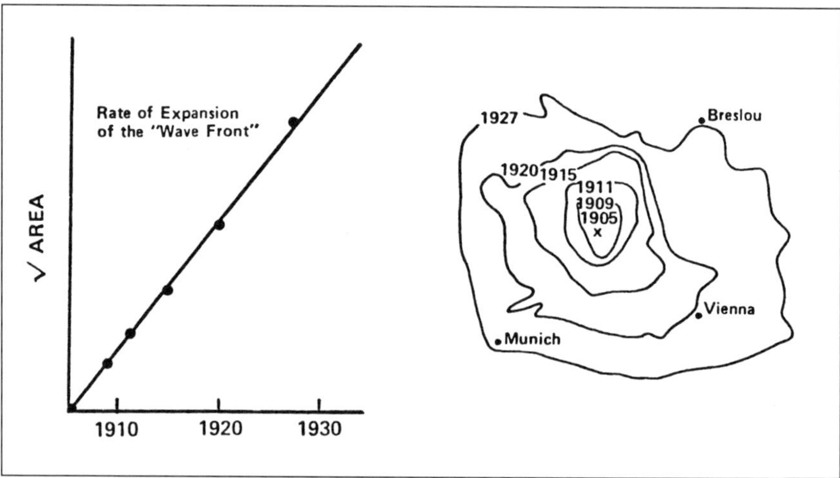

Fig. 1.2. The spread of the muskrat in central Europe. Left: areas in which the muskrats were found in central Europe in different years following their introduction in 1905. Right: the square root of the area is plotted against time to yield the rate of expansion of the wave front.

had discredited this line of explanation and cast it in a negative light (Anthony 1997). A new approach, which looked inward and placed value on the autonomous development of a society, was now in fashion. And, if one had to deal with a case of diffusion, it was generally agreed that the cultural mode was the one that was involved. The extent of the acceptance of this position is reflected by Edmonson's "Neolithic Diffusion Rates" (1961), which was one of the few studies on the question in the literature at the time. While those offering comments on his article in *Current Anthropology* were critical of many aspects of Edmonson's study, no one really questioned that what was being measured, if one attempted to make such a measurement, was a rate of cultural diffusion. In short, there was the need to introduce the concept of demic diffusion and to borrow the wave of advance model from the field of population biology, and to apply them to an archaeological problem.

In 1971, I presented the model at the Sheffield meeting on the explanation of culture change organized by Colin Renfrew. On one hand, fashion was now on our side. There was an enthusiasm for formal treatments of this kind in archaeology at the time that the model was first introduced. On the other hand, almost no one in prehistory then could actually read the mathematics of the model. Indeed, at a conference that focused on models in archaeology, most of us were still neophytes at model building. Upon completing my thesis at London, I joined Cavalli-Sforza at Stanford University in 1972, where we worked in close collaboration for the next five years. The studies would include taking a closer look at the individual components of the model, the measurement of the rate of spread with a much larger database, the development of simulation studies that focused on settlement systems, and the examination of the genetic implications of the spread.[10]

On a separate front, in 1974, I was invited by the new University of Calabria to organize an archaeological survey in the region that forms the toe of Italy. I had recently visited the excavations on the Aldenhoven Platte in Germany, where the full distribution of long houses of Linear Pottery age on the landscape was being brought to light, and I was duly impressed (Kuper et al. 1974). Studies of this kind on Neolithic settlement patterns were completely lacking in the western Mediterranean at the time. When I arrived in Calabria, not much was known about the Neolithic period (one leading archaeologist indeed had once claimed that there were no Neolithic sites to be found in Calabria), and three of the four sites where excavations had been conducted so far were cave sites. This would all change rapidly with the survey, where our specific interest was in finding Mesolithic and Neolithic sites. The repeated, intensive coverage of the Acconia area produced from one field season to the next richer and richer patterns of Stentinello or impressed ware Neolithic habitation (Ammerman 1985). And the excavations at the site of Piana di Curinga yielded a different picture for the early Neolithic than the one obtained before from the investigation of cave sites in Calabria—the first good series of wattle and daub houses in the western Mediterranean (Ammerman et al. 1988, 122–3). Moreover, at Acconia, some Paleolithic remains were recovered from the land surface but nothing was found on the Mesolithic side in either the survey or the excavations. In short, at least in the areas where we had done extensive fieldwork in Calabria (Acconia, Nicotera, and Le Castella), two conclusions could be drawn: (1) there was no evidence for continuity in settlement patterns over the Neolithic transition, and (2) there was good evidence for local population growth in different parts of Calabria with the appearance of the Neolithic.[11] Of no less importance for the development of my own thinking was the growing realization that the household—as a common locus for residence, for the production and consumption of food, and for kinship and personal identity—represented one of the keys to the study of economic and social life in the Neolithic period (Ammerman et al. 1988, 136–9; Ammerman 1989b).

There was a need to bring together the various lines of investigation that Cavalli-Sforza and I had done over the years and shape them into a comprehensive statement. Now living on different coasts (I began teaching in New York in 1977) and each with a busy schedule, the book would take a number of years to complete. It finally came out in 1984—that is, at a time when indigenism was at its peak. No attempt will be made here to review the various arguments put forward in the book, which are familiar to those interested in the question. A few comments, however, may be in order on the reception of the book. This was mixed as one might expect. In general, the book was greeted positively by those in the fields of human biology and genetics. On the side of Neolithic studies, there was the common misunderstanding that our model was somehow a historical claim that there had been constant, uniform, and continuous movement of first farmers on the ground (not something envisioned, by the way, even in Skellam's original application of the model to the spread of the muskrat in Central Europe). And there were other misrepresentations of what we had said in the book as well.[12] But *pazienza,* as they say in Italy—sooner or later it will all come out in the wash. The misrepresentations can be viewed, in retrospect, as part of the natural resistance that the book faced, since its main argument ran clearly against the tide of indegenism at the time.

The Question of Indigenism

Indigenism, as mentioned earlier, is the position in anthropology and archaeology that regards the developments in the life of a given society or population as a self-contained affair. It can be interpreted as part of the necessary reaction to the excesses of the "old" diffusionism. The "new" indigenism, however, is not without serious limitations of its own. We first used the term in the preface to our book (Ammerman and Cavalli-Sforza 1984, xiv). We did not try to explore the intellectual roots of indigenism as such. For want of a better term, indigenism served simply to define the counter position to our own in the ongoing debate over the Neolithic transition in Europe.

It was only some years later—in looking back from a greater distance in 1996—that I began to think about the question of indigenism in greater depth. There are a number of factors that contribute directly and indirectly to the attraction of this notion in our time. To begin with, indigenism represents the working principle behind the majority of studies in cultural and social anthropology since the Second World War. In this field, a lone anthropologist would set out with a knapsack full of notebooks to chart the rules of organization of a remote, indigenous society. Notably, indigenism is at the heart of the school of thought known as structuralism. Structuralism was a term coined by the anthropologist Claude Lévi-Strauss to describe a method of applying models of linguistic structure to the study of society as a whole, in particular to customs and myths. Here emphasis is placed on cultural difference and the need to respect it. Societies, in this view, are structures that have relatively stable relations among their elements. Change over time is not a leading concern and, if it does occur, it is considered to have its source within a given society. Thus, structuralism, without going into the reasons for why it arose in France after the war (e.g., Lilla 1998), spoke of each culture as autonomous.

Another strand of indigenism is connected with nationalism. In the 1960s and 1970s, there was the widespread belief that each nation in Europe was entitled—by a right of chauvinism, as it were—to agricultural origins of its own. Accordingly, one had a series of claims for local domestication in Romania, in Italy, in France, and elsewhere—almost none of which has withstood the test of time.[13] In Europe, archaeology is organized essentially along national lines. This carries with it the understandable aspiration of each national archaeology to advance its own nation's identity. In effect, each national archaeology has the charge of tracing its own country's distinctive and ideally independent course of development over time. Again, emphasis is placed on the autonomous, on the autochthonous, and on cultural difference.

A third strand at work involves the apprehensiveness in the post-Holocaust world that we have all come to associate with the displacement of people. The Holocaust has redefined the way in which we think about the question of the movement of people. Indeed, the 20th century has already witnessed enough of this. The forced displacement of people in Kosovo and in East Timor, as we saw in 1999, runs completely counter to the sensibility of the post-modern world. However, the idea of indigenous people also contains its own paradox. Today many groups that would like to identify themselves in this way have actually experienced geographical displacement in recent historical times, and thus they are no longer autochthonous as such (Beteille 1998).

Still another way of looking at indigenism would be as a post-colonial involution, where the essential aim is to rewrite the past. Indigenism seems to be most pronounced in those nations that have had a long colonial experience. In adjusting to a world that has changed, there is a need to realign one's identity. We live in an era of empires giving way to the conscious rediscovery of ethnic identity and to the natural right of self-determination. In this context, there is a strong urge to affirm that one is in control of one's own native past (Ammerman 1989a, 165). By way of involution, when it comes time to look outward again and consider other societies, the archaeologist tends to project indigenism on the rest of the world as well.

In combination, these four strands yield a complex of values and attitudes that has a firm hold on our time. Thus, there is far more at stake, when it comes to the Neolithic transition in Europe, than simply the knowledge of how early farming began in this part of the world. If our formulation of the question is correct, then the Neolithic transition poses a fundamental challenge to the paradigm of indigenism. If this important chapter in the prehistory of Europe—the shift to a new form of production—can be explained in some other way, then the paradigm itself runs the risk of losing power. The door is open for moving beyond indigenism.

The full realization of what was at stake began to dawn on me when I was working on a quite different problem.[14] In studying the cultural history of the many different attempts to reconstruct the ancient city of Rome, there was the distance to see the debate over the Neolithic transition in a new light. I now recognized the wider potential significance that our work on the Neolithic transition had in the sense of taking one of the first exploratory steps beyond indigenism. We did not set out to challenge the intellectual fashion of the day; we simply followed the lead of our own curiosity in trying to solve a complex problem. The irony is that indigenism, as a way of viewing the world, is clearly out of step when it comes to the political situation in Europe today. Looking forward, the transition now in progress—hopefully a more rapid one than the Neolithic transition—points toward more open frontiers and the freer circulation of people and information in Europe. From this perspective, indigenism is an anachronism: something from a former age (a way of rendering the past bound to the Cold War years just after the Second World War) that is incongruous in terms of where Europe is headed today. Perhaps it is time to move beyond indigenism in the study of the past as well as in our preparation for the future.

Conclusions

In 1990, I had the chance to travel to Irian Jaya, the western part of New Guinea that belongs to Indonesia, and experience life among two groups with a stone age technology: the Jali in the Highlands, who live on the basis of gardening, and a small group on the Wildeman River in the southern Lowlands, who live for the most part by foraging.[15] My experience in Irian Jaya forced me to think about some things for the first time: in the case of the Jali, who follow essentially a sedentary way of life, I was struck by the fluidity of residence—the open circulation of people between settlements in the community. The dynamics of residence (i.e., where the individuals in a given community actually choose to reside at various times in their lives) has not received proper attention in the anthropological literature. The real challenge, of course, is to keep track of all of the movements and relocations that the people in a community experience over the course of a generation. The point here is that a sedentary way of life should not be equated with immobility. There is a tendency in Neolithic studies to equate the life history of an individual with the history of a given site (Ammerman and Cavalli-Sforza

1979, 288). In fact, in the case of small-scale societies of the kind observed in New Guinea, the vital events in an individual's life—birth, marriage, and death—commonly take place not at one and the same site but at different places. Thus, demographic events themselves are distributed over the landscape. And the fluidity of residence in the case of the Jali only adds to the level of mobility within the society. The suggestion here is that our common notions about the settlement systems and the demography of early farmers are far too immobile.

Five points emerge from this opportunity to reflect upon my own experience in the study of the Neolithic transition. The first concerns the collaboration between archaeology and human genetics that Cavalli-Sforza and I were able to pioneer. We had the courage to accept this challenge at a time when it was far from clear where the research would lead. In retrospect, the collaboration between Cavalli-Sforza and myself now seems like the obvious and natural thing to do. In fact, our collaboration was seen as an unconventional and even radical move in the early 1970s. There was at that time no tradition of collaboration between our two fields of study. Much of our initial effort was spent in trying to bridge the differences in our backgrounds. Looking back, we now see that this effort was well spent.[16]

The second involves the idea of demic diffusion, which is now widely accepted and used in the literature. There was a clear need, as explained earlier, to coin this new term. At the time that we proposed the concept, diffusion in any form was out of fashion. In the 1990s, the pendulum began to swing back the other way, and the spread of agriculture has once again emerged as a question of major interest.[17] The concept of demic diffusion helped to fill a major gap in terms of how we think about the movement of people in prehistory (e.g., fig. 1.3).[18]

The third point relates to the wave of advance model, which made it possible to bring local population growth, site relocation, and the rate of spread of early farming together in a quantitative framework for the first time. In retrospect, the wave of advance can be seen today as a first-generation model, a conceptual tool at the general level that helped in reformulating the question of the Neolithic transition. In more recent studies, others have gone on to develop second-generation models that treat site relocation in a more complex way. For example, if one takes a closer look at the situation on the ground in a given region of Europe, one often finds that the expansion was

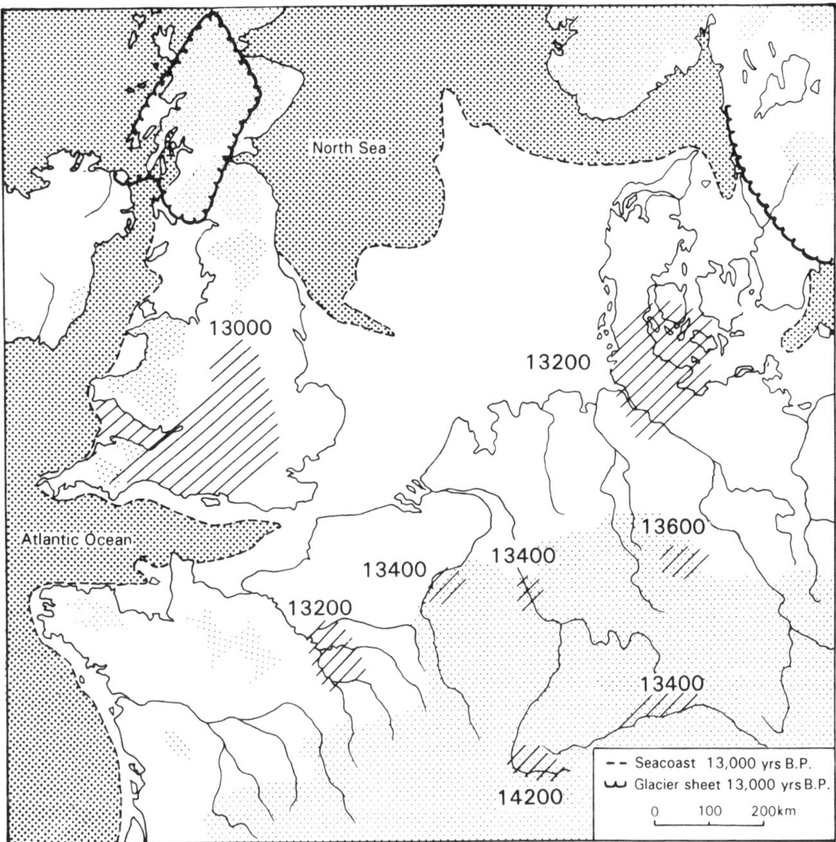

Fig. 1.3. Map showing the timing of the start of the pioneer phase of reoccupation in different areas of northwest Europe at the end of the ice age. (Housley et al. 1997, fig. 14)

less continuous in space than the general model predicts. Clusters of early farming sites are separated by open spaces on the map. What this implies is that not all environments were perceived as the same by the first farmers. Some places on the landscape were more attractive than others. In short, the wave of advance model was a point of departure. It has led to the development of second-generation models such as the one proposed by Van Andel and Runnels (1995) for Greece, where alluvial plains with arable soils are seen as the preferred places and comprise the main stepping stones in the initial spread there. The wave of advance model now serves as a yardstick when it comes to the comparison of competing models that attempt to explain the Neolithic transition.

The fourth point concerns the fieldwork that I did on Neolithic settlement patterns in Calabria, which offered a chance to translate method and theory into practice. In turn, the results from the field showed how narrow our ideas about Neolithic economics had been—essentially the reduction of the economy to subsistence—and the need to pay more attention to exchange systems and the household. In other words, there was the opportunity for a dialogue between ideas and practice in the study of Neolithic settlement patterns in Calabria.

The fifth point concerns indigenism, which continues to cast its long shadow over how we think about the Neolithic transition in Europe. It is now time to move beyond indigenism. Over the long run, one of the main contributions of our work on the Neolithic transition may well turn out to be that of focusing attention on this very problem for the study of the remote past in our time.

REFERENCES

Ammerman, A.J. 1985. *The Acconia Survey: Neolithic Settlement and the Obsidian Trade. Occasional Publication* 10. Institute of Archaeology, University of London, London.

———. 1989a. "On the Neolithic Transition in Europe: A Comment on Zvelebil and Zvelebil (1988)." *Antiquity* 63:162–5.

———. 1989b. "Toward the Study of Neolithic Households." *Origini* 14:73–82.

———. 1996. "Reconstructing the Ancient City." *Center* 16:48–51.

Ammerman, A.J., and L.L. Cavalli-Sforza. 1971. "Measuring the Rate of Spread of Early Farming in Europe." *Man* 6:674–88.

———. 1973. "A Population Model for the Diffusion of Early Farming in Europe." In *The Explanation of Culture Change*, edited by C. Renfrew, 343–57. London.

———. 1979. "The Wave of Advance Model for the Spread of Early Farming." In *Transformations: Mathematical Approaches to Culture Change*, edited by C. Renfrew and K.L. Cooke, 275–93. New York.

———. 1984. *The Neolithic Transition and the Genetics of Populations in Europe*. Princeton.

Ammerman, A.J., L.L. Cavalli-Sforza, and D.K. Wagener. 1976. "Towards the Estimation of Population Growth in Old World Prehistory." In *Demographic Anthropology: Quantitative Approaches*, edited by E. Zubrow, 27–61. Albuquerque.

Ammerman, A.J., G. Shaffer, and N. Hartmann. 1988. "A Neolithic Household at Piana di Curinga, Italy." *Journal of Field Archaeology* 15:121–40.

Anthony, D. 1997. "Prehistoric Migration as Social Process." In *Migrations and Invasions in Archaeological Explanation*, edited by J. Chapman and H. Hamerow, 21–32. British Archaeological Reports, International Series 664. Oxford.

Barker, G.W. 1985. *Prehistoric Farming in Europe*. Cambridge.

Bernabeu, J., V. Villaverde, E. Badal, and Y.R. Martínez. 1999. "En torno a la neolitización del Mediterráneo peninsular: Valoración de los procesos postdeposicionales de la Cova de les Cendres." In *Geoarqueologia I Quaternari Litoral*, 69–81. Valencia.

Beteille, A. 1998. "The Idea of Indigenous People." *Current Anthropology* 39:187–91.

Cavalli-Sforza, L.L., P. Menozzi, and A. Piazza. 1994. *The History and Geography of Human Genes*. Princeton.

Clarke, D.L. 1968. *Analytical Archaeology*. London.

Costantini, L. 1989. "Plant Exploitation at Grotta dell'Uzzo, Sicily: New Evidence for the Transition from Mesolithic to Neolithic Subsistence in Southern Europe." In *Foraging and Farming: The Evolution of Plant Domestication*, edited by D.R. Harris and G.C. Hillman, 197–206. London.

Dennell, R.W. 1992. "The Origins of Crop Agriculture in Europe." In *The Origins of Agriculture*, edited by C.W. Cowan and P.J. Watson, 71–100. Washington, D.C.

Diamond, J. 1997. *Guns, Germs, and Steel: The Fates of Human Societies*. New York.

Edmonson, M.S. 1961. "Neolithic Diffusion Rates." *Current Anthropology* 2:71–102.

Fisher, R.A. 1937. "The Wave of Advance of Advantageous Genes." *Annals of Eugenics London* 7:355–69.

Fort, J., and V. Mendez. 1999. "Time-Delayed Theory of the Neolithic Transition in Europe." *Physical Review Letters* 82:867–70.

Gkiasta, M., T. Russell, S. Shennan, and J. Steele. 2003. "Neolithic Transition in Europe: The Radiocarbon Record Revisited." *Antiquity* 77:45–62.

Harris, D.R., ed. 1996. *Origins and Spread of Agriculture and Pastoralism in Eurasia*. Washington, D.C.

Hodson, F.R., D.G. Kendall, and P. Tautu, eds. 1971. *Mathematics in the Archaeological and Historical Sciences*. Edinburgh.

Heun, M., R. Schafer-Pregl, D. Klawan, R. Castagna, M. Accerbi, B. Borghi, and F. Salamini. 1997. "Site of Einkorn Wheat Domestication Identified by DNA Fingerprinting." *Science* 278:1312–4.

Housley, R.A., C.S. Gamble, M. Street, and P. Pettitt. 1997. "Radiocarbon Evidence for the Lateglacial Human Recolonisation of Northern Europe." *Proceedings of the Prehistoric Society* 67:25–54.

Kuper, R., H. Löhr, J. Lüning, and P. Stehli. 1974. "Untersuchungen zur neolithischen Besiedlung der Aldenhovener Platte IV." *Bonner Jahrbucher* 174:424–508.

Landes, D.S. 1969. *The Unbound Prometheus*. Cambridge.

Lilla, M. 1998. "The Politics of Jacques Derrida." *The New York Review of Books* 45 (11):36–41.

Moore, A.M.T., G.C. Hillman, and A.J. Legge. 2000. *Village on the Euphrates*. New York.

Özdoĝan, M., and N. Basgelen. 1999. *Neolithic in Turkey*. Istanbul.

Peltenburg, E., S. Colledge, P. Croft, A. Jackson, C. McCartney, and M.A. Murray. 2000. "Agro-Pastoralist Colonization of Cyprus in the 10th Millennium BP: Initial Assessment." *Antiquity* 74:844–53.

Price, T.D. 2000. "Europe's First Farmers: An Introduction." In *Europe's First Farmers*, edited by T.D. Price, 1–18. Cambridge.

Semino, O., et al. 2000. "The Genetic Legacy of Paleolithic *Homo sapiens sapiens* in Extant

Europeans: A Y-Chromosome Perspective." *Science* 290:1155–9.

Skellam, J. 1951. "Random Dispersal in Theoretical Populations." *Biometrika* 38:196–218.

Ucko, P.G., and G.W. Dimbleby, eds. 1969. *The Domestication and Exploitation of Plants and Animals*. London.

Van Andel, T.H., and C.N. Runnels. 1995. "The Earliest Farmers in Europe." *Antiquity* 69:481–500.

Whittle, A. 1996. *Europe in the Neolithic*. Cambridge.

Zvelebil, M. 1986. "Mesolithic Prelude and Neolithic Revolution." In *Hunters in Transition: Mesolithic Societies of Temperate Eurasia and Their Transition to Farming*, edited by M. Zvelebil, 5–15. Cambridge.

———. 1996. "The Agricultural Frontier and the Transition to Agriculture in the Circum-Baltic Region." In *The Origins and Spread of Agriculture and Pastoralism in Eurasia*, edited by D.R. Harris, 323–35. London.

Zvelebil, M., and K.V. Zvelebil. 1988. "Agricultural Transition and Indo-European Dispersals." *Antiquity* 62:574–83.

NOTES

[1] A preliminary version of my paper, with the same title as this chapter, was sent out to the other participants six months before the conference in order to give them a more tangible sense of the meeting's aim. This example was meant to encourage the participants to look back on their own experience in the preparation of their own papers.

[2] The first person to achieve the great Renaissance ambition of reconstructing the ancient city of Rome was Pirro Ligorio, whose *Imago*, a bird's-eye-view map of the entire city, was printed in 1561.

[3] The AMS method involves counting directly the number of ^{12}C and ^{14}C atoms in a sample. Previously, the traditional method was based on counting beta decay events in a sample and thus called for samples of much larger size (that is, larger than individual seeds). In particular, the AMS method now provides control over the ages of seeds found out of stratigraphic context at cave sites as a result of taphonomic processes. This had been a major source of confusion in the literature prior to the 1980s.

[4] There are three stages in Zevelebil's availability model: the availability phase at the start (where domesticated plants and animals contribute less than 10% of the subsistence economy), the substitution phase (where farming continues to rise slowly to 50%), and the consolidation phase. Recent studies in Denmark do not support such a gradual shift even in the case of some regions of northwest Europe (see ch. 14 in this volume).

[5] For the formal model, see Ammerman and Cavalli-Sforza (1984, 67–71). Recently, Fort and Méndez (1999, 867–70), two physicists in Catalonia, have refined the mathematical treatment of the migratory component in the model through the use of Einstein's approach to Fickian diffusion. This now makes it possible to handle in a better way the occurrence of time delays in the diffusionary process (i.e., the time that passes between site-relocation

events). Note that we had earlier discussed the discrete treatment of time and space in the context of different kinds of settlement systems; see Ammerman and Cavalli-Sforza (1979, 285–8). It is worth adding that the value for the initial growth rate used by Fort and Méndez (1999, 869) is probably too high. It is unrealistic to expect such a high rate of growth (3% per year; near the maximum for a human population) in the frontier zone generation after generation. When a lower value for the initial growth rate (say 2%) is used in combination with a value of 15 years for the average site-relocation time, the classical wave of advance model, based on parabolic reaction-diffusion (PRD) equations, yields much the same rate of advance as the new model using the hyperbolic reaction-diffusion (HRD) equations. I wish to thank Fort and Mendez for making this further comparison. At the conceptual level, their new HRD treatment is the preferred one.

[6] Recall that David L. Clarke (1968) had only recently published his seminal work, *Analytical Archaeology*, which borrowed heavily from the "new" geography and which encouraged the use of models in prehistory.

[7] See the mix up, e.g., in Zvelebil and Zvelebil (1988, 577), who misread our position as that of "the wave of advance hypothesis." As I noted in my reply to them (Ammerman 1989a, 162), we have never held such a position. Instead, we have written about the demic hypothesis and about the wave of advance model.

[8] There has been a long tradition of digging at cave sites in countries like Spain, France, and Italy. This was a productive field strategy when the primary goal was that of working out chronological sequences and tracing the initial outlines of prehistory. Since such sites do not normally represent the main form of Neolithic habitation, they comprise a biased sample with regard to the proper study of social and economic questions in the Neolithic period. This may well apply to the Mesolithic period as well. It will still take another decade or more to break away from this old habit in the Mediterranean countries. A positive sign here is the work done by Zilhão (see ch. 11) and the recent study of post-depositional processes at the cave site of Cova de les Cendres in Spain (Bernabeu et al. 1999).

[9] As part of one of the sessions at the Venice meeting, we reviewed the radiocarbon dates that are currently available for different parts of Europe. This confirmed the overall pattern for the spread at the macro level and the average rate for Europe as a whole (see now Gkiasta et al. 2003, who find the same basic trends). However, the current evidence documents more variation in the rate of dispersal at the regional level. In particular, there is good evidence for the spread slowing down as it reached the northwestern periphery of Europe, where conditions at high latitudes were less favorable for the practice of early farming. In the case of the Mediterranean, there is now a clearer picture as well. The average rate in the eastern Mediterranean was comparatively slow, but it accelerated in the west. There are now good dates for the arrival of food production as a package in Cyprus (already by 8000 B.C. cal.; see the chapter by Guilaine in this volume; Peltenburg et al. 2000), at Knossos on the island of Crete (around 7000 B.C., where the recent excavations by Nikos Efstratiou and dates run at Oxford now confirm the previous results of John D. Evans), and southern Italy (around 6200 B.C.; see the chapters by Biagi and Skeates in this volume). This gives an average rate between Cyprus and southern Italy of about 1 km per year. In contrast, early farming reached the east coast of the Iberian Peninsula by around 5500 B.C., indicating a

higher average rate (in the range of 3 to 5 km per year) in the west Mediterranean (see ch. 11 by Zilhão). The faster tempo in the west may be connected with improvements in seafaring technology over time. Note that, in light of the statistical nature of radiocarbon determinations, one has to be cautious about estimating the rate of spread between two regions that are close to one another (given the short distances involved, even small differences in the estimated arrival times can affect the rate measurement).

[10] In one study, we explored several methods for estimating the rate of population growth from archaeological evidence (Ammerman et al. 1976). In another, there was a further consideration of more formal aspects of the wave of advance model (including the consideration of geographic barriers) and the simulation of local population growth and site relocation in the context of a settlement system in archaeology (Ammerman and Cavalli-Sforza 1979).

[11] On the question of continuity over the transition, see the Preface. There is, of course, the possibility that some Mesolithic sites located on the coastline at the time may have been lost as a result of the transgression in the first half of the Holocene. Note, however, that this would have affected early Neolithic sites on the coastline as well. In comparing the distributions of late Mesolithic and early Neolithic sites in a given area, the coastline was, in effect, much the same for these two times. The lack of Mesolithic sites in the area immediately behind the coast today (in all three survey areas) indicates a different kind of settlement system than the one that operated in the Neolithic period.

[12] E.g., there was the assertion by some archaeologists that we took the position that a high rate of population growth was needed in order to make the spread happen. In the book, we had shown just the opposite: that a modest rate of local population growth in the frontier zone (as low as 1% per year) was sufficient to generate the observed spread (Ammerman and Cavalli-Sforza 1984, 80–2). The claim was even made by one reviewer that there was no evidence for the spread of early farming in Europe on the basis of the radiocarbon dates that were available at the time (clearly a counterfactual statement in the mid 1980s).

[13] One of the more remarkable cases is that of the Grotta dell'Uzzo in Sicily, which is still sometimes cited in support of early cereal domestication and plant use in the central Mediterranean by those less familiar with the situation in Italy. At the Venice meeting, the following points came up in the discussion. There are problems with the stratigraphy of the part of the cave sequence that yielded the various "Mesolithic" seed remains; this was pointed out by Biagi, who participated in the fieldwork at the site. There are technical problems with those few ^{14}C dates from site in the literature (see the chapter by Skeates in this volume). No attempt has been made so far to date any of the seeds in question by the AMS method (notwithstanding the recommendation to do so by myself and others over the last decade). AMS dating could show that the seed remains are out of stratigraphic context. The claim for early seeds at the Grotta dell'Uzzo (Costantini 1989) uncritically entered the literature in *Foraging and Farming: The Evolution of Plant Exploitation*; it has unfortunately spread since then (Dennell 1992, 82; Whittle 1996, 289–91; Price 2000, 6). All of this illustrates the ease with which such a questionable claim can spread in the literature in the wishful context of indigenism.

[14] Since 1984, most of my fieldwork has been done at archaeological sites in the cities of Rome, Athens, and Venice. The investigations in Rome led to my interest in studying the reconstructions of the ancient city. The main focus of our research in the three cities is on environmental archaeology. This is what I had in mind when I first went to study with G. Dimbleby at the Institute of Archaeology in London. Thus, the study of the reconstructions of ancient Rome has brought me back full cycle to the Neolithic transition (the problem that I had switched to in my first year at London). Such are the cycles of doing research.

[15] In the case of those living on the banks of the Wildeman River, contact with those working for lumber companies in recent years has introduced infectious diseases, which have decimated the native population. Recall the approach to native societies in New Guinea taken by Diamond (1997), which is more open and more historically oriented than the one usually adopted by the anthropologist. There are still a few places in the world where one can experience life among such groups, and the prehistorian should take advantage of this opportunity while it still exists.

[16] It led, e.g., to the worldwide synthesis on the geographic distribution of the classical polymorphisms in human genetics, which showed that 27% of the overall variation observed in extant European populations can be linked with the spread of early farming in Europe (Cavalli-Sforza et al. 1994, 255–301). This result is now confirmed by the more recent analysis of an independent body of genetic data, the markers of the Y chromosome (Semino et al. 2000). The study of mitachondrial DNA, a third genetic system, gives much the same result; in this case, about one-fifth of the variability in European populations is now thought to be connected with the spread of early farming (see the chapter by Sykes in this volume).

[17] This shift is reflected by the contributions to the volume edited by Harris (1996).

[18] Housley et al. (1997) have recently shown that northwest Europe witnessed its own major episode of demic diffusion some 13,000 to 12,000 years ago. This involved the reoccupation of land newly opened up by the retreating glacier sheet at the end of the last ice age. It is not without irony that a precedent for a demic expansion is now offered by the late Paleolithic hunters and gatherers of northern Europe—the region that represents the cradle of indigenism in the case of Neolithic studies.

PART II

The Origins of Agriculture

— 2 —

Investigating Agricultural Transitions: A Comparative Perspective

Patty Jo Watson

Although unfamiliar with the details of agricultural transitions in Europe, I have had experience with them elsewhere, first in 1955 as a participant in the last field season at Jarmo in northern Iraq, and continuing with research in four different world areas: western Asia (Iran and Turkey), eastern North America (Kentucky), southwestern North America (Arizona and New Mexico), and North China (Henan Province). The intensities of those experiences were quite different, however, being greatest for western Asia and eastern North America, less for the southwest, and superficial for North China.

As a first-year student in the University of Chicago Anthropology graduate program, I was drawn to Near Eastern prehistory because the work of Professor Robert J. Braidwood and his team seemed so compelling. Robert and Linda Braidwood had recently returned from a very successful season (1950–1951) in northern Iraq where they had been excavating the site of Jarmo, then the world's oldest farming village. They were planning to go back with experts who could help them retrieve physical remains of the earliest domesticated plants and animals and who would aid them in describing the ancient Jarmo landscape: topography, flora, and fauna. I was fortunate to be included as a field assistant when the Oriental Institute, University of Chicago, Iraq-Jarmo Project returned to Iraq with National Science Foundation funding (the first ever awarded for an Old World archaeological proposal) in the fall of 1954 (Braidwood and Howe 1960; Watson 1991, 1–2).

I was the most junior and in every way the least experienced and least knowledgeable staff member, but through daily association with Braidwood himself, Vivian Browman, Bruce Howe, Frederick Matson, Charles Reed, Herbert Wright, and—for several weeks in the spring—Hans Helbaek, I could not have been provided a better opportunity to learn all about the great rewards and the major pitfalls of interdisciplinary research on the origins of agropastoralist economies.

That first field experience was followed by one in Iran (with the Iranian Prehistoric Project, 1959–1960), two fall seasons in Turkey (with the Istanbul-Chicago Prehistoric Project, 1968 and 1970), two summers in New Mexico (Cibola Archeological Project, 1972 and 1973), and five brief consulting visits to Neolithic sites in the Yellow River drainage of north China (1990s). During the 1960s, my own geographic focus had shifted from western Asia to eastern North America, where I became centrally involved in the Cave Research Foundation Archeological Project (Carstens and Watson 1996; Crothers et al. forthcoming; Kennedy and Watson 1997; Watson 1969, 1974, 1997), and the Shell Mound Archaeological Project (Marquardt and Watson 1983, forthcoming)—closely related efforts in western central Kentucky. The cave work began in 1963 and has continued to the present with several particularistic goals of no concern here but also with the general objective of helping to elucidate the transition to agriculture in the Eastern Woodlands of North America. The Shell Mound Project was designed to pursue the same problem. This is not the place to provide detailed summaries, but a brief overview is helpful before discussing the archaeological investigation of agricultural transitions in general.

Two Projects on Agricultural Transitions in Precolumbian North America

As the result of persistence and hard work over the past two decades, a small group of dedicated paleoethnobotanists and archaeologists has succeeded in establishing eastern North America as one of the seven or eight world areas where an agricultural system was developed indigenously (i.e., where there was an agricultural transition independent of those elsewhere, notably in this case, independent of Mesoamerica: Decker-Walters et al. 1993; Fritz 1990, 1991, 1994c; Gremillion 1994; Newsom et al. 1993; Smith 1989, 1992; Watson 1985, 1989, 1995; Yarnell 1994). The earliest version of this agricultural complex includes two gourds *(Cucurbita pepo ovifera* and *Lagenaria siceraria;* the gourds from both these plants were used as containers and probably also as net floats and rattles), two starchy seeded food plants *(Chenopodium bushianum* and *Phalaris caroliniana),* and two oily seeded food plants *(Helianthus annuus* and *Iva annua).* This set of indigenous crops is documented archaeobotanically by 3000 b.p. (uncalibrated) but individual cultigens, most notably the *pepo* gourd, appear much earlier: ca. 5500 b.p. in Maine and Pennsylvania (Petersen and Sidell 1996; Hart and Sidell 1997) and perhaps by 7000 b.p. in west-central Illinois (Asch 1993). Domesticated sunflower and sumpweed are dated to approximately 4000 b.p. (in central Tennessee and west-central Illinois, respectively), domestic chenopod (southern Ohio and eastern Kentucky) to 3500 b.p.

Like that of several other archaeologists, my role in this research has been to retrieve empirical evidence pertaining to the origins and early development of the agricultural economy in question and to collaborate as fully as possible with appropriate specialists. Some of the prehistoric farmers in what is now the midwestern United States were also cavers who intensively mined many kilometers of passageways in the Mammoth Cave System (west central Kentucky) leaving abundant archaeobotanical remains as they did so. The most important category in this 2,000 to 3,000 year old archaeological record is that of human paleofecal deposits. Thousands of these invaluable data repositories are well preserved in the dry cave passages; they have yielded some of the most detailed evidence available anywhere for an early agrarian diet. So far, however, we have not found data in the Mammmoth Cave System that document antecedents to the 3000–2000 b.p. agricultural complex. Hence, for several years we have been engaged in complementary research at aboveground sites near Mammoth Cave and especially downstream on the middle Green River, where several dozen Archaic shell middens chronologically precede and overlap the cave material. In spite of the abundance of charred plant remains in these middens and our persistence in retrieving and analyzing it, we have not succeeded in documenting Late Archaic incipient agriculture along this stretch of the Green River (probably because it was not there, at least not evidenced in riverine shell mounds). We did find charred *Cucurbita pepo* gourd rind that was for a short period in the early 1980s the oldest AMS-dated cultigen north of Mexico. But there are by now several *pepo* gourd occurrences as old and older. The best evidence currently available for the earliest version of an Eastern Agricultural Complex (sunflower, sumpweed, chenopod, *pepo* gourd) is from rockshelters in eastern Kentucky and northwestern Arkansas at ca. 3000 b.p. (Fritz 1990, 1994c, 1997; Fritz and Watson 1993; Gremillion 1994, 1997). Although *pepo* gourd was domesticated by 5500 b.p., if not 7000 b.p., and sunflower and sumpweed by 4000 b.p., it appears that transitions to agriculture in the midwest and midsouth took place in the first half of the fourth millennium b.p.

Lessons from Archaeological Research on Early Agriculture

Methods

One of the first things I noticed about interdisciplinary research as exemplified by Braidwood's behavior and that of most of his field and analytical collaborators is the critical importance of continuous communication among all personnel. This is entirely obvious, and easy to say, but maintaining the communication network was extremely difficult in that pioneer, pre-email era when the international research team was scattered geographically and when every individual except Braidwood himself was bootlegging most of the research time necessary to retrieve and analyze data (i.e., they were fitting the archaeological research into cracks and crannies in and around their real jobs). In that early period, the appropriate personnel were all self-taught, ad hoc experts; there were no formally trained archaeobotanists, paleoethnobotanists, zooarchaeologists, or geoarchaeologists. At Jarmo in 1955, Helbaek and Reed were creating nearly all the necessary research protocols as they went along. And there were many surprises on both sides—the natural scientific and the archaeological ones—regarding essential materials and procedures.

Although communication media among collaborators are now much improved, there are still very serious potential and real difficulties in maintaining the necessary liaisons between and among all the different specialists. Both the glories and the failures of interdisciplinary archaeological investigation of agropastoralist transitions flow directly or indirectly from this single source: coordination and communication. Braidwood's dictum that all the experts must participate wholeheartedly in the field as well as in the lab has been confirmed repeatedly. The other axiom just noted, that good coordination and a strong communication network are absolutely essential, is equally well-established and just as difficult to practice at present as it was at the dawn of modem interdisciplinary archaeology. Given all that, what do we now know about when, where, how, and why pristine agriculture or agropastoralism happened and how these economies moved into regions that lacked them?

Results and Explanations

It is well established that hominid and human groups everywhere throughout the several million year history of our kind were skillful at individually and corporately learning, retaining, and passing onto their offspring immense amounts of detail about relevant physical and social environments. Major differences in technological development since the late Pleistocene, including the creation and diffusion of agropastoral economies, are not owing to biological variation among human populations in various locales. What then does explain the differential fates of human societies? Jared Diamond (1997) argues at length that physical environment—global position, landforms, flora, fauna—is the ultimate explanation for the striking diversity among human cultures in various world areas, enabling some to expand at the expense of others (Diamond 1997).

West Asia

As regards the specific case of West Asian agropastoral origins, scholars of the early 20th century also focused primarily upon the physical environment, especially climatic change, as the major causal factor in the emergence of early food producing communities. Much of this initial environmental emphasis result from geologist Rafael Pumpelly's interdisciplinary archaeological work in Central Asia during the first decade of the last century. At any rate, V. Gordon Childe's suggestions about the beginnings of plant and animal domestication highlight the importance of lowland riverine and oasis locales during presumed Early Holocene desiccation, as did Pumpelly's discussion (Childe 1936, ch. 5; Pumpelly 1908, chs. 4–5). In contrast, Braidwood emphasized differential climatic regimes in the upland ("the hilly flanks of the Fertile Crescent") and lowland zones of West Asia in his first explanatory models during the early 1950s (Braidwood 1952,1953; Watson 1991, 3–4; 1995). He rejected the suggestion of large-scale climatic change as a primary shaper of human society and of cultural innovations such as food production, because results from Herbert Wright's initial geomorphological fieldwork in northern Iraq (1950–1951) indicated no major early Holocene environmental fluctuations. In the 1960s and 1970s, however, as Wright and others carried out palynological research in West Asia and the eastern Mediterranean that earlier assessment proved wrong (i.e., the end Pleistocene-early Holocene climate of Western Asia *was* markedly different—much colder and drier—from the modem one) and climatologically-oriented proximate explanations for agricultural origins in the Near East again received considerable attention. As palynological and paleoclimatic data accumulated for West Asia, environmental emphases increased, especially as regards the Levant and early small grain domestication (McCorriston and Hole 1991; Moore and Hillman 1992; Wright 1983, 1993).

During this same period, however, counter-currents of another sort were competing with the environmental/ecological inferential frameworks of the 1970s and 1980s. By the beginning of the 1990s, considerable credence was being given to the suggestion that relations between humans and their biological and geological surroundings are so intricately entwined and thoroughly recursive for any human society—past or present—that environmental forces cannot be disentangled from their cultural matrices. Very different arguments from this basic premise were advanced by David Rindos (1984), for example, and by archaeologists such as Bender (1978), Hayden (1996), and Hodder (1990). Social and cultural imperatives were accorded considerable weight or given central place in such expositions. Even for the West Asian paradigm case of cultural ecological, environmental explanation for agricultural origins, some of the most authoritative recent formulations (e.g., Bar-Yosef and Belfer-Cohen 1991) seem to give social imperatives a significant role in accounting for the origin of agriculture in a specific region (the Southern Levant) at a specific time (the early Holocene). The sociocultural (or sociopolitical) imperatives adduced in this case and in others range from various forms of intracommunity social action by dominant groups (with concomitant effects upon ideological and techno-economic systems) to initiation and maintenance of external alliance and exchange networks stimulating production of certain crops as prestige boosters and/or trade goods.

Europe, Africa, and South Asia

In other parts of the Old World, archaeological data are very sparse to nearly nonexistent, but it does seem that elements of the West Asian food producing economy diffused both west (to Europe and northeastern Africa) and east (to South Asia). Sheep, goats, wheat, and barley were added by 7,000 years ago to the fishing-gathering-hunting subsistence system of dwellers along the Nile. Cattle, however, were probably domesticated ca. 9000 b.p. in the eastern Sahara *from* indigenous wild herds occupying what was then a relatively moist, savanna-like environment. A variety of small grains (sorghum, finger millet, African rice, tef, noog) were prehistorically domesticated in sub-Saharan Africa, as was enset (a plant in the same family as the banana but with a large edible corm) and possibly various tubers. However, chronologies and other information are virtually completely lacking (Harlan 1989, 1992; Marshall 1998; Marshall and Hildebrand 2002; Wetterstrom 1993, 1998).

Data for the northerly parts of South Asia are somewhat better, sufficient to show that agricultural transitions there were long, complex processes eventually combining winter crops (wheat and barley) with summer monsoon crops such as African and East Asian millets, perhaps also an indigenous millet and rice (presumably from East Asia). Wheat, barley, sheep, and goats were apparently introduced from West Asia but cattle (of the humped *Bos indicus* type) are thought to have been indigenously domesticated ca. 7000 b.p. (Meadow 1996, 1998).

East Asia, Southeast Asia, and the Islands of the Southwestern Pacific

A considerable amount is known about many aspects of the later Neolithic of northern China but very little information is available concerning its early stages. Nor are there any detailed data for origins and development of the southern Chinese Neolithic (Cohen 1996; Crawford 1992). Recent research on the origins of rice agriculture in the Yangtze River drainage is continuing but is still in its early stages (Crawford and Chen 1998; Zhao 1998).

Besides rice, Southeast Asian agriculture, as well as that of New Guinea and of islands in the southwestern Pacific, includes an important array of tuber, root, and tree crops, virtually unknown archaeologically. In New Guinea, there is early landscape modification (thought by some to be early Holocene in age), perhaps for the growing of native yams and taro, but so far no accompanying archaeobotanical remains have been found. Hence, the nature and chronology of indigenous New Guinea agriculture is still being debated. This discussion is complicated by the fact that, as just noted, there is no detailed information available for the origins and development of mainland Southeast Asian agriculture either (Bayliss-Smith 1996; Bellwood 1996; Golson 1989; Hather 1996; Spriggs 1996).

The Americas

For eastern North America, there is by now a good archaeobotanical database but still plenty of room for explanatory formulations highlighting either environmental or sociocultural imperatives. The most influential view at present is a Rindosian co-evolutionary model based upon sound but broad-scale paleoenvironmental assumptions (Smith 1992). Although internally coherent and well-crafted, this floodplain argument does not square well with the currently available empirical data (Fritz and Watson 1993; Gremillion 1993, 1994; Watson 1985).

Agricultural transitions in southwestern North America are variable and different from those of eastern North America. The Southwest may be more comparable to Europe in some important regards because, it is believed, "real" agriculture (first maize and squash, with beans, cotton, and bottle gourds added later) entered this region from Mexico beginning in the second millennium b.c., so the main story is one of agricultural acquisition rather than invention. On the other hand, there were several indigenous cultigens in the Southwest (including little barley, goosefoot, agave; possibly panic grass, amaranth, and tepary beans). And some of these may predate maize and squash (Cordell 1997, 269-80; Huckell 1995). After decades of almost total neglect, pre-maize and squash plant use is now being actively investigated throughout this region and is beginning to yield results somewhat at variance with the traditional maize-centric formulations. In the Phoenix and Tucson basins, where relatively intensive and extensive paleoethnobotanical research has been carried out in conjunction with federally-mandated salvage archaeology, prehistoric cultivation of agave and perhaps of other indigenous plants as well as the systematic use of several more (e.g., saguaro, cholla) is indicated, together with maize agriculture in river valleys or with the aid of irrigation canals (Cordell 1997, 269–80; Huckell 1995).

New information from a site in northern Mexico, Cerro Juanaquena, has important implications for our understanding of agricultural transitions in the southwestern U.S. and northwestern Mexico region. It is apparently very large (some 10 ha) and includes an impressive series of terraces on a hill overlooking a small river, yet it is also quite old (ca. 2900 b.p.). Maize is present in the archaeobotanical assemblage, as are the remains of a number of other plants, including domestic amaranth (Fritz et al. 1999; Smith 1998, 205).

Very important archaeological and archaeobotanical work was carried out farther south in Mexico at several sites in the states of Tamaulipas and Puebla during the 1960s (McClung de Tapia 1992). No further excavations at these sites have taken place, but the older data are currently being reassessed and reevaluated (Fritz 1994a, 1994b, 1999). By 5000 b.p., maize was present, having been created by gatherer-hunters from the wild ancestor, teosinte *(Zea mays parviglumis),* somewhere in the Balsas River drainage of western Mexico (Fritz 1994a, 1994b, Smith 1998, ch. 7). Unfortunately, political conditions have hindered systematic research there for the past few decades, so the regional archaeology is not well known.

Evidence bearing on early use and domestication of plants other than maize (squashes, bottle gourd, chili peppers, beans) is available from the 1960s Tamaulipas and Puebla excavations. And some specimens have recently been AMS dated. *Cucurbita pepo* was apparently domesticated by 9000 b.p., C. *argyrosperma* and C. *moshata* by ca. 5000–4000 b.p. (Smith 1997a, 1997b), and common beans *(Phaseolus vulgaris)* much later (2500 B.P. at the earliest; Kaplan and Lynch 1999).

In contrast to the situation in Mexico, field research on agricultural transitions in Central and South America is being actively pursued especially in lowland areas. Tubers and root crops (e.g., yams, manioc, arrowroot), plus an array of other plants (e.g., chili peppers, cotton, bottle gourd) as well as maize, squashes, and beans are among the species being investigated via macrobotanical and microbotanical (pollen, phytoliths) remains. Preliminary results are said to indicate early (early to mid-Holocene) varied and dynamic subsistence systems incorporating a melange of wild, propagated, cultivated, and domestic plants, but without standardized, expansive, food producing economies prior to ca. 7000 b.p. (Bray 2000; Pearsall 1992, 1995; Piperno et al. 2000; Piperno and Pearsall 1999).

Conclusions

In the half-century since Braidwood built upon Pumpelly's approach to fieldwork, and Childe's and Peake and Fleure's theorizing, to create a powerful mode of interdisciplinary empirical research (Braidwood and Howe 1960; Peake and Fleure 1927), many methodological obstacles have been overcome. Several new subdisciplines have been developed and models explaining agricultural transitions have become increasingly sophisticated. As indicated in the regional sketches briefly presented above, there is now sufficient momentum for major breakthroughs in proximal explanations for agricultural transitions in global prehistory and perhaps for a new generation of ultimate explanatory models as well.

Future directions for the areas I know most about are fairly obvious: (1) refinement of proximate explanatory formulations and models to incorporate combinations of social and environmental imperatives and to instigate collection of new, detailed empirical data relevant to how and why specific agricultural transitions took place; (2) testing of the basic assumptions underlying parts of ultimate causation models such as Jared Diamond's. For example, if, as he says, a maize-based agricultural system is inherently superior to the eastern North American pre-maize agricultural complex, then why was intensive maize (plus beans and squash) agriculture practiced at many small Fort Ancient sites in the Ohio River valley but not at the approximately contemporary, biggest site complex north of Mexico—greater Cahokia in the middle Mississippi River valley? The archaeobotany of Cahokia includes some maize and a little squash (no beans) with the indigenous small-seeded crops but does not reveal a heavily maize/squash/beans centered economy like that at several Fort Ancient communities of about the same age.

As to agricultural transitions in Europe, the most striking explicit and implicit theme at the Venice conference itself was that of a package-transmission mode for the West Asian agropastoral economy into Europe. That is, most participants seemed to assume (at least for purposes of argument) that wheat, barley, legumes, sheep, goats, pigs, and cattle moved into Europe from Anatolia as a viable economic unit (perhaps with some minor variations in components here and there). Package-transmission was also once believed to explain New World agricultural transitions north of Mexico but we now know that the postulated package (maize, beans, and squash) never existed for eastern North America, and probably not for most of the Southwest either. Throughout the Holocene, there was great local variation in plant use, both north and south of the Mexican border, with differing combinations of wild, cultivated, and domestic species evidenced in different locales across all of North America. It is doubtless convenient to begin with the simplifying assumption that agropastoralism spread *en bloc* into Europe, and such a formulation may be justified for certain very broad-scale purposes but it seems unlikely that this is actually how transitions were effected in most places. Even if the West Asian species did move into some parts of Europe as a unit, the mere arrival of the unit (and its human handlers) does not, of course, explain why the package was accepted or rejected. I think that the more we learn about early and middle Holocene plant and animal use in both Western Asia and eastern Europe, the more diversity we will find. Archaeobotanical and zooarchaeological remains must be systematically recovered, dated accurately, and carefully analyzed before the empirical patterning can be clearly defined but that patterning will probably be more diverse and mosaic-like than monolithic and package-like in most regions.

To illuminate agricultural transitions in Europe adequately, it seems that considerable chronological refinement is necessary in all regions with special attention to AMS radiocarbon determinations on critical plant and animal remains. More widespread and more systematic attention to archaeobotany, zooarchaeology, and geoarchaeology of relevant sites is also required if something approaching a comprehensive account is to be obtained. The methods and techniques are available and the broad outline of the story is known, but for most European regions the dynamic and fascinating details have yet to be revealed.

REFERENCES

Asch, D.L. 1993. "Aboriginal Specialty-Plant Cultivation in Eastern North America: Illinois Prehistory and a Post-Contact Perspective." In *Agricultural Origins and Development in the Midcontinent,* edited by W. Green, 25–86. Iowa City.

Bayliss-Smith, T. 1996. "People-Plant Interactions in the New Guinea Highlands: Agricultural Hearthland or Horticultural Backwater?" In *The Origins and Spread of Agriculture and Pastoralism in Eurasia,* edited by D.R. Harris, 499–523. London.

Bellwood, P. 1996. "The Origins and Spread of Agriculture in the Indo-Pacific Region: Gradualism and Diffusion or Revolution and Colonization?" In *The Origins and Spread of Agriculture and Pastoralism in Eurasia*, edited by D.R. Harris, 465–98. London.

Bar-Yosef, O., and A. Belfer-Cohen. 1991. "From Sedentary Hunter-Gatherers to Territorial Farmers in the Levant." In *Between Bands and States,* edited by S. Gregg, 181–202. Center for Archaeological Investigations, Occasional Papers No. 9, Southern Illinois University. Carbondale, Ill.

Bender, B. 1978. "From Gatherer-Hunter to Farmer: A Social Perspective." *World Archaeology* 10:204–22.

Braidwood, R.J. 1952. *The Near East and the Foundations for Civilization.* Eugene, Oreg.

———. 1953. *Prehistoric Men.* 3rd ed. Chicago Natural History Museum, Anthropology Series No. 37. Chicago.

Braidwood, R.J., and B. Howe. 1960. "The General Problem." In *Prehistoric Investigations in Iraqi Kurdistan,* edited by R.J. Braidwood and B. Howe, 1–8. Oriental Institute, University of Chicago, Studies in Ancient Oriental Civilization No. 31. Chicago.

Bray, W. 2000. "Ancient Food for Thought." *Nature* 408:145–6.

Carstens, K., and P.J. Watson, eds. 1996. *Of Caves and Shell Mounds.* Tuscaloosa, Ala.

Childe, V.G. 1936. *Man Makes Himself.* London.

Cohen, D. 1996. "The Origins of Domesticated Cereals and the Pleistocene-Holocene Transition in East Asia." In *The Transition to Agriculture in the Old World,* edited by O. Bar-Yosef. Special issue of *The Review of Archaeology* 19:22–9.

Cordell, L. 1997. *Archaeology of the Southwest.* San Diego.

Crawford, G.W. 1992. "Prehistoric Plant Domestication in East Asia." In *The Origins of Agriculture, An International Perspective,* edited by C.W. Cowan and P.J. Watson, 7–38. Washington, D.C.

Crawford, G. W., and S. Chen. 1998. "The Origins of Rice Agriculture: Recent Progress in East Asia." *Antiquity* 72:858–66.

Crothers, G., C.H. Faulkner, J. Simek, P.J. Watson, and P. Willey. Forthcoming. "Woodland Cave Archaeology." In *The Woodland Southeast,* edited by D. Anderson and R. Mainfort. Tuscaloosa, Ala.

Decker-Walters, D., T.W. Walters, C.W. Cowan, and B.C. Smith. 1993. "Isozymic Characterization of Wild Populations of Cucurbita pepo." *Journal of Ethnobiology* 13:55–72.

Diamond, J. 1997. *Guns, Germs, and Steel: The Fates of Human Societies.* New York.

Fritz, G.J. 1990. "Multiple Pathways to Farming in Prehistoric Eastern North America." *Journal of World Prehistory* 4: 387–435.

———. 1991. "New Dates and Data on Prehistoric Agriculture: The Legacy of Complex Hunter-Gatherers." *Annals of the Missouri Botanical Garden* 82:3–15.

———. 1994a. "Are the First American Farmers Getting Younger?" *Current Anthropology* 4:387–435.

———. 1994b. "Reply." *Current Anthropology* 35:639–43.

———. 1994c. "In Color and In Time: Prehistoric Ozark Agriculture." In *Agricultural Origins and Development in the Midcontinent,* edited by W. Green, 105–26. Iowa City.

———. 1997. "A Three-Thousand-Year-Old Cache of Crop Seeds from Marble Bluff, Arkansas." In *People, Plants, and Landscapes: Studies in Paleoethnobotany,* edited by K. Gremillion, 42–62. Tuscaloosa, Ala.

———. 1999. "Gender and the Early Cultivation of Gourds in Eastern North America." *American Antiquity* 64:417–29.

Fritz, G.J., K. Adams, R. Hard, and J. Roney. 1999. "Evidence for Cultivation of *Amaranthus* 3000 Years Ago at Cerro Juanaquena, Chichuahua." *Abstracts and Directory.* 22nd Annual Meeting of the Society for Ethnobiology, Oaxaca, 11–12. Oaxaca.

Fritz, G.J., and P.J. Watson. 1993. "Terraces, Coves, Caves, and Hollows." *SEAC Bulletin* 36:4–9.

Golson, J. 1989. "The Origins and Development of New Guinea Agriculture." In *Foraging and Farming: The Evolution of Plant Exploitation,* edited by D.R. Harris and G.C. Hillman, 678–87. London.

Gremillion, K. 1993. "Coevolution, Storage, and Early Farming in Eastern Kentucky." *SEAC Bulletin* 36:21.

———. 1994. "Evidence of Plant Domestication from Kentucky Caves and Rockshelters." In *Agricultural Origins and Development in the Midcontinent,* edited by W. Green, 87–103. Iowa City.

———. 1997. "New Perspectives on the Paleothnobotany of the Newt Kash Shelter." In *People, Plants, and Landscapes: Studies in Paleooethnobotany,* edited by K. Gremillion, 23–41. Tuscaloosa, Ala.

Harlan, J .R. 1989. "Wild-Grass Seed Harvesting in the Sahara and Sub-Sahara of Africa." In *Foraging and Farming,* edited by D.R. Harris and G.C. Hillman, 79–98. London.

———. 1992. "Indigenous African Agriculture." In *The Origins of Agriculture: An International Perspective,* edited by C.W. Cowan and P.J. Watson, 59–70. Washington, D.C.

Hart, J., and N. Siddell. 1997. "Additional Evidence for Early Cucurbit Use in the Nothern Eastern Woodlands East of the Allegheny Front." *American Antiquity* 62:523–37.

Hather, J.G. 1996. "The Origins of Tropical Vegeculture: Zingiberaceae, Araceae, and Dioscoreaceae in Southeast Asia." In *The Origins and Spread of Agriculture and Pastoralism in Eurasia,* edited by D.R. Harris, 538–50. London.

Hayden, B. 1996. "Pathways to Power: Principles of Creating Socioeconomic Inequalities." In *Foundations of Social Inequality,* edited by T.D. Price and G.M. Feinman, 15–78. New York.

Hodder, I. 1990. *The Domestication of Europe.* Oxford.

Huckell, B.B. 1995. *Of Marshes and Maize: Preceramic Agricultural Settlements in the Cienega Valley, Southeastern Arizona.* University of Arizona Anthropological Papers No. 59. Tucson.

Kaplan, L., and T.F. Lynch. 1999. "Pphaseolus (Fabaceae) in Archaeology: AMS Radiocarbon Dates the Their Significance for Pre-Columbian Agriculture." *Economic Botany* 53:261–72.

Kennedy, M., and P.J. Watson. 1997. "The Chronology of Early Agriculture and Intensive Mineral Mining in the Salts Cave and Mammoth Cave Region, Mammoth Cave National Park, Kentucky." *Journal of Cave and Karst Studies, National Speleological Society Bulletin* 59:5–9.

Marquardt, W.H., and P.J. Watson. 1983. "The Shell Mound Archaic of Western Kentucky." In *Archaic Hunters of the American Midwest,* edited by J. Phillips and J. Brown, 323–39. New York.

Marquardt, W.H., and P.J. Watson, with the editorial assistance of M.C. Kennedy. Forthcoming. *Archaeology of the Middle Green River, Kentucky.*

Marshall, F.B. 1998. "Early Food Production in Africa." In *The Transition to Agriculture in The Old World,* edited by O.Bar-Yosef. *The Review of Archaeology* 19 (2):47–58.

Marshall, F.B., and E. Hildebrand. 2002. "Cattle before Crops: The Beginnings of Food Production in Africa." *Journal of World Prehistory* 16:99–143.

McClung De Tapia, E. 1992. "The Origins of Agriculture in Mesoamerica and Central America." In *The Origins of Agriculture: An International Perspective,* edited by C.W. Cowan and P.J. Watson, 143–71. Washington, D.C.

McCorriston, J., and F. Hole. 1991. "The Ecology of Seasonal Stress and the Origins of

Agriculture in the Near East." *American Anthropologist* 93:46–69.

Meadow, R.H. 1996. "The Origins and Spread of Agriculture and Pastoralism in Northwestern South Asia." In *The Origins and Spread of Agriculture and Pastoralism in Eurasia,* edited by D.R. Harris, 390–412. London.

———. 1998. "Pre- and Proto-Historic Agricultural and Pastoral Transformations in Northwestern South Asia." In The *Transition* to *Agriculture* in *the Old World,* edited by O. Bar-Yosef. *The Review of Archaeology* 19 (2):12–21.

Moore, A.M.T., and G.C. Hillman. 1992. "The Pleistocene to Holocene Transition and Human Economy in Southwest Asia: The Impact of the Younger Dryas." *American Antiquity* 57:482–94.

Newsom, L.A., S.D. Webb, and J.S. Dunbar. 1993. "Historical and Geographic Distribution of *Cucurbita pepo* Gourds in Florida." *Journal of Ethnobiology* 13:75–97.

Peake, H.J.E., and H.J. Fleure. 1927. *Peasants and Potters.* Oxford.

Pearsall, D.M. 1992. "The Origins of Plant Cultivation in South America." In *The Origins of Agriculture: An International Perspective,* edited by C.W. Cowan and P.J. Watson, 173–205. Washington, D.C.

———. 1995. "Domestication and Agriculture in the New World Tropics." In *Last Hunters, First Farmers: New Perspectives on the Prehistoric Transition to Agriculture,* edited by T.D. Price and A.T. Gebauer, 157–92. Santa Fe.

Petersen, J.B., and N.A. Sidell. 1996. "Mid-Holocene Evidence of *Cucurbita* sp. from Central Maine." *American Antiquity* 61:685–98.

Piperno, D.R., and D.M. Pearsall. 1998. *The Origins of Agriculture in the American Neotropics.* San Diego.

Piperno, D.R., A.J. Ranere, I. Holst, and P. Hansell. 2000. "Starch Grains Reveal Early Root Crop Horticulture In the Panamanian Tropical Forest." *Nature* 40:894–97.

Pumpelly, R., ed. 1908. *Explorations in Turkestan: Expedition of 1904. Prehistoric Civilization of Anau: Origins, Growth and Influence of Environment.* Carnegie Institute Publication 73. Washington, D.C.

Rindos, D. 1984. *The Origins of Agriculture: An Evolutionary Perspective.* New York.

Smith, B.D. 1989. "Origins of Agriculture in Eastern North America." *Science* 246:1566–71.

———. 1992. *Rivers of Change: Essays on Early Agriculture.* Washington, D.C.

———. 1997a. "The Initial Domestication of *Cucurbita Pepo* in the Americas 10,000 Years Ago." *Science* 276:932–4.

———. 1997b. "Reconsidering the Ocampo Caves and the Era of Incipent Cultivation in Mesoamerica." *Latin American Antiquity* 8:342–83.

———. 1998. *The Emergence of Agriculture.* 2nd ed. New York.

Spriggs, M. 1996. "Early Agriculture and What Went Before in Island Melanesia: Continuity or Intrusion?" In the *Origins and Spread of Agriculture and Pastoralism in Eurasia,* edited by D.R. Harris, 524–37. London.

Watson, P.J., ed. 1969. *The Prehistory of Salts Cave, Kentucky.* Illinois State Museum Reports of Investigations No. 16. Springfield, Ill.

———, ed. 1974. *Archaeology of the Mammoth Cave Area.* New York.

———. 1985. "The Impact of Early Horticulture in the Upland Drainages of the Midwest and Midsouth." In *Prehistoric Food Production in North America,* edited by R.I. Ford, 99–147. University of Michigan, Museum of Anthropology, Anthropological Papers 75. Ann Arbor.

———. 1989. "Early Plant Cultivation in the Eastern Woodlands of North America." In *Foraging and Farming: The Evolution of Plant Exploitation,* edited by D.R. Harris and G.C. Hillman, 555–70. London.

———. 1991. "The Origins of Food Production in West Asia and Eastern North America: A Consideration of Interdisciplinary Research in Anthropology and Archaeology." In *Quaternary Landscapes,* edited by L. Shane and E. Cushing, 1–37. Minneapolis.

———. 1995. "Explaining the Transition to Agriculture." In *Last Hunters, First Farmers: New Perspectives on the Prehistoric Transition to Agriculture,* edited by T.D. Price and A.B. Gebauer, 21–37. Santa Fe.

———, ed. 1997. *Archaeology of the Mammoth Cave Area.* Rev. ed. St. Louis.

Wetterstrom, W. 1993. "Foraging and Farming In Egypt: The Transition from Hunting and Gathering to Horticulture in the Nile Valley." In *The Archaeology of Africa: Food, Metals, and Towns,* edited by T. Shaw, P. Sinclair, B. Andah, and A. Akpoko, 163–226. London.

———. 1998. "The Origins of Agriculture in Africa: With Particular Reference to Sorghum and Pearl Millet." In *The Transition to Agriculture in the Old World,* edited by O. Bar-Yosef. In *The Review of Archaeology* 19 (2):30–46.

Wright, H.E., Jr. 1983. "Climatic Change in the Zagros Mountains—Revisited." In *Prehistoric Archeology along the Zagros Flanks,* edited by L. Braidwood, R. Braidwood, B. Howe, C. Reed, and P. Watson, 505–9. Oriental Institute Publication 105. Chicago.

———. 1993. "Environmental Determinism in Near Eastern Prehistory." *Current Anthropology* 34:458–69.

Yarnell, R.A. 1994. "Plant Husbandry in Eastern North America: A Brief and Reasonably True Account." In *Agricultural Origins and Development in the Midcontinent,* edited by W. Green, 7–24. Iowa City.

Zhao, Z. 1998. "The Middle Yangtze Region in China is One Place Where Rice was Domesticated: Phytolith Evidence from the Diaotonghuan Cave, Northern Jiangxi." *Antiquity* 72:885–97.

— 3 —

Paradigms and Transitions: Reflections on the Study of the Origins and Spread of Agriculture

David R. Harris

The conference in Venice in October 1998 on "The Neolithic Transition in Europe" offered an unusual twofold challenge. First, participants were invited to look back over 25 years of research on the Neolithic transition in Europe and forward to further research that we would like to see undertaken over the next decade. And secondly, we were encouraged to reflect on how, individually, we came to be involved in the subject and how our ideas subsequently evolved. Few scholarly meetings explicitly acknowledge how the life experiences of participants—as well as the prevailing zeitgeist of their times—influence the intellectual positions they adopt, and still fewer meetings aim to contribute in this way to the making of intellectual history. In response to this challenge and to the spirit of the conference, my contribution is therefore cast in more personal terms than is usual for a scientific paper.

In selecting participants for the meeting, the organizers cast their net wide. Some of us were invited not for our knowledge of European prehistory but in the hope that we would bring to the discussions a comparative view from the world beyond Europe. I was one of that group: my archaeological and ecological fieldwork on early agriculture having been conducted mainly outside Europe in subtropical and tropical areas of Western and Central Asia, northern Australia and New Guinea, and Mesoamerica. This chapter of *The Widening Harvest* relates to my involvement, over four decades, in research on the origins and spread of agriculture to the broader intellectual currents that have shaped interpretations of the Neolithic transition in Europe.

The 1950s: Agricultural Origins and Dispersals

The title of this section encapsulates, for me at least, how what we now tend to label the transition to agriculture was conceptualized in the 1950s, when my interest in the subject was first aroused. In 1952 the American cultural geographer Carl O. Sauer published a slim volume with that title. It was the text of five lectures that he gave that year to the American Geographical Society in New York, and it presented in a mere 110 pages a speculative and alluringly well written interpretation of plant and animal domestication and the beginnings of agriculture worldwide, which, Sauer (1952, iii) himself said, was not a "well-polished abstract of accepted learning as much as a prospectus of what is not securely within our grasp." His provocative synthesis was dismissed as an untestable "exercise in imagination" in a highly critical review by the Harvard botanist Paul Mangelsdorf (1953, 89), who stressed the lack of evidence for the pattern of origins and dispersals that Sauer postulated and dismissed the "old argument between diffusion and independent invention" as "essentially futile since there is no doubt that both occur and that both have played a part in man's cultural history."

Mangelsdorf's view can be seen in retrospect as an early expression of a difference of emphasis on the role of diffusion that has characterized interpretations of agricultural origins in the Americas and the Old World since the 1950s with a greater role attributed to diffusion in the latter, particularly within Eurasia. No doubt this stems in part from the strong influence on European archaeologists of Gordon Childe's famous hypothesis of the Neolithic Revolution in the Near East, which he had propounded in the 1920s and 1930s and which included the assumption that agriculture had been brought to Europe by Neolithic colonists from the Near East (Childe 1928, 1934, 1936). But it may also reflect actual differences in how agriculture evolved in the two hemispheres.

Another major strand in the intellectual framework of the 1950s was an ecological viewpoint that emphasized the role of human-environment interactions in the origins and spread of agriculture. Gordon Childe had proposed climatic change toward greater aridity ("desiccation") as the primary cause of the Neolithic Revolution in the Near East, but he did not show any awareness of how knowledge of the ecology and behavior of the earliest crops and domestic animals could aid understanding of the process of domestication. It seems more likely that he was simply uninterested in, rather than unaware of, ecological concepts (Harris 1994, 30) because when he was formulating, in London, his ideas about the Neolithic Revolution, the British botanist A.G. Tansley (1920, 1935) was developing and publishing his views on plant succession, climax vegetation, and the ecosystem.

It was not until after the Second World War that ecological thinking began to have a significant impact on interpretations of the origins of agriculture and, more generally, of prehistoric human subsistence. In Britain, Grahame Clark, then a young lecturer in the Department of Archaeology at Cambridge, was mainly responsible for bringing an ecological perspective to bear on European prehistory—particularly in his seminal book *Europe: The Economic Basis* published in1952; while in the same year, in the United States, Sauer's *Agricultural Origins and Dispersals* revealed how deeply his thinking was imbued with ecological awareness, as the title of the first chapter, "Man—Ecologic Dominant," implies.

The publication in 1952 of these two otherwise very different books can be seen in retrospect to mark the beginning of the ecological (some would say, less aptly in my view, environmental) approach to the study of the transition to agriculture, which has continued, although with its strength somewhat attenuated, to the present. My own lifelong interest in the origins and spread of agriculture began in the mid 1950s, when I was a graduate student at the University of California in Berkeley. I read both Sauer and Clark soon after arriving in Berkeley, and it was then that I first explored the literature on the European Mesolithic and Neolithic. I was already familiar with Childe's ideas on the spread of agriculture into Europe. But now I discovered that much earlier in the 20th century Gradmann (1901, 1906) had suggested that the loess and other light soils of central and western Europe were devoid of forest during the Neolithic and that the first farmers chose to settle there because (he assumed) they did not have the technical capacity to clear forest or woodland. I also learned of Godwin's and Iversen's palynological research in the 1930s and 1940s, which had demonstrated that the light soils were by no means treeless during the Sub-Boreal period and that Neolithic farmers had used fire to clear patches of woodland for their crops by the system of shifting cultivation or Brandwirtschaft (Godwin 1934, 1944; Iversen 1941, 1949). I was to return much later, after observing shifting (swidden) cultivation at first hand in the South American tropics (Harris 1971), to the relationship between swidden and settlement in Neolithic Europe as well as in the tropics (Harris 1972).

In the intellectual climate of Berkeley in the late 1950s, as elsewhere in Anglo-American academia, explanations of cultural change, including the origins of agriculture, continued to be expressed in terms of culture-historical processes of diffusion, migration, and colonization. They were generally regarded as self-evident processes, and little attention was paid to how they actually functioned to bring about change. Diffusion was accepted as a norm, but how innovations spread, and what might account for whether they were adopted or rejected, was largely ignored. Thus, for example, it was tacitly assumed that agriculture, once it arose in the Near East, would "naturally" have spread to Europe, because it was seen as a way of life inherently superior to that of the hunter-gatherers it progressively replaced. However, the lack of interest in just how major cultural changes, such as the transition to agriculture, took place, was soon to give way, in the 1960s, to new theoretical concerns with processes of cultural change that introduced into the study of "agricultural origins and dispersals" the novel concept of building and testing explanatory models. Culture history, which had long been regarded as sufficient explanation in itself, began to give way to a search for explanatory variables couched in terms of systems analysis.

The 1960s and 1970s: Model Building and Anti-Diffusionism

The 1960s witnessed a paradigm shift that transformed the social sciences. A major element in this transformation was the conceptualization of human societies as systems and the introduction of concepts used in systems analysis such as equilibrium and negative and positive feedback (Maruyama 1963). The possibility of analyzing societies in terms of sets of interacting variables with predictable outcomes seemed to offer new and exciting prospects of explaining cultural change, and the systemic approach soon attracted theoretically innovative anthropologists, archaeologists, and geographers.

In the United States, Lewis Binford (1968; Binford and Binford 1968) and Kent Flannery (1968) emerged as champions of the "new archaeology," while in Britain it was the "new geography" of Richard Chorley and Peter Haggett (1967) that led the way, soon to be followed by David Harvey (1969), and, in archaeology, by David Clarke (1972) and Colin Renfrew

(1973). These innovators argued for a more "scientific" approach to the explanation of economic and social change through the specification of causal variables in "processual" models that would, ideally, be capable of being tested against "real-world" data. Soon after I returned to Britain from California in 1958, I became aware that this model-building revolution was under way and began to see how it might lead to better understanding of the origins of agriculture.

While undertaking fieldwork in the western United States, Mexico, and the Caribbean in the 1950s, I had already developed a strongly ecological outlook, and the "systems thinking" of the new geography and new archaeology attracted me, especially as a means of modeling the conditions under which agriculture might have originated. I hoped it might be possible to formulate a general explanatory model applicable worldwide to the study of how agriculture arose in pristine situations (as opposed to the adoption of agriculture following contact with pre-existing areas of origin). The "Research Seminar in Archaeology and Related Subjects" that took place in London in 1968 stimulated these endeavors.

This event, organized by Peter Ucko and Geoffrey Dimbleby, brought together a remarkably broad spectrum of archaeologists, anthropologists, biologists, and geographers united by their interest in how humans had exploited and domesticated plants and animals through the ages. Although the seminar did not strongly reflect the new archaeology or geography, one contribution in particular, by Flannery (1969), did so. In his "Origins and Ecological Effects of Early Domestication in Iran and the Near East" he built on the equilibrium model of post-Pleistocene human adaptations put forward by Binford (1968), and, using ethnographic data on hunter-gatherer groups, he proposed a model of population pressure and disequilibrium relative to environmental carrying capacity to explain the origins and early development of agriculture in Southwest Asia.

My contribution to the seminar (Harris 1969), entitled "Agricultural Systems, Ecosystems and the Origins of Agriculture," was general rather than regionally specific like Flannery's. But it also sought to model the conditions under which hunter-gatherers were likely to have begun to cultivate and, in due course domesticate, plants. It was in this paper, too, that I first outlined a hypothesis concerned with the spread of agriculture, which proposed that agricultural systems based on seed-crop cultivation were inherently more likely to spread—for ecological and nutritional reasons—than those based mainly on root and tuber crops. I elaborated this hypothesis in papers written for the next two Seminars in Archaeology and Related Subjects, which were held in London in 1970 and Sheffield in 1971 (Harris 1972, 1973), and in the first of these I briefly discussed the topic that pre-occupied us in Venice 28 years later—the spread of agriculture in Europe during the Neolithic.

The seminar at Sheffield in 1971 reflected more strongly than its predecessors the new processual, model-building orientation that was now in the ascendant in much anthropological, archaeological, and geographical discourse (Renfrew 1973). As its subtitle, "Models in Prehistory," made clear, there was now an explicit concern with the development of heuristic models that could guide research design and help archaeology to become a more scientific discipline. It was at the Sheffield seminar that Albert Ammerman's and Luca Cavalli-Sforza's model for the spread of farming in Europe was first introduced to many of us, although an earlier version had been published that year in the journal *Man* (Ammerman and Cavalli-Sforza 1971). Six years previously Grahame Clark (1965) had drawn attention to the consistently time-transgressive pattern of radiocarbon dates for early farming in Europe, which suggested that it took some 5,000 years for agriculture to spread from the Levant (Jericho) to Britain. But Ammerman and Cavalli-Sforza were the first to propose, in their "wave of advance" model drawn from population biology, how the process could have taken place, and also to relate it to the radiocarbon chronology.

By contrasting "demic" with "cultural" diffusion—the former involving population growth and displacement and resembling what anthropologists had previously referred to as "primary" diffusion and the latter involving the transmission of cultural traits without significant population displacement ("secondary" and "stimulus" diffusion)—Ammerman and Cavalli-Sforza introduced into the archaeological literature on early agriculture an important conceptual distinction that continues to guide debates about how agriculture spread from its areas of origin. However, to demonstrate from archaeological data in any given situation the relative importance of these two modes of diffusion remains a complex and difficult task.

In retrospect, there is a neat irony in the fact that the first attempt (by Ammerman and Cavalli-Sforza) to model how prehistoric agriculture might have spread coincided with the widespread rejection by archaeologists of diffusion in favor of "independent origins." It was always likely that attempts to build general heuristic models, divorced from particular regional and chronological contexts, would engender the belief that agriculture could have arisen independently in many different parts of the world. The title of my main contribution to the model building of this time—"Alternative pathways toward agriculture"—can be taken to imply that view, although that was not my intention and the paper did not address the question of many versus few independent areas of origin (Harris 1977a).[1] For whatever reasons, and I do not doubt that they include changes in societal attitudes outside academia, the tide of opinion during the 1970s and into the 1980s was strongly anti-diffusionist. It became the fashion, wherever archaeologists were investigating Mesolithic-Neolithic (or equivalent) transitions, to look for evidence of independent—and preferably very early—domestication and agriculture, or at least to interpret such evidence as there was in those terms.[2] Europe came to be regarded as a special case of diffusion, although even there arguments were advanced for independent domestication and indigenous food production (e.g., Dennell 1983, 155–68).

In these debates on diffusion versus independent origins, my position was that the issue was of less immediate interest than attempts to model how agriculture arose in pristine situations, for even the most extreme diffusionist had to admit that it did so at least once somewhere in the world! I did not dismiss diffusionary processes as unimportant—both history and everyday experience show them to be an integral part of group and individual interaction—but I was more concerned to elucidate how agriculture could have begun in the first place.

The 1980s: From "Origins" to "Spread"

By the mid 1970s, it had become apparent that although the general models that several of us had proposed, largely on the basis of ecological and ethnographic data, had undoubted heuristic value, they tended to be too general to be tested, in their entirety, against independently generated, especially archaeological, evidence. They could be dissected, and certain parts of them, could, in some areas at least, be judged against archaeological evidence. An example of this procedure (equivalent to Binford's "middle-range theory"; Binford 1977, 7) in which an attempt was made to test the proposition that sedentary settlement preceded the emergence of agriculture was an analysis of the plant remains from the site of Tell Abu Hureyra in Syria, where we were able to demonstrate with a high degree of probability that the site was occupied year-round before agriculture (in the sense of the cultivation of domesticated cereals) was practiced there (Hillman et al. 1989). This conclusion did not preclude the possibility of the "pre-domestication cultivation" of wild cereals having preceded agriculture, and it also underlined the need for precise definitions of such general terms as cultivation, domestication, and agriculture.[3]

It was also in the mid 1970s that I became convinced that if we could investigate more precisely how foragers and farmers interacted in the "ethnographic present," as well as in the past, we might get closer to understanding the process by which hunter-gatherers did, or did not, adopt agriculture. Some European prehistorians had long been interested in how late Mesolithic and early Neolithic populations had interacted, and in the 1980s there were attempts, for example by Douglas Price and by Marek Zvelebil and Peter Rowley-Conwy (1984, 1986), to develop models of such interaction and test them against the relatively rich archaeological record of, in particular, Scandinavia.

Following the publication by Ammerman and Cavalli-Sforza (1984) of their fully developed model for the spread of farming across Europe, the debate became more polarized between proponents of demic diffusion as the primary process responsible for the Neolithic transition in Europe and those prehistorians who saw the process more as one that involved the selective adoption of components of the Southwest Asian Neolithic "package" by resident Mesolithic populations—some of whom were now portrayed as culturally complex "affluent foragers" (Price 1985). No doubt, as Ammerman suggested in his "position paper" before the Venice meeting, opposition among prehistorians to his and Cavalli-Sforza's 1984 book was also part of a wider sociopolitical mind set which, in the post-colonial era that stressed nationalism and minority rights, favored what he calls indigenism.

While this debate was focused on Europe, I looked elsewhere, at a great remove geographically from the European scene, for a context in which to study more recent forager-farmer interaction ethnographically and historically as a means of improving models of such interaction in the prehistoric past. This search led to Torres Strait, which, since my days as a graduate student in Berkeley, I had regarded as an interesting and somewhat puzzling "frontier" between Australia, the continent of hunter-gatherers, and neighboring New Guinea, a continental island many of whose inhabitants practiced agriculture. Back in the 1950s, we used to wonder why agriculture had never spread across this narrow, island-studded strait into Australia. And even though this naïve question had long since been replaced by the more universal one of why agriculture had ever begun anywhere, Torres Strait still seemed to be a promising locale in which to examine how hunter-gatherers and cultivators had interacted in the recent past.

The attraction of the Strait for such a case study rested not only on the geographical contiguity of northeastern Australia and southern Papua New Guinea but also on the fact that historical accounts existed of the beginning, in the 1840s, of sustained European contact with the Torres Strait islanders. In addition to these historical sources, there were the six volumes of the *Reports of the Cambridge Anthropological Expedition to Torres Straits* (Haddon 1901–1935), which provided uniquely detailed ethnographic information on the life of the islanders in 1898. Furthermore, by the end of the 1970s, archaeological investigations in the Highlands of Papua New Guinea, led by Jack Golson (1977), had shown convincingly that cultivation there dated back at least 6,000 years, suggesting that the Torres Strait divide between hunter-gatherers and agriculturalists might be of great antiquity.

In 1974, I began ethnobotanical fieldwork in the Strait (Harris 1977b), and this led, over the next decade, to archival research and trial excavations on the western islands and adjacent mainland to the north, which demonstrated that, in the mid 19th century, a north–south gradient from agriculture to foraging, and from sedentary to seasonally mobile settlement, had existed across the Strait. At that time, a complex pattern of socioeconomic interaction, which involved both inter-island and trans-strait trade, linked the communities of the islands and the mainlands to north and south. I proposed a "trade-and-horticulture hypothesis" to explain the adoption of crop cultivation on particular islands within an economic system in which the exploitation of wild plants and marine fish, turtle, and dugong predominated (Harris 1979). I also suggested that this represented a self-equilibrating system of resource use capable of sustaining the populations of the western islands over the long term rather than being a prelude to full dependence on agriculture. It proved impossible, however, to determine conclusively from the archaeological data when crop cultivation began or for how long this system of interdependent wild-resource procurement and small-scale cultivation had existed prior to the mid 19th century (Barham and Harris 1985; Harris 1994).[4] I did not at this time draw attention to how this case study might aid the interpretation of prehistoric forager-farmer interaction in other parts of the world, such as Europe, but it is worth noting now the general similarity that exists between my trade-and-horticulture hypothesis and, for example, Zvelebil's interpretation of forager-farmer interaction in the circum-Baltic region (Zvelebil 1996).

The 1990s: Diffusion Reevaluated

By the end of the 1980s, anti-diffusionism was on the wane and the priority accorded earlier to the search for agricultural "origins" had given way to a growing interest, principally among archaeologists working in the Old World, in the processes by which agricultural systems developed, spread, and were adopted—whether as packages or piecemeal.

During the 1980s and early 1990s, impressive advances were made in our understanding of early agriculture and pastoralism in southwest Asia, and there was renewed interest in how agro-pastoralism spread west into Europe and east into central and southern Asia. In 1993, I organized a Prehistoric Society conference in London on this theme, and one of the striking features of the resulting book (Harris 1996b) is the acceptance by most of the contributors of the major role that diffusion (demic or cultural) seems to have played in the establishment of agriculture throughout Eurasia. That the spread of agro-pastoralism into much of eastern and central Europe involved demic diffusion ("colonization" or "migration"), represented most clearly by the archaeological signature of the Linear Pottery or Linearbandkeramic (LBK) culture, is now widely accepted.[5] So too is the evidence that around the margins of this area, especially in northern and northwestern Europe, the spread took place more slowly, and full dependence on agriculture was preceded by long periods of interaction between foragers and farmers, with the former selectively appropriating elements of material culture, including domestic animals and crops, from the latter. Thus, the processes by which agro-pastoralism became established in Europe as a whole are now seen as having been much more complex and regionally varied than a demic-diffusion model alone would predict,[6] and interest is now turning to smaller-scale analysis of how environmental, demographic, and technological variables may have interacted to promote or delay the process of spread (e.g., Collard and Shennan 2000; Pétrequin et al. 1998).

Increasing interest in how agro-pastoralism spread from the core area of its origin in southwest Asia was also behind my own most recent field project—an investigation, with British, Russian, and Turkmenian colleagues, of the early Neolithic site of Jeitun (Djeitun) in southern Turkmenistan. From 1989 to 1994, six seasons of excavation at the site and ecological surveys in the surrounding desert, piedmont, and mountain environments have shown conclusively that agro-pastoralism based on the cultivation of barley and einkorn wheat and the herding of domestic goats and sheep was established at Jeitun by 5000 b.c. (6000 B.C. cal.) (Harris et al. 1993, 1996). There are compelling reasons for interpreting the "Jeitun Culture" sites on the piedmont of southern Turkmenistan as the result of diffusion (probably mainly demic) from southwest Asia (Harris and Gosden 1996; Harris 1998a), and comparison with the spread of agro-pastoralism westward across Anatolia and into southeastern Europe suggests that the rates of diffusion may have been similar (Harris 1998a, 79). A clearer picture also is emerging of how agro-pastoralism originated and spread *within* the "Fertile Crescent" of southwest Asia.[7]

In looking back over my own 40-year involvement in the study of "agricultural origins and dispersals," through the paradigm shifts that have so strongly shaped how research has been undertaken, I find that two overarching conclusions can be drawn. First, that agriculture originated independently very seldom in very few regions of the world at times when particular environmental, demographic, and socioeconomic circumstances combined to initiate, and sustain, a gradual transition from dependence on wild foods to dependence on crop cultivation, and, in even fewer areas, also on the raising of domestic animals. Secondly, and following inevitably from the first conclusion, that it was through processes of diffusion (probably mainly demic) that agriculture eventually became established throughout most of the planet's cultivable lands. Expressing it another way, historical contingency, rather than normative "laws" of human behavior, explains why our forager ancestors ever became farmers.

Postscript: Looking to the Future

Over the next decade, we can expect to obtain more archaeological, genetic, and paleoenvironmental data with which to test and refine our understanding of the Neolithic transition in Europe. Many types of evidence can contribute to this improved understanding, and I will conclude by stressing the potential value of new data from six domains of enquiry that seem to me to offer particular promise.

1. Molecular genetics of modern and prehistoric human populations in Europe and adjacent regions, including the analysis (when feasible) of ancient DNA from accurately dated Mesolithic and Neolithic skeletal samples
2. Isotopic analysis of accurately dated and provenienced human bones as a source of information on Mesolithic and Neolithic diet
3. Paleopathological investigation of Mesolithic and Neolithic human skeletal samples to advance knowledge of the role of disease in the transition to agriculture
4. Radiocarbon dating by accelerator mass spectrometry (AMS) of accurately identified samples, from known archaeological contexts, of the crops and domestic animals that were introduced from southwest Asia and spread across Europe during the Neolithic: principally, barley, wheat, goat, and sheep
5. Ecological modeling of how the crops of southwest Asian origin adapted, as they spread through central and western Europe, to differences in day length, soils, vegetation, hydrology, and the seasonal distribution of rainfall and temperature
6. Ecological modeling of the impact on temperate European forests and woodlands of such modes of exploitation as burning, browsing, grazing, leaf harvesting, coppicing, and shifting (swidden) cultivation as an aid to interpretation of the Mesolithic and Neolithic pollen data.

REFERENCES

Ammerman, A.J., and L.L. Cavalli-Sforza. 1971. "Measuring the Rate of Spread of Early Farming in Europe." *Man* 6:674–88.

———. 1973. "A Population Model for the Diffusion of Early Farming in Europe." In *The Explanation of Culture Change: Models in Prehistory,* edited by C. Renfrew, 343–57. London.

———. 1984. *The Neolithic Transition and the Genetics of Populations in Europe.* Princeton.

Barham, A.J., and D.R. Harris. 1985. "Relict Field Systems in the Torres Strait Region." In *Prehistoric Intensive Agriculture in the Tropics,* edited by I.S. Farrington, 247–83. *BAR-IS* 232 (i). Oxford.

Bar-Yosef, O., and R.H. Meadow. 1995. "The Origins of Agriculture in the Near East." In *Last Hunters-First Farmers: New Perspectives on the Prehistoric Transition to Agriculture,* edited by T.D. Price and A.B. Gebauer, 39–94. Santa Fe.

Binford, L.R. 1968. "Post-Pleistocene Adaptations." In *New Perspectives in Archeology,* edited by S.R. Binford and L.R. Binford, 313–41. Chicago.

———. 1977. Introduction to *For Theory Building in Archaeology*, edited by L.R. Binford, 1–10. New York.

Binford, S.R., and L.R. Binford, eds. 1968. *New Perspectives in Archeology.* Chicago.

Bogucki, P. 1996. "The Spread of Early Farming in Europe." *American Scientist* 84:242–53.

———. 2000. "How Agriculture Came to North-Central Europe." In *Europe's First Farmers,* edited by T.D. Price, 197–218. Cambridge.

Childe, V.G. 1928. *The Most Ancient East: The Oriental Prelude to European Prehistory.* London.

———. 1934. *New Light on the Most Ancient East.* London.

———. 1936. *Man Makes Himself.* London.

Chorley, R.J., and P. Haggett, eds. 1967. *Models in Geography.* London.

Clark, J.G.D. 1952. *Prehistoric Europe: The Economic Basis.* London.

———. 1965. "Radiocarbon Dating and the Spread of Farming Economy." *Antiquity* 39:45–8.

Clarke, D.L. 1972. *Models in Archaeology.* London.

Collard, M., and S. Shennan. 2000. "Processes of Culture Change in Prehistory: A Case Study from the European Neolithic." In *Archaeogenetics: DNA and the Population Prehistory of Europe*, edited by C. Renfrew and K. Boyle, 89–97. Cambridge.

Dennell, R.W. 1983. *European Economic Prehistory: A New Approach.* London.

Flannery, K.V. 1968. "Archeological Systems Theory and Early Mesoamerica." In *Anthropological Archeology in the Americas,* edited by B.J. Meggers, 67–87. Washington, D.C.

———. 1969. "Origins and Ecological Effects of Early Domestication in Iran and the Near East." In *The Domestication and Exploitation of Plants and Animals,* edited by P.J. Ucko and G.W. Dimbleby, 73–100. London.

Godwin, H. 1934. "Pollen Analysis: An Outline of the Problems and Potentialities of the Method." *New Phytologist* 33:278–305, 325–58.

———. 1944. "Neolithic Forest Clearance." *Nature* 153:511.

Golson, J. 1977. "No Room at the Top: Agricultural Intensification in the New Guinea Highlands." In *Sunda and Sahul: Prehistoric Studies in Southeast Asia, Melanesia, and Australia*, edited by J. Allen, J. Golson, and R. Jones, 601–38. London.

Gradmann, R. 1901. "Das mittelseuropäische Landschaftsbild nach seiner geschichtlichen Entwicklung." *Geographische Zeitschrift* 7:361–77, 435–47.

———. 1906. "Beziehungen zwischen Plfanzengeographie und Siedlungsgeschichte." *Geographische Zeitschrift* 12:305–25.

Haddon, A.C. 1901–1935. *Reports of the Cambridge Anthropological Expedition to Torres Straits.* 6 vols. Cambridge.

Harris, D.R. 1969. "Agricultural Systems, Ecosystems and the Origins of Agriculture." In *The Domestication and Exploitation of Plants and Animals*, edited by P.J. Ucko and G.W. Dimbleby, 3–15. London.

———. 1971. "The Ecology of Swidden Cultivation in the Upper Orinoco Rain Forest, Venezuela." *Geographical Review* 61:475–95.

———. 1972. "Swidden Systems and Settlement." In *Man, Settlement and Urbanism,* edited by P.J. Ucko, R. Tringham, and G.W. Dimbleby, 245–62. London.

———. 1973. "The Prehistory of Tropical Agriculture: An Ethnoecological Model." In *The Explanation of Culture Change: Models in Prehistory*, edited by C. Renfrew, 391–417. London.

———. 1977a. "Alternative Pathways toward Agriculture." In *Origins of Agriculture,* edited by C.A. Reed, 179–243. The Hague.

———. 1977b. "Subsistence Strategies across Torres Strait." In *Sunda and Sahul: Prehistoric Studies in Southeast Asia, Melanesia and Australia,* edited by J. Allen, J. Golson, and R. Jones, 421–63. London.

———. 1979. "Foragers and Farmers in the Western Torres Strait Islands: An Historical Analysis of Economic, Demographic and Spatial Differentiation." In *Social and Ecological Systems,* edited by P.C. Burnham and R.F. Ellen, 75–109. London.

———. 1989. "An Evolutionary Continuum of People-Plant Interaction." In *Foraging and Farming: The Evolution of Plant Exploitation,* edited by D.R. Harris and G.C. Hillman, 11–26. London.

———. 1990. *Settling Down and Breaking Ground: Rethinking the Neolithic Revolution.* Amsterdam.

———, ed. 1994. *The Archaeology of V. Gordon Childe: Contemporary Perspectives.* London.

———. 1995. "Early Agriculture in New Guinea and the Torres Strait Divide." In *Transitions: Pleistocene to Holocene in Australia and Papua New Guinea*, edited by J. Allen and J.F. O'Connell, 848–54. *Antiquity* 69, Special Number 265.

———. 1996a. "Domesticatory Relationships of People, Plants and Animals." In *Redefining Nature: Ecology, Culture and Domestication,* edited by R. Ellen and K. Fukui, 437–63. Oxford.

———, ed. 1996b. *The Origins and Spread of Agriculture and Pastoralism in Eurasia.* London.

———. 1998a. "The Spread of Neolithic Agriculture from the Levant to Western Central Asia." In *Origins of Agriculture and Crop Domestication,* edited by A.B. Damania, J. Valkoun, G. Willcox, and C.O. Qualset, 63–80. Aleppo.

———. 1998b. "The Origins of Agriculture in Southwest Asia." *The Review of Archaeology* 19:5–11.

———. Forthcoming a. "Climatic Change and the Beginnings of Agriculture: The Case of the Younger Dryas." In *Evolution on Planet Earth*, edited by A. Lister and L. Rothschild. New York.

———. Forthcoming b. "Development of the Agro-Pastoral Economy in the Fertile Crescent during the Pre-Pottery Neolithic B (PPNB)." In *The Transition from Foraging to Farming in Southwest Asia,* edited by U. Baruch, S. Bottema, and R. Cappers. Berlin.

Harris, D.R., and C. Gosden. 1996. "The Beginnings of Agriculture in Western Central Asia." In *The Origins and Spread of Agriculture and Pastoralism in Eurasia,* edited by D.R. Harris, 370–89. London.

Harris, D.R., C. Gosden, and M.P. Charles. 1996. "Jeitun: Excavations at an Early Neolithic Site in Southern Turkmenistan." *Proceedings of the Prehistoric Society* 62:423–42.

Harris, D.R., V.M. Masson, Y.E. Berezkin, M.P. Charles, C. Gosden, G.C. Hillman, A.K. Kasparov, G.F. Korobkova, K. Kurbansakhatov, A.J. Legge, and S. Limbrey. 1993. "Investigating Early Agriculture in Central Asia: New Research at Jeitun, Turkmenistan." *Antiquity* 67:324–38.

Harvey, D. 1969. *Explanation in Geography.* London.

Higgs, E.S., ed. 1972. *Papers in Economic Prehistory.* Cambridge.

———. 1975. *Palaeoeconomy*. Cambridge.

Higgs, E.S., and M.R. Jarman. 1969. "The Origins of Agriculture: A Reconsideration." *Antiquity* 43:31–41.

Hillman, G.C., S.M. Colledge, and D.R. Harris. 1989. "Plant-Food Economy during the Epipalaeolithic Period at Tell Abu Hureyra, Syria: Dietary Diversity, Seasonality and Modes of Exploitation." In *Foraging and Farming: The Evolution of Plant Exploitation,* edited by D.R. Harris and G.C. Hillman, 240–68. London.

Iversen, J. 1941. "Landnam i Danmarks Stenalder." *Danmarks geologiske Undersogelse,* Series 2, 66:1–68.

———. 1949. "The Influence of Prehistoric Man on Vegetation." *Danmarks geologiske Undersogelse,* Series 4, 3 (6):1–25.

Lahr, M.M., R.A. Foley, and R. Pinhasi. 2000. "Expected Regional Patterns of Mesolithic-Neolithic Human Population Admixture in Europe Based on Archaeological Evidence." In *Archaeogenetics: DNA and the Population Prehistory of Europe*, edited by C. Renfrew and K. Boyle, 81–8. Cambridge.

Mangelsdorf, P.C. 1953. Review of *Agricultural Origins and Dispersals,* by C.O. Sauer. 1952. *American Antiquity* 19:87–90.

Maruyama, M. 1963. "The Second Cybernetics: Deviation-Amplifying Mutual Causal Processes." *American Scientist* 51:164–79.

Pétrequin, P., R-M Arbogast, C. Bourquin-Mignot, C. Lavier, and A. Viellet. 1998. "Demographic Growth, Environmental Changes and Technical Adaptations: Responses of an Agricultural Community from the 32nd to the 30th Centuries BC." *World Archaeology* 30:181–92.

Price, T.D. 1985. "Affluent Foragers of Mesolithic Southern Scandinavia." In *Prehistoric Hunter-Gatherers: The Emergence of Cultural Complexity*, edited by T.D. Price and J.A. Brown, 341–63. Orlando.

———. 1987. "The Mesolithic of Europe." *Journal of World Prehistory* 2:225–305.

———, ed. 2000. *Europe's First Farmers.* Cambridge.

Renfrew, C., ed. 1973. *The Explanation of Culture Change: Models in Prehistory*. London.

Renfrew, C., and K. Boyle, eds. 2000. *Archaeogenetics: DNA and the Population Prehistory of Europe*. Cambridge.

Sauer, C.O. 1952. *Agricultural Origins and Dispersals.* New York.

Solheim, W.G., II. 1972. "An Earlier Agricultural Revolution." *Scientific American* 226 (4):34–41.

Tansley, A.G. 1920. "The Classification of Vegetation and the Concept of Development." *Journal of Ecology* 8:118–44.

———. 1935. "The Use and Abuse of Vegetational Concepts and Terms." *Ecology* 16:284–307.

Zilhão, J. 1993. "The Spread of Agro-Pastoral Economies across Mediterranean Europe: A View from the Far West." *Journal of Mediterranean Archaeology* 6:5–63.

Zvelebil, M. 1996. "The Agricultural Frontier and the Transition to Farming in the Circum-Baltic Region." In *The Origins and Spread of Agriculture and Pastoralism in Eurasia,* edited by D.R. Harris, 323–45. London.

———. 2000. "The Social Context of the Agricultural Transition in Europe." In *Archaeogenetics: DNA and the Population Prehistory of Europe*, edited by C. Renfrew and K. Boyle, 57–79. Cambridge.

Zvelebil, M., and M. Lillie. 2000. "Transition to Agriculture in Eastern Europe." In *Europe's First Farmers*, edited by T.D. Price, 57–92. Cambridge.

Zvelebil, M., and P. Rowley-Conwy. 1984. "Transition to Farming in Northern Europe: A Hunter-Gatherer Perspective." *Norwegian Archaeological Review* 17:104–28.

———. 1986. "Foragers and Farmers in Atlantic Europe." In *Hunters in Transition: Mesolithic Societies of Temperate Eurasia and Their Transition to Farming,* edited by M. Zvelebil, 67–93. Cambridge.

NOTES

[1] This paper was presented at a conference in Chicago in 1973 but not published until 1977.

[2] See, e.g., Solheim (1972), and, more generally, the work of Eric Higgs and his students at Cambridge, who rejected the conventional dichotomy between "hunter-gatherer" and "agricultural" economies and argued that the domestication of plants and animals had begun before the Neolithic; see Higgs and Jarman (1969) and Higgs (1972, 1975).

[3] As I argued in a series of papers in which I developed models of people-plant and people-animal interaction; see Harris (1989, 1990, 1996a).

[4] The earliest date for cultivation that we obtained was 780 ± 70 b.p. (uncal.)

[5] See, e.g., Bogucki (1996, 2000) and ch. 13 of this volume; Zvelebil and Lillie (2000, fig. 3.1).

[6] See the chapters in this volume and in Price (2000); also those by Zvelebil and by Lahr et al. in Renfrew and Boyle (2000).

[7] See, e.g., Bar-Yosef and Meadow (1995), Harris (1998b), and chs. 4 and 5 in this volume.

— 4 —

The Abu Hureyra Project: Investigating the Beginning of Farming in Western Asia

Andrew M.T. Moore

Present evidence indicates that farming began earlier in Western Asia than in any other region across the globe. That alone would make the area of prime importance to students of the human past. But there are other reasons why we should examine the transition from foraging to farming there with particular attention. For the moment, Western Asia is the only region where we can trace a continuous sequence of development from the foraging societies of the late Pleistocene through the farming communities of the early Holocene to the first civilizations. Furthermore, the farming villages of the earlier Neolithic in Western Asia were often spectacular in their great size, in the variety of their artifacts, and in their richly developed ideologies. They thus provide striking testimony of the dramatic immediate consequences of the development of a farming way of life. The agricultural economy—based on domesticated wheat, barley, rye, pulses, sheep, goats, cattle, and pigs—that developed in Western Asia later spread to Europe. Thus, the development of agriculture in Western Asia is crucial to our understanding of the formation of farming communities across the European continent.

My aim in this paper is to provide some insights into how my collaborators and I have pursued our research on the beginning of agriculture at Abu Hureyra, an early village in the valley of the Euphrates River in northern Syria (fig. 4.1). In accordance with the themes of the Venice meeting, I shall explain something of the history of the project, and the unique set of opportunities,

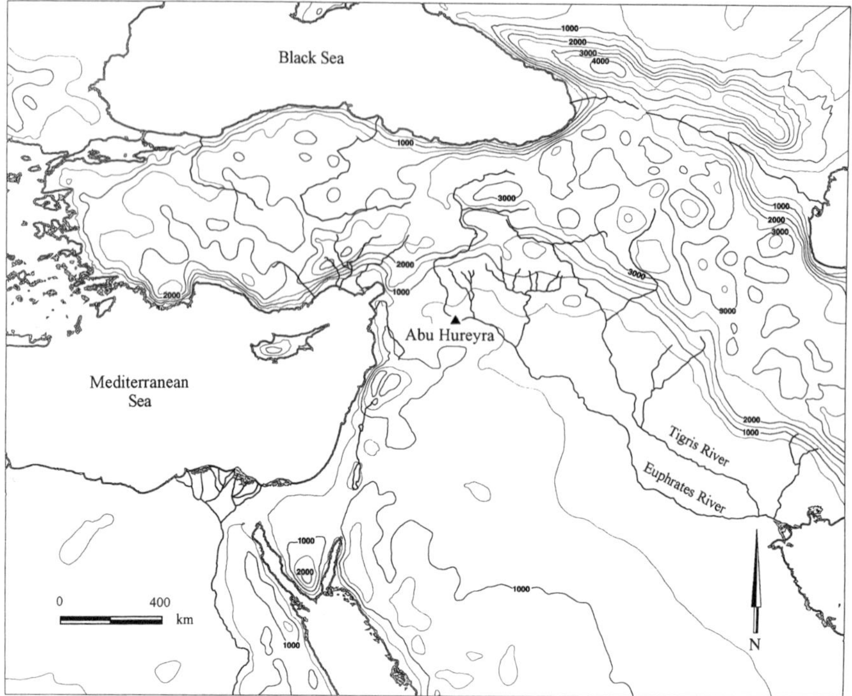

Fig. 4.1. The location of Abu Hureyra in Western Asia

discoveries, scientific advances, and personal interactions that have enabled us, after many years of research, to explain how farming developed there.

Abu Hureyra was inhabited from ca. 11,500 to 7,000 b.p. (uncal.), a sweep of years that spanned the transition from hunting and gathering to farming across Western Asia. The Abu Hureyra project began in 1971 when we conducted two brief surveys of the section of the Euphrates Valley in which the site was located. I was then a postgraduate student of 26 at Oxford preparing a doctoral thesis on the Neolithic of the Levant. In tracing the evolution of the project, it will be helpful to recall what we knew then about the beginning of agriculture in Western Asia.

Data were first systematically collected in the 1950s during the excavations by Kenyon (1957) at Jericho and the work by Braidwood and Howe (1960) in the foothills of the Zagros Mountains. Only in the 1960s were extensive excavations undertaken in Anatolia (Mellaart 1975). By 1970, we had a reasonably continuous record of cultural development from the

Epipaleolithic through the Neolithic in the Levant and Mesopotamia, but the sequence was still incomplete in Anatolia. It appeared that these three regions had played a significant role in the development of farming and village life, yet nothing was known of the connections between them because Syria, the region that linked the others, had hardly been explored.

What of the domestication of plants and animals? By 1970, the widely accepted view was that cereals and pulses had been domesticated around the Fertile Crescent. It was thought that Palestine had been an early center of cereal domestication and cultivation (Vaux 1966), while sheep and possibly goats were probably domesticated first in the Zagros (Mellaart 1967). Pigs and cattle could have been domesticated anywhere within or even beyond the core area of agricultural development. The general view was that farming had developed slowly in the early to mid Holocene and that many human groups had maintained a hunting and gathering way of life long after their neighbors had adopted farming.

These tentative interpretations rested on scanty evidence. Most of the early villages that had been investigated had yielded some animal bones. But these had rarely been collected systematically, and the samples were of modest size. The quantities of plant remains recovered from all investigated sites together were minuscule because flotation was just coming into use.

During the 1960s, most prehistoric archaeologists in Western Asia were preoccupied with cultural classification, as were many still in Europe. Kenyon (1979, chs. 2–3) developed a cultural sequence based on her work at Jericho that has since been widely adopted in the Levant. Braidwood and Howe (1960) did the same for Mesopotamia and the Zagros. This preoccupation with cultural classification has remained an important theme in archaeological research on the development of Epipaleolithic and Neolithic societies in Western Asia down to the present (see the *Tübinger Atlas des Vorderen Orients).* It has provided a useful framework for articulating the archaeological evidence, but has little to say about how an agricultural way of life may have developed.

The hypothesis put forward by Childe (1928) for explaining the Neolithic Revolution depended heavily on the notion that the climatic changes that had marked the transition from Pleistocene to Holocene in higher latitudes had had significant effects in Western Asia. Research in Western Asia after the Second World War, notably by Braidwood and Howe (1960, 181) and their colleagues, did not support this idea. For a long period, then, environmental change fell out of favor as a contributing factor in the development of agriculture. Braidwood himself preferred to invoke cultural influences to account for it. More significantly, he drew the attention of archaeologists and others to the existence of a "nuclear zone" in the "hilly flanks of the Fertile Crescent" where the wild ancestors of the domesticates were to be found today (Braidwood and Howe 1960). Assuming no significant climatic shifts in the transition from Pleistocene to Holocene, Braidwood thought that plants and animals would have been domesticated in this nuclear zone, and thus that we should seek the first farming villages there.

But in the late 1960s, new ideas were in the air. The publication of David Clarke's *Analytical Archaeology* in 1968 had a profound impact on students like myself who were then engaged in postgraduate study because it opened up new possibilities for interpreting the archaeological record of the human past. We were even more dissatisfied than students in their mid 20s usually are with the prevailing consensus on the development and spread of farming imparted to us by the senior generation of scholars. We younger archaeologists setting out for Western Asia from Oxford, London, and other universities were impatient to demonstrate that we could do new and different things, and ultimately arrive at fresh, more informative conclusions. We had yet to learn how long it would all take.

Much of Western Asia was open to archaeologists then, in contrast with the present day. We could travel in nearly every country and reach remote regions that had only recently become easily accessible through the construction of modern roads. Thus, it was possible to initiate research in regions that had hitherto not been systematically explored.

Archaeologists were introducing new fieldwork techniques and beginning to work with new analytical perspectives. A notable example was the British Academy Major Research Project in the Early History of Agriculture. My two founder collaborators in the Abu Hureyra Project, Tony Legge and Gordon Hillman, were both associated with the British Academy venture. Tony and other members of the British Academy team were developing new flotation techniques that were to be of fundamental importance in our work at Abu Hureyra. A few archaeologists like David French (1972, 181) at the British Institute of Archaeology in Ankara were experimenting with the concept of total recovery of artifacts and organic remains from archaeological sites, an approach that commended itself to us.

Much of this came together at Abu Hureyra. We planned to carry out a large-scale excavation there and to maintain meticulous records of stratification and context. We aimed for total recovery of artifacts and animal bones, and the largest possible samples of plant remains. The results have fully justified our efforts. We recovered the samples we needed to answer many of the questions that arose early in the investigation, and have been able to go back to the collections again and again to answer new questions as our research has proceeded.

The Abu Hureyra Project

Abu Hureyra is a very large (11.5 ha) site located on the Middle Euphrates bend, far outside Braidwood's nuclear zone. When we first saw it in 1971, the surface finds indicated that the site had been inhabited during the early Neolithic. Its size and considerable depth of deposit promised a long sequence of occupation stretching farther back in time.

We had gone to the Euphrates Valley at the invitation of the Syrian authorities to take part in the campaign of salvage archaeology that preceded the completion of a new dam. This was the first invitation that had been issued to a British team in over a decade, and we were acutely aware of the need to conduct the project in a manner that accorded fully with Syrian administrative norms. Our work at Abu Hureyra was a salvage exercise, and this imposed severe constraints on what we could accomplish. We managed just two seasons of work in 1972 and 1973, a total of six months' excavation, before the dam was finished and the site disappeared under water early in 1974. We planned at the outset to dig throughout each season and to recover as much material as we could in the brief time available. Detailed study of our excavated material and records would have to be postponed until after the excavation. We had to concentrate on our main objectives during the excavation, knowing that we could not return to the site later. The onset of the October War during the second season of excavation threatened to end our project before we had met even these limited goals. We were far from the front line, however, and so in no danger. With the support of the Syrian authorities, we continued to work, albeit with considerable difficulty, until we had completed our season as planned, the only project in the entire Middle East that remained in the field during that troubled time.

The project was truly collaborative and interdisciplinary from the beginning. Gordon Hillman as archaeobotanist and Tony Legge as archaeozoologist agreed to participate well before the excavation began. We have continued to work together on the Abu Hureyra project over the succeeding 29 years and are joint authors of the book describing the results, *Village on the Euphrates* (Moore et al. 2000). Such a sustained and productive collaboration over so long a period is perhaps unique in archaeology.

Gordon Hillman, Tony Legge, and I firmly believed in the analytical and interpretive value of an ecological perspective. Accordingly, we intended to study the site in its environmental context to understand what had happened there and, by extension, elsewhere during the period in which agriculture developed. This has remained a consistent theme throughout our work. We had two main aims from the beginning: to determine the cultural sequence of occupation on the site and to establish the changes that the inhabitants made in their economy. The first would be resolved through excavation and the study of the artifacts we recovered. The second required the collection and analysis of large quantities of plant remains and animal bones.

Our first aim was speedily accomplished by digging a series of trenches across the site that penetrated to the subsoil beneath. In the last few days of excavation at the end of the first season, we discovered that at the northern end of the site there was an earlier settlement of Mesolithic or Epipaleolithic affinities, Abu Hureyra 1, that underlay the Neolithic village of Abu Hureyra 2. This gave the project a new and special significance because it raised the possibility that we could compare conditions before and after the introduction of farming at the same site. The stratigraphic and cultural framework was essential to our understanding of the site but, once established, it ceased to be the main focus of our work.

There were details of the stratigraphic sequence, however, that have taken a long time to establish with finality. In trench E, the one trench in which we found superimposed deposits of Abu Hureyra 1 and 2, it appeared that there had been a hiatus between the two episodes of occupation. This seemed to be confirmed by the initial series of 13 radiocarbon dates on charcoal that we obtained in the late 1970s from the British Museum Laboratory (fig. 4.2). These gave a general chronology for the sequence of occupation and confirmed that the site was first settled before 11,000 B.P., but the dates could not tell us precisely how long the two settlements had been occupied. Moreover, they seemed to suggest that the hiatus in occupation between Abu Hureyra 1 and 2 could have lasted as much as 1,400 years. We had no more charcoal samples large enough to date using the conventional method so there

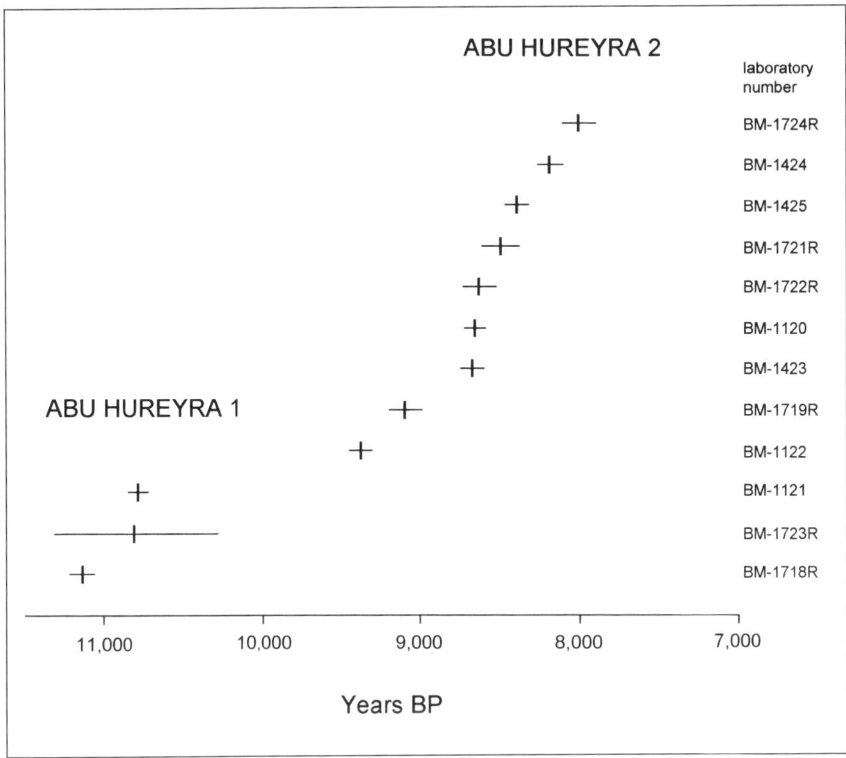

Fig. 4.2. Radiocarbon dates for Abu Hureyra from the British Museum Laboratory in chronological order

the matter had to rest. We accepted that, on the evidence available to us, there had probably been a gap in occupation between the two settlements.

This interpretation changed with the advent of accelerator mass spectrometry (AMS) radiocarbon dating. The Research Laboratory for Archaeology and the History of Art at Oxford was a pioneer in the development of this new technique. They approached us in the early 1980s to see if we could provide samples for dating, at first on an experimental basis. We had plenty of charred seeds and charcoal from our flotation samples as well as animal bones that could not be dated by conventional means but were appropriate for AMS dating. The results were immediately encouraging. Where we could compare the Oxford AMS dates directly with those from the British Museum laboratory, they matched.

We have since obtained over 50 AMS dates for Abu Hureyra in a dating program that has continued to the present, transforming our understanding

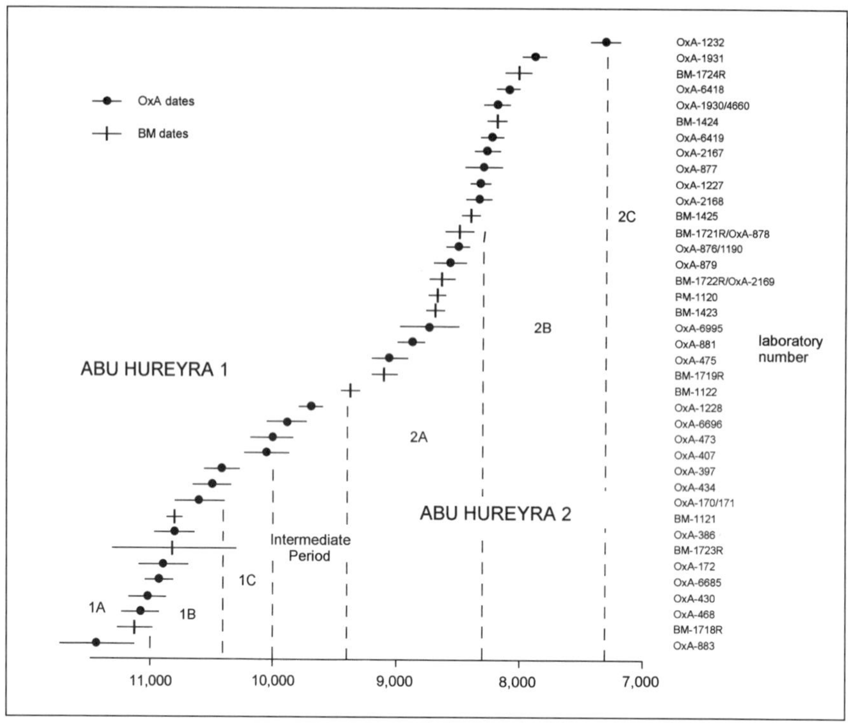

Fig. 4.3. Selected radiocarbon dates for Abu Hureyra from the British Museum (BM) and Oxford (OxA) laboratories in chronological order

of the chronology of the site (Moore 1992). The AMS dates have confirmed the long sequence of occupation, beginning about 11,500 B.P. and ending some time after 7000 B.P. (fig. 4.3). Most importantly, they have also demonstrated that occupation was continuous with no hiatus between Abu Hureyra 1 and 2. Abu Hureyra 1 was inhabited from ca. 11,500 to 10,000 B.P. There followed an intermediate episode of occupation from ca. 10,000 to 9400 B.P. Then the village of Abu Hureyra 2 developed from ca. 9400 to after 7000 B.P. We had found traces of occupation deposit at the bottom of trench G that contained organic material. Fortunately, we had passed all of it through the flotation machines. In the deposit we found domestic cereal grains, other charred seeds, and also some twig charcoal. The date from this twig charcoal is 9680 ± 90 B.P. (OxA-1228), confirming that we did actually excavate a little deposit from the Intermediate period. Of course, much more probably lay nearby in areas that we did not have time to dig.

The AMS dates have not only confirmed the continuity of occupation at Abu Hureyra across the Pleistocene-Holocene boundary and from Epipaleolithic to Neolithic, but have also demonstrated an uninterrupted sequence from hunting and gathering through the early stages of farming to a developed farming way of life. None of this could have been anticipated at the outset of the work, since we could not have known then that the site contained such a record. And, indeed, we have been able to establish these facts only through prolonged study of the material we recovered from the site, using every methodological and technical advance as it has become available.

The key to all this lay in our decision to aim for near total recovery of animal bones, plant remains, and other organic material from the site during the excavation. We dry sieved all the excavated soil, an exercise that yielded over two metric tons of animal bones as well as huge quantities of small artifacts. And we passed large samples of soil, as much as four wheelbarrow loads (200 lt), from each level dug from the main trenches through flotation machines. Such a large-scale application of flotation was unprecedented. It yielded well over 500 lt of carbonized plant material for later study, an extraordinary amount, certainly, but we have since learned barely enough to answer all the questions that our research has raised.

The Economy of Abu Hureyra

At the conclusion of the excavation in 1973, we thought that the settlement of Abu Hureyra 1 had probably been inhabited by hunters and gatherers, while Abu Hureyra 2 was a village of farmers. But Gordon had already found a few grains of wild einkorn and other cereals in the flotation samples from Abu Hureyra 1 and identified many more as his first serious examination of the plant remains got under way in 1974. These cereals were far outside their present-day distribution, raising the possibility that they might have been cultivated. We mentioned this in our preliminary report (Moore 1975). If confirmed, it would push the beginning of farming back into the Epipaleolithic and so into the late Pleistocene.

The question of whether or not we had evidence for early farming in Abu Hureyra 1 has become the most important issue that has preoccupied us throughout our ensuing research. The arguments have gone back and forth among us as our investigations have proceeded and new evidence has come to light, both from the Abu Hureyra samples themselves and ancillary field studies. But it is only in the last few years that the matter has been resolved definitively.

Gordon Hillman presents the latest results of our analysis of the plant remains in more detail in his contribution to this volume (ch. 5), but it will be useful to summarize the economic sequence at Abu Hureyra here (fig. 4.4). The environment of the Euphrates Valley and adjacent undulating plain was highly favorable for hunters and gatherers toward the end of the Pleistocene. The land beyond the valley supported rich open woodland and grassland steppe, while the valley itself was filled with a dense forest of trees, vines, and marsh plants. Both the valley and the steppe supported an abundance of game. A group of hunter-gatherers founded the sedentary settlement of Abu Hureyra 1 about 11,500 B.P., choosing a spot that was on a gazelle migration route. They killed these animals in large numbers during the spring migration and kept much of the meat for consumption later in the year (Legge and Rowley-Conwy 1987). This continued until well into Abu Hureyra 2 times. The bulk of their food, however, came from a multitude of wild plants, some of which were staples.

A fundamental change took place ca. 11,000 B.P. with the onset of the Younger Dryas. This cool, dry climatic episode, which lasted 1,000 years, had a dramatic effect on the environment across Western Asia just as it did elsewhere in the world. It disrupted the gathering activities of the inhabitants of the village of Abu Hureyra 1 and was a major factor in their decision to adopt farming ca. 11,000 B.P.

How did we make the connection between the Younger Dryas and the vegetation record from Abu Hureyra? Gordon had now published an extensive analysis of the plant remains from Abu Hureyra 1 (Hillman et al. 1989, fig. 14.1). His diagram clearly indicated that there had been a major change in vegetation soon after 11,000 B.P. with a decline in open woodland species, especially trees, wild cereals, and wild legumes. We discussed this at some length but could not for the time being provide a satisfactory explanation for it.

In the same year I attended the Valbonne conference on the late Epipaleolithic of the Levant and saw briefly Baruch and Bottema's slide of a pollen diagram they had recently prepared from a new core made in the Huleh Basin. Later, in 1991 while writing the chapter on the environment of Abu Hureyra for our own book, I asked the editors of the Valbonne volume (Bar-Yosef and Valla 1991), then in press, for a copy of this pollen

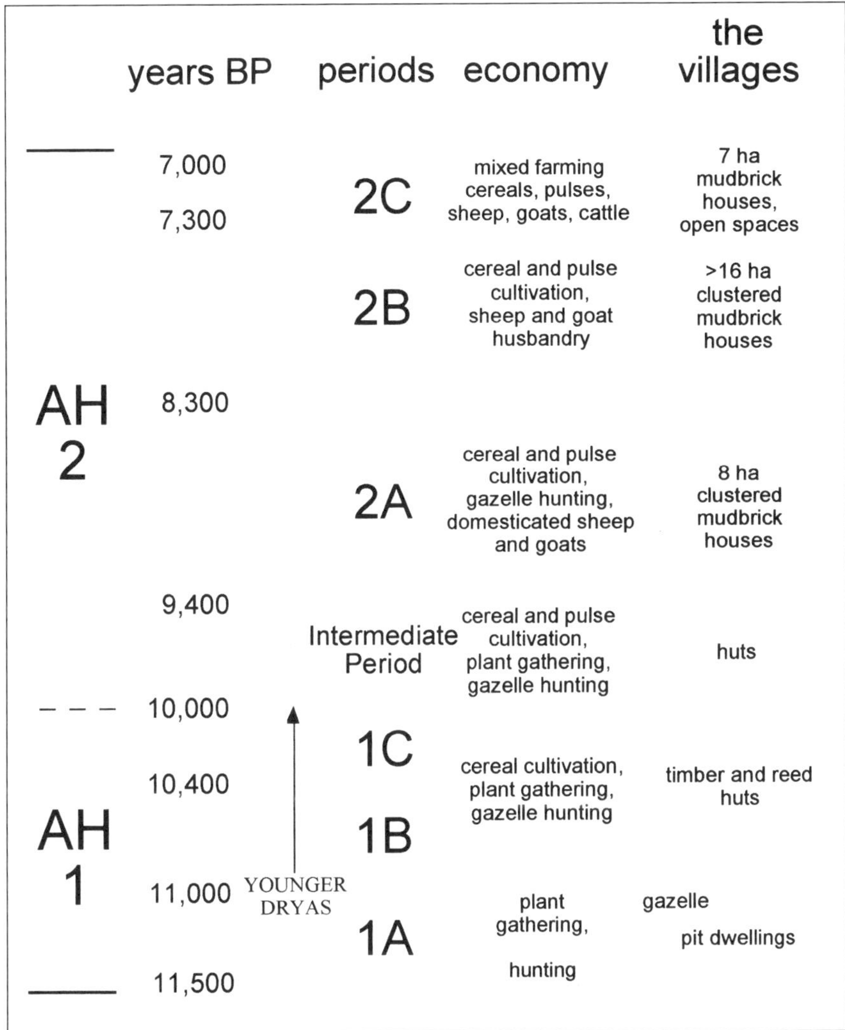

years BP
periods
economy
the villages
7,000
7,300
2C
mixed farming cereals, pulses, sheep, goats, cattle
7 ha mudbrick houses, open spaces
2B
cereal and pulse cultivation, sheep and goat husbandry
>16 ha clustered mudbrick houses
AH 2
8,300
2A
cereal and pulse cultivation, gazelle hunting, domesticated sheep and goats
8 ha clustered mudbrick houses
9,400
Intermediate Period
cereal and pulse cultivation, plant gathering, gazelle hunting
huts
10,000
1C
10,400
cereal cultivation, plant gathering, gazelle hunting
timber and reed huts
AH 1
1B
11,000
YOUNGER DRYAS
1A
plant gathering,
gazelle
hunting
pit dwellings
11,500

Fig. 4.4. The sequence of occupation at Abu Hureyra showing the principal characteristics of the settlement and its economy in successive periods

diagram. When I received the diagram, I could see at once that it showed a sharp retreat of forest that had lasted a little over a millennium. The beginning of this retreat was dated to ca. 11,500 B.P., but it clearly marked the same episode as the one Gordon had identified in the vegetation sequence at Abu Hureyra. Scientific journals in the early 1990s were full of references to the worldwide impact of the Younger Dryas, and it became clear to Gordon and me, as well as to Baruch and Bottema (1991, 17), that this climatic reversal had caused the abrupt changes in vegetation to be seen in the Abu Hureyra plant remains and the Huleh core.

Gordon and I published our interpretation of this evidence in 1992 (Moore and Hillman 1992), and linked it to the beginning of farming in Western Asia. But we still needed confirming evidence from Abu Hureyra that agriculture really had begun there coincident with the onset of the Younger Dryas. That came later.

Gordon had identified a very few domestic-looking cereal grains from Abu Hureyra 1 early on in the research, but there were doubts over the identifications. And he was fully aware that such grains might have filtered down from overlying Abu Hureyra 2 levels. Then in the early 1990s, he and Susan Colledge found a few more. By that time, we knew that we could date such grains directly by AMS and so prove whether or not they were really as early as the deposits in which they were found. Beginning in 1995, we asked Susan to go back through every one of the flotation samples from Abu Hureyra 1, some of which had not been sampled before, and to select out any grains that looked domestic—a huge task that took her many months. The total firmly identified to date (March 2001) stands at over 45 grains of domestic wheat, barley, and rye. A few of these were large enough to be dated by the Oxford Laboratory. In March 1997, we learned that the dates for two of the grains, both of domestic rye, were early. One grain, 10,930 ± 120 B.P. (OxA-6685), dated from the onset of the Younger Dryas and was contemporary with the vegetation changes we had already noted in the Abu Hureyra 1 sequence. The other, 9860 ± 220 B.P. (OxA-6996), dated from the Intermediate Period that followed Abu Hureyra 1, at the beginning of the climatic amelioration that followed the Younger Dryas. Taken together, and in the light of much other supporting evidence that Gordon illuminates in his chapter, these two dates indicate that the cultivation of domestic rye began very early at Abu Hureyra (during the late Pleistocene), and that the catalyst for the adoption of cultivation was probably the climatic changes brought on by the Younger Dryas.

The Oxford Laboratory has since obtained additional confirming dates on grains of domestic rye from Abu Hureyra 1 levels (Moore et al. 2000, app. 1). Evidence now suggests that the cultivation of rye and perhaps other cereals began ca. 11,000 B.P., early in the occupation of Abu Hureyra 1. Thereafter, the cultivation of domestic cereals and pulses contributed increasingly to the food supply in the later centuries of Abu Hureyra 1, through the Intermediate period, and into Abu Hureyra 2.

The inhabitants of the settlement kept small flocks of domestic sheep and goats at least as early as the beginning of Abu Hureyra 2, ca. 9400 B.P. Thus, these two species were present in domesticated form almost from the start of the Holocene in North Syria. Then about 8300 B.P., a date we have determined quite precisely, the villagers sharply reduced their dependence on hunted gazelle and expanded the size of their flocks of sheep and goats. We believe that this change was probably the result of an abrupt decline in the size of the gazelle herds following over hunting by communities along their migration route. Thereafter, during the eighth millennium B.P., a mature Neolithic economy developed at Abu Hureyra based on the cultivation of a variety of cereals and legumes and the herding, not only of sheep and goats, but also of cattle and pigs.

This sequence was marked by sudden changes followed by long periods of adjustment. This step-by-step process was dynamic, yet the full Neolithic economy took a long time to develop. The evidence from Abu Hureyra demonstrates that the beginning of agriculture had its roots well back in the hunting and gathering practices of Epipaleolithic communities and that the adoption of farming was itself a late Pleistocene event. The sequence of development from foraging to a mature farming way of life was continuous at Abu Hureyra and perhaps at a few other sites in Western Asia, for example, Jericho. Elsewhere, the onset of the Younger Dryas and the later spread of farming throughout the region during the early Neolithic led to widespread disruption in patterns of settlement (Moore and Hillman 1992, 491; Moore 1985, 60). It is important to recognize, too, that not only were the first cereals domesticated by 11,000 B.P., but animals were domesticated by the 10th millennium B.P. in Northern Syria, a quite different scenario than the one envisaged when we started work. And, finally, it is clear that environmental change, and specifically the rapid fluctuations that accompanied the transition from Pleistocene to Holocene, were key factors in triggering the switch from foraging to farming. There are lessons here for our colleagues who are tracing the spread of farming into Europe and are seeking to understand the processes that underlay it. We have learned, too, that local factors such as the setting of a site and shifts in vegetation zones were often of compelling significance in the course of events, an observation that surely resonates with our colleagues working on the European Mesolithic and Neolithic.

The Partnership

The three of us—Tony, Gordon, and myself—have been working together on the Abu Hureyra project since its inception in 1971. Many others have joined us as collaborators along the way. The title page of *Village on the Euphrates* carries nine additional names, and many more people have helped carry the research forward at different times. It is but rarely that collaboration in a single archaeological or indeed any other project can be sustained over so long a period, and we have been asked how it has been accomplished.

The most important single factor, I would argue, is the unique importance of the site and its record. It is this that has repeatedly brought us back to the task of analysis and interpretation, even as other distractions, professional and personal, have intervened. It has also been the most compelling factor for our many other collaborators.

But there are other, more personal, influences that help to explain how Tony, Gordon, and I have continued to work together for so long. The answer is not to be found in any uniquely harmonious blend of personalities, for we each possess very different temperaments. But, as with partners in a successful marriage, we have come to understand each other extraordinarily well, have built good, flexible working relationships and, above all, have learned to be patient. We are, in short, good friends, but not the best of friends. Thus, each of us has always been able to work with the other two.

Of growing importance over the years has been our increased respect for each other's scientific contributions to the project. Tony's detailed delineation of the stages of animal domestication based on a massive study of the huge sample of animal bones carried out over many years has transformed our understanding of this process, not only at Abu Hureyra but across Western Asia. Tony's discovery, with Peter Rowley-Conwy, of the special significance of gazelle at Abu Hureyra has been of great interpretive value. Knowing that the inhabitants over many generations depended for much of their meat supply on the mass slaughter of these animals during their spring migration, and that this pattern persisted long after the transition from foraging to farming, has enabled us to think about domestication and agricultural development in new ways. Gordon's remarkable reconstructions of changing vegetation patterns around Abu Hureyra from the Late Pleistocene to the present have provided the environmental context for much that happened there, and have made extraordinary contributions to our understanding of the site. And his discovery of evidence for cultivation of domestic rye at such an early date has at last settled one of the most difficult questions we have faced, and furthermore has fundamentally altered our conception of the beginning of agriculture in Western Asia.

Finally, because the project has been a truly interdisciplinary one, our research has been interdependent. We have each needed the contributions of the other two to reach our conclusions. None of us could have carried the project alone.

Looking Forward

What might we hope to achieve in future work on the problems of agricultural development in Western Asia? Serious gaps remain in our knowledge of the sequence of development from Epipaleolithic to Neolithic in, for example, southeast Turkey, where the recent excavations by Rosenberg (1994) have demonstrated the potential significance of this region. Similarly, we need to find out much more about what was going on in Northern Mesopotamia, Syria outside the salvage areas where work has been concentrated in recent years, and in Anatolia.

A second, and surely obvious, point is that if we wish to examine the transition from foraging to farming, then we need to recover appropriate evidence: that is, remains of the plants and animals that were the fundamental constituents of both ways of life. This requires saving very large quantities of such material from excavations—much bigger samples than are normally retained.

And, finally, given that environmental change was so important in the transition from foraging to farming, it follows that an ecological approach to the problem, informed by research on the accompanying social changes, will continue to yield the most productive results. We hope that those extensive areas of Western Asia at present closed to archaeological research may open once again so that field investigations of this fundamentally important subject may continue.

REFERENCES

Baruch, U., and S. Bottema. 1991. "Palynological Evidence for Climatic Changes in the Levant ca. 17,000–9,000 B.P." In *The Natufian Culture in the Levant*, edited by O. Bar-

Yosef and F.R. Valla, 11–20. Ann Arbor.

Bar-Yosef, O., and F.R. Valla, eds. 1991. *The Natufian Culture in the Levant.* Ann Arbor.

Braidwood, R.J., and B. Howe. 1960. *Prehistoric Investigations in Iraqi Kurdistan. Studies in Ancient Oriental Civilization* 31. Chicago.

Childe, V.G. 1928. *The Most Ancient East.* London.

French, D.H. 1972. "Excavations at Can Hasan III 1969–1970." In *Papers in Economic Prehistory,* edited by E.S. Higgs, 181–90. Cambridge.

Hillman, G.C., S.M. Colledge, and D.R. Harris. 1989. "Plant-Food Economy during the Epipalaeolithic Period at Tell Abu Hureyra, Syria: Dietary Diversity, Seasonality, and Modes of Exploitation." In *Foraging and Farming,* edited by D.R. Harris and G.C. Hillman, 240–68. London.

Kenyon, K.M. 1957. *Digging up Jericho.* London.

———. 1979. *Archaeology in the Holy Land.* 4th ed. London.

Legge, A.J., and P.A. Rowley-Conwy. 1987. "Gazelle Killing in Stone Age Syria." *Scientific American* 255 (8):88–95.

Mellaart, J. 1967. "The Earliest Settlements in Western Asia." *Cambridge Ancient History* 1. Rev. ed. Fascicle 59. Cambridge.

———. 1975. *The Neolithic of the Near East.* London.

Moore, A.M.T. 1975. "The Excavation of Tell Abu Hureyra in Syria: A Preliminary Report." *Proceedings of the Prehistoric Society* 41:50–77.

———. 1985. "The Development of Neolithic Societies in the Near East." *Advances in World Archaeology* 4:1–69.

———. 1992. "The Impact of Accelerator Dating at the Early Village of Abu Hureyra on the Euphrates." *Radiocarbon* 34:850–8.

Moore, A.M.T., and G.C. Hillman. 1992. "The Pleistocene to Holocene Transition and Human Economy in Southwest Asia: The Impact of the Younger Dryas." *American Antiquity* 57:482–94.

Moore, A.M.T., G.C. Hillman, and A.J. Legge. 2000. *Village on the Euphrates.* New York.

Rosenberg, M. 1994. "Hallan Çemi Tepesi: Some Further Observations Concerning Stratigraphy and Material Culture." *Anatolica* 20:121–40.

Vaux, R. de, O.P. 1966. "Palestine during the Neolithic and Chalcolithic Periods." *Cambridge Ancient History* 1. Rev. ed. Fascicle 47. Cambridge.

— 5 —

Investigating the Start of Cultivation in Western Eurasia: Studies of Plant Remains from Abu Hureyra on the Euphrates

Gordon Hillman

Curiosity over gaps in our understanding of the evolution of wheat was, perhaps, a somewhat trivial reason for choosing to undertake research on a subject as far-reaching in its implications for human history as that of the origins of agriculture. But such was my initial prompting.

As students in the Department of Agricultural Botany at the University of Reading (U.K.) during the 1960s, our teachers had inspired us with the sheer beauty of evolutionary dynamics as expounded by giants such as Stebbins (1950, 1974) and Grant (1958, 1963), with the tragically unfinished work of Vavilov, and with the elegance of Riley's genomic analyses of wild wheats in search of the elusive donor of the B genome (Riley 1965). But in studies of the evolution of wheat, there was much that could not be understood without more information from archaeology concerning, for example, the precise ecological and agronomic circumstances associated with cereal domestication in Southwestern Asia. Our teachers had introduced us to the many advances in this area achieved by Hans Helbaek, the renowned Danish archaeobotanist, but it was clear that much remained to be done. And for a young agricultural botanist keen to research the evolution of wheat and with interests in prehistory, the possibility of collaborating with archaeologists in further studies of ancient remains of crops from early agrarian sites in Southwest Asia was extremely attractive.

A few years earlier the Head of Department, Hugh Bunting, had arranged for Hans Helbaek to be awarded an honorary doctorate and encouraged him (in vain) to join the department in the hope that this would allow him to devote more time to studying plant remains from archaeological sites in Southwestern Asia. It was similarly thanks to Hugh's energy and encouragement that, in 1969, I embarked on a doctoral program loftily devoted to "the origins of agriculture in Anatolia." This was to involve a study of plant remains meticulously recovered in huge quantities by David French from his excavations at the Aceramic Neolithic site of Can Hasan III in central Turkey, using his revolutionary system of large-scale flush-flotation. The work on the plant remains would be undertaken at the British Institute of Archaeology at Ankara (BIAA), following a preliminary year of research training in archaeobotany with Maria Hopf in Mainz, Germany, who was then working on the plant remains from Jericho (Hopf 1983). Now, 30 years later, it is clear that, without the resulting five years living and working in Turkey, it would have been difficult to lay proper foundations for those background studies of Southwest Asian plant ecology that eventually allowed us to identify the changes in environment seemingly responsible for triggering the start of cereal cultivation at Epipaleolithic Abu Hureyra.

In embarking on the studies of the Can Hasan III plant remains, however, it was apparent that no serious analyses could begin until there was a comprehensive reference collection of modern seeds, including a wide range of forms of the archaic crops that were still under cultivation. En route to Ankara in 1969, I therefore visited every Turkish institution likely to house seed reference collections. By the time I reached Ankara, it was apparent that I would have to collect my own. What I did not then realize was how many years it would take.

It had been clear from the outset that recognition of the ecological significance of the many non-cultigens present in the remains would require an intimate knowledge of the local flora and its ecology. I already had a good knowledge of the British flora from a childhood spent running wild in the woodlands, marshes, and downlands of the Sussex Weald, and from teenage years spent actively exploring these and other habitats prior to working as a field assistant on the U.K. Nature Conservancy's high moorland reserve in the Northern Pennines. I was also closely familiar with the flora of several parts of mainland Europe from having worked for almost five years in the European Herbarium of London's Natural History Museum under

Alexander Melderis, the Latvian grass taxonomist. But while these experiences provided useful foundations, more specific knowledge was clearly needed.

It was also a matter of embarrassment to me that, although I had been raised in arable and dairy farming and in my family's nursery-gardening firm, I had no specific knowledge of traditional systems of agriculture and horticulture in arid-zone Southwestern Asia. It would clearly be presumptuous to attempt to interpret plant remains in terms of ancient agricultural systems without some familiarity with survivors of such systems in the present day. In my naïveté, I assumed that all archaeologists involved in the study of early agriculture routinely began by mastering such knowledge.

That same summer spent as part of the remarkable Aşvan Project (French 1972, 1973)[1] allowed me to begin rectifying all three of these deficiencies immediately, as well as to start learning a little of archaeological excavation and French's uniquely efficient system of flush-flotation (French 1971). The village of Aşvan, on a tributary of the Upper Euphrates, was paradise on earth, and several summers spent there allowed the background studies to proceed apace. Firstly, the surrounding area had a rich and diverse flora, so it was possible to start assembling a seed reference collection and simultaneously to familiarize myself with parts of the Turkish flora and its ecology, including the local crops (most of which were complex land-races) and the stands of wild cereals that I discovered growing up in the nearby Munzur Mountains. Secondly, agriculture at Aşvan and in neighboring Kurdish villages was totally nonmechanized. Most of the land-race elements and the technology applied to them were probably millennia old, as, seemingly, was the village architecture and perhaps even aspects of the social organization. The hospitable villagers were also more than happy to share their expertise with an ignorant stranger, even one who had only just begun learning Turkish.

Living in this present-day Iron Age village of Aşvan was my real introduction to archaeology. Seeing the myriad activities of daily life and the resulting accumulation of tell-tale debris in the living areas and store rooms, in the courtyards and sheep pens, on the threshing yards, around and within the diverse forms of bread oven, on the ash middens, beside the eroding mudbrick walls, and entangled in the piles of stored straw, I blithely assumed that observations of living archaeology of this sort were what provided all archaeologists with their means of interpreting excavated remains in terms of possible ancient lifeways. So I immediately started devising strategies for assembling an explicitly defined interpretive model (of the sort I assumed to be standard throughout archaeology) to allow me to interpret my ancient remains of crops and their weeds in terms of past agrarian activities—on a testable and scientifically repeatable basis.

All that was necessary was to get the villagers to explain the sequence of operations they applied to each crop; record details of the activities involved in each operation; take samples of the products and by-products at each step; analyze their composition; and explore the pattern of correlation between the composition of products and the operations that had generated them. If systematic correlations were found, then, given archaeological samples of equivalent composition, I would be able to suggest (as a working hypothesis) that they had been generated by similar activities applied in the past. Perhaps it would also be possible to deduce something of the distribution of these inferred past activities relative to excavated structures, and for some of these structures, even to suggest some of their past functions. Four seasons of such studies at Aşvan and in nearby villages produced a wealth of data and the beginnings of a coherent ethnographic model (Hillman 1972, 1973).

It was only after years of living and working with archaeologists that I finally realized that, although most archaeological interpretation was dependent ultimately on some sort of ethnographic analogy, it rarely used explicitly defined ethnographic models applied in a testable and scientifically repeatable fashion, and that, in Old World archaeobotany, not only were no such models in use, it appeared that none had ever been formulated. The impressive and very innovative interpretive systems of Robin Dennell (1972, 1974) were, I eventually learned, based on elegant sieving experiments combined with extrapolations from assumed functions of excavated structures, not on ethnographic observation. Follow-up studies in other areas of Anatolia, especially in areas still growing glume wheats such as emmer, eventually allowed the formulation of models that now offer a partial basis for tracing the emergence of specific forms of husbandry and crop processing (Hillman 1981, 1984), including perhaps some of those applied during the early phases of cultivation (Harris 1984). The work also served to encourage equivalent studies elsewhere, with Glynis Jones, in particular, taking the subject into altogether higher realms of statistical sophistication that allowed more explicit hypothesis testing and greater scientific repeatability.[2]

Although the Aceramic Neolithic site of Can Hasan III was insufficiently early to provide evidence of the first phases of cultivation in the Near East, work on its plant remains provided the impetus to tackle a number of problems that would later beset us (in even more extreme form) at yet earlier sites such as Abu Hureyra. Principal among these problems were the inadequacies of the criteria used to identify remains of ancient food plants, particularly in taxonomically critical groups such as the wheats. It quickly became clear that many of the criteria used by archaeobotanists were based on inadequate research or on modern reference specimens obtained from botanic gardens where many of the cereals had been misidentified, mislabeled, or allowed to hybridize. We therefore had to start the very time-consuming process of isolating new criteria, using specimens collected direct from the field and identified with due rigor.[3] It was equally clear that, when identifying badly fragmented remains of seeds and fruits from very early sites, we would have to use features of *internal anatomy*. However, most such features were (and remain) largely unknown to most archaeobotanists, and most of them would need to be researched from scratch. Overall, the work would be very time consuming. But it could not be ducked. If plant remains are misidentified, even the most elegant interpretations based on them are worthless.[4]

Each of these areas of background study, together with assembling a comprehensive seed reference collection, had to be completed before research could begin in earnest on the plant remains from Can Hasan III. There was no option but to stay in Turkey some years longer than first planned, find funding for the extra years, and put the doctoral studies on hold. The five years spent in Turkey also ensured exposure to a wealth of archaeological ideas through David French and the (literally) hundreds of archaeologists and anthropologists who visited the BIAA. With French's evangelical zeal for the "new archaeology," every meal, every tea break, every excursion, turned into a seminar, often lasting deep into the night, and sometimes continuing for days.

During this period, a winter sojourn at the BIAA by Andrew and Barbara Moore had led to my agreeing to join their planned excavations at Abu Hureyra and to analyze the seed remains. Andrew ensured that we had ample opportunity to agree on sampling, recovery, and other on-site strategies during the year preceding the start of excavation in 1972, including strategies such as preliminary on-the-spot analyses of flots to assess sample richness and allow immediate adjustment of sample size. Would that prior negotiations of this sort were standard practice!

As Andrew Moore describes in his contribution to this volume (ch. 4), he encountered Epipaleolithic levels in the final hours of the first season. Examination of the first of the resulting flots back in Ankara revealed that wild-type einkorn wheat (or something closely resembling it) was present, just as at the nearby site of Mureybit (van Zeist and Bottema 1968; van Zeist and Bakker-Heeres 1986). We therefore faced precisely the same dilemma as van Zeist and his colleagues, namely that the wild-type einkorn could have come either from foraging wild stands (whether local or distant) or from "pre-domestication cultivation." But even to begin to resolve the matter, it would be necessary to know (a) how close to Abu Hureyra any stands of wild einkorn could have grown, and (b) how long the process of domestication would have taken (i.e., whether prolonged periods of pre-domestication cultivation were possible or even probable). The relevant studies had never been undertaken, so in 1973, I started field experiments in stands of wild einkorn in east and central Turkey to measure domestication rates. The 1973 season at Abu Hureyra also saw the search for habitats within ca. 50 km of Abu Hureyra capable of supporting wild stands of einkorn. We found no trace of them.

Meanwhile, analysis of some of the Epipaleolithic samples from Abu Hureyra had revealed a broad spectrum of wild food plants and possible weeds of cultivation (segetals) whose ecology had never been studied adequately. The field season at Abu Hureyra in 1973 was also used to continue the background studies started in 1972. In particular, we targeted the ecology of these segetals and wild food-plants both in the Euphrates Valley and in the steppe hinterland right down to the frontier with Iraq. Before I left Turkey in 1975, I completed a series of extensive field studies at locations across the Anatolian Plateau focusing especially on the ecology of the steppic feather-grasses (*Stipa* spp.) and several of the other wild food plants and weeds found in large numbers at Abu Hureyra. These and subsequent ecological studies have proved invaluable during the ensuing 24 years of research.

After leaving Turkey and joining the Department of Botany at University College Cardiff (Wales), teaching commitments in crop evolution and in plant taxonomy, combined with local research on Welsh archaeobotany, ensured that progress on the analysis of the Abu Hureyra remains was piecemeal. Nevertheless, it proved possible to complete a study on criteria for identifying wild and domestic ryes (Hillman 1978), which subsequently proved crucial for the work at Abu Hureyra (Hillman et al. 1993).

A return to full-time research on the Abu Hureyra plant remains came only when David Harris suggested that I bring the botanical side of the Abu Hureyra Project to his Department of Human Environment at the Institute of Archaeology in London in 1981, and that we apply for state funding to employ a research assistant for three years. We duly appointed Sue Colledge, an experienced archaeobotanist. The funding also allowed an SEM-based project, undertaken by Sue Colledge, to explore the potential of alternative indicators of domestication (discussed below), and another major round of ecological fieldwork in Syria in 1983 (done together with Sue Colledge, David Harris, Tony Legge, and Peter Rowley-Conwy, which targeted specific questions that had arisen from the analysis of the Abu Hureyra remains during the preceding decade).

For many of the wild food plants found in the Epipaleolithic levels at Abu Hureyra, this new phase of the work also allowed us to expand our field experiments on (a) methods of harvesting the various edible seeds, roots, and so forth; (b) the problems of bulk-processing them, especially those likely to have served as caloric staples and/or requiring detoxification; and (c) ways of preparing them as food. The starting point for most of our experiments was ethnographic accounts of equivalent practices among hunter-gatherers who used similar resources in recent times. We also organized analyses of the nutrient status of a number of these wild foods, together with the study of any toxins present both before and after detoxification procedures and their preparation as food.[5]

Each aspect of these studies produced new information, which required us repeatedly to re-evaluate our ideas on early subsistence at Abu Hureyra. Much of the new information impinged on the question of whether cultivation of cereals and legumes had already started during the Epipaleolithic. It is therefore appropriate to outline how and why our ideas on this central subject have changed over the past 25 years.

Changing Our Minds about Epipaleolithic Cultivation

From the outset of the Abu Hureyra Project, we were aware of the possibility that the wild-type cereals, at least, might already have been under pre-domestication cultivation from some point during the Epipaleolithic. More specifically, we had offered this as one of three possible explanations for the presence of morphologically wild-type cereals at both Abu Hureyra and Tell

Mureybit where the then climatic evidence suggested that natural stands of wild cereals could not have grown locally in quantities that would have allowed them to make a significant contribution to caloric need (Moore et al. 1975).

But even if morphologically wild-type cereals were already under cultivation, it was still uncertain precisely how long it would have taken for them to have become domesticated (i.e., how many crop generations would have been required for selective pressures—resulting from cultivation—to advantage domestic-type mutants such that they eventually came to dominate the crops). This uncertainty had far-reaching implications. For example, if domestication had been very rapid, then we would be unlikely to encounter any cases of pre-domestication cultivation in archaeological remains, and this would suggest that the wild-type cereals from Epipaleolithic Abu Hureyra had probably been gathered from wild stands. If, however, domestication required, for example, 1,000 years, then the emergence of morphologically fully-domesticated cereals by 9800–9600 B.P. at Tell Aswad, Jericho, and in the Neolithic levels at Abu Hureyra would have required cultivation to have started by ca.10,800 B.P., and there would be a chance of eventually encountering archaeological remains of wild-type (or partially wild-type) crops from some point during the prolonged period of pre-domestication cultivation at one or more sites in the region. In that case, Epipaleolithic Abu Hureyra could theoretically have been one such settlement where cultivation of wild-type cereals was already under way. This was clearly a hypothesis that we had to test exhaustively.

The only way to resolve these uncertainties was to measure domestication rates experimentally. In essence, this would involve field measurements (both in wild stands and in sown crops of wild-types) of the selection pressures that are produced when wild-type crops are exposed to those husbandry practices likely to have been applied by the first farmers.[6] I completed the first round of field trials in Turkey in 1973–1974, and the second one with the geneticist Stuart Davies in 1979–1980 (in Wales at University College Cardiff's Department of Botany, where both of us then worked and where the necessary large glasshouse facilities were available). We then incorporated our measurements of selection pressure into mathematical models describing the frequencies of recessive alleles in a series of populations with different levels of outbreeding. Much to our surprise, the models indicated that, once cultivators started using sickles or uprooting to harvest the wild cereals, the crop was likely to have become dominated by domesticates with non-shattering ears in about 200 years, and even within 25 years, given less probable circumstances (Hillman and Davies 1990, 1992). Thus, the appearance of domesticates early in the PPNB (as at Tell Aswad and Jericho; van Zeist and Bakker-Heeres 1979; Hopf 1983) did not, of itself, necessarily make it inevitable that cultivation in the region had started back in the Epipaleolithic. Indeed, prolonged cultivation involving the harvesting regime specified was unlikely ever to have taken place without morphologically recognizable domesticates soon becoming manifest in the remains.

While these domestication experiments were under way, we simultaneously launched a series of studies to isolate much more subtle markers of the domestication syndrome, again with a view to resolving the question of whether cultivation of caloric staples was initiated already during the Epipaleolithic. More specifically, we hoped to isolate heritable features that would have emerged under cultivation even more rapidly than features such as non-shattering ears or fat grains, and that would therefore allow us to detect the start of cultivation well before the manifestation of these two classic markers of domestication. I initiated a legume-based study with one of my Cardiff students (Stocks 1979) to look for SEM-observable anatomical features linked to loss of seed-dormancy, which theoretically should have become manifest in the crop population within a year of the start of cultivation. If such anatomical features survived in charred archaeological specimens, they could, in theory, provide evidence of cultivation in the remains of seeds harvested from the very first crops ever sown by the settlement concerned. Thereafter, this work on legumes (and parallel work on cereals) continued at the Institute of Archaeology in the capable hands of Ann Butler, Sue Colledge, and eventually Frances McLaren and Michelle Cave. Within a few years, we had explored exhaustively the potential of a wide range of both histological and chemical markers that we hoped might distinguish cultivated wild-types from truly wild forms gathered from wild stands (Butler 1989, 1990, 1991; Cave 1989, 2000; Colledge 1988; Hillman et al. 1982, 1989a, 1993; McLaren et al. 1990).

But even before we began these studies of the anatomy and chemistry of the domestication syndrome, we also had been studying the potential of indirect indicators of cultivation, particularly the use of remains of "weed" seeds.[7] The presence of a number of these segetal marker species among the plant remains from Abu Hureyra had already tentatively hinted at some form of Epipaleolithic cultivation. More specifically, many of the wild seed species I had found in the initial analyses of the charred remains were the very species, which I had already found to thrive as weeds of cultivated cereals under traditional systems of husbandry in Syria and southeastern Turkey.[8] I therefore undertook additional field studies in 1972–1975 of patterns of weed infestation in areas of steppe in Syria and central Turkey newly taken into cultivation, as well as in fields which had been cultivated for 5 to 10 years. The results reinforced the conclusion that certain components of the seed assemblages from Epipaleolithic Abu Hureyra were possibly indicative of cultivation (Moore 1979).

Some years later, however, new ecological data led us to reevaluate this evidence. We eventually concluded that the wild cereals had probably been gathered entirely from wild stands, after all, and that the occupants of all three Epipaleolithic phases were probably hunter-gatherers. More specifically, in 1983, a further round of funding for the analysis of plant and animal remains from Abu Hureyra allowed us to undertake detailed studies in central Syria of the patterns of association of a range of wild food plants with "weeds" and various ecological marker-species in remnants of woodland-steppe and in relatively pristine, moist, treeless steppe, following an unusually wet winter. The study revealed that, under these periodically moist conditions, most of the plants which are today typical weeds of traditional cultivation can (and do) grow successfully in pristine steppe,[9] and that they achieve particularly high densities in shallow depressions, wadi systems, and in other areas of natural disturbance such as mountain screes. Therefore, seeds from such plants identified in the charred remains from Abu Hureyra theoretically could have been gathered from uncultivated steppe and not just from weeds under cultivation, and so their presence could not be used uncritically as an unequivocal indicator of cultivation after all (Hillman et al. 1989a). This inevitably made earlier arguments for possible cultivation decidedly less secure (Moore 1979).

Furthermore, we now had other, more positive evidence for the wild-type cereals not being under cultivation. Throughout the Epipaleolithic occupation we had encountered a consistent pattern of association: the charred grain remains of wild-type einkorn and wild-type annual rye (both of them potential cultigens), found together with grains of a second species of rye, wild mountain-rye *(Secale montanum* Guss). This pattern suggested that all three cereals were growing together and were harvested and processed as a mixture. And because wild mountain-rye is a perennial and does not normally survive under cultivation,[10] this association suggested that all three cereals had been growing together in wild stands, not in sown crops. We therefore concluded that the two potentially cultivable cereals (wild-type einkorn and wild-type annual rye) were probably not under cultivation after all. This clearly tipped the argument firmly in favor of the occupants of Abu Hureyra having remained hunter-gatherers throughout the Epipaleolithic (Hillman et al. 1989a).

For several years this remained our conclusion, and despite deficiencies in our data and some inconsistencies, we were happy to accept that there was probably no local initiation of the cultivation of caloric staples, no on-the-spot domestication, and that the agriculture we find in the overlying Neolithic levels arrived fully-fledged from elsewhere. I have to admit that I found (and continue to find) hunter-gatherer subsistence vastly more interesting than that of farmers, so I was correspondingly pleased that we now had more than a millennium's worth of data that, we hoped, could yield useful information on local foraging strategies that were still seemingly unsullied by any systematic cultivation of staples.

It was not long, however, before we started to unearth several lines of additional evidence which now, a decade later, have finally forced us to conclude (in my case rather reluctantly) that the Epipaleolithic occupants were, in fact, cultivating one or more of their caloric staples from early in phase 2 of the Epipaleolithic, around 11,000 b.p. uncal. (ca. 13,000 calibrated B.P.; Stuiver et al. 1998). In short, we would have to eat our words.

The New Evidence

The new evidence includes (a) data from additional analyses of the plant remains, (b) further field studies of the present-day ecology of the wild plants represented by the remains, and (c) further field experiments on the harvesting and processing of a wide range of wild plants that appear to have served as food. Equally important was the recalculation of the entirety of our accumulated data in the form of percentages rather than the original absolute numbers per unit volume of source deposit. This immediately revealed distinct trends of which we had hitherto been unaware.

At the same time, we also achieved a major breakthrough in our understanding of the Epipaleolithic distribution of Abu Hureyra's vegetal resources. This breakthrough came just five years ago when time allowed me to devote two months to modeling the distribution of potential, present-day vegetation in the region, as it would be in the absence of deforestation, cultivation, and grazing by domestic animals.[11] Again, I used my own accumulated plant-ecological data, together with geological data, rainfall data, the few detailed publications on southwest Asian vegetation, and old maps of the region which, because they indicated even the smallest settlements and farmsteads and predated the widespread adoption of water pumps for irrigation, could be used to delineate the former limits of traditional rain-fed cultivation.[12] The results surprised us. Without deforestation, cultivation, and heavy grazing, the vegetation of the northwest quadrant of the Fertile Crescent would be vastly more mesic today than any of us had ever dared suggest before. Moreover, when we used published pollen data to adjust this map to the situation that probably prevailed at the start of the Epipaleolithic occupation at Abu Hureyra, we discovered that wild cereals such as einkorn (and probably a form of wild annual rye) would have grown within a radius of 1–2 km of the site. At last, the presence of these and other food plants in the remains from Abu Hureyra started to make sense, and many other seeming inconsistencies that had worried us for two decades now found ready resolution.

Combined, all these new data reveal that the people who first occupied Abu Hureyra around 11,600 B.P. had access to an abundance and diversity of wild plant foods that allowed at least some of them to occupy the site year-round, just as we had argued in 1989. The data further indicated that, at first, they were, indeed, strictly hunter-gatherers and there was no hint of any systematic cultivation of caloric staples.[13]

Our new data also indicate that, after four or so centuries of successful subsistence as hunter-gatherers, foraging was severely affected by advancing desiccation. The first wild foods to disappear from the Abu Hureyra remains were the fruits and seeds of drought-sensitive plants of oak-dominated park-woodland. Next, the supply of wild lentils and other large-seeded legumes appears to have collapsed, followed soon afterward by drastic reductions in grain supplies of the wild wheats and ryes. Thereafter, we see a collapse: first in the still more drought-resistant feather-grasses, and finally in food-seeds from the extremely drought-tolerant shrubby chenopods. By this time, our evidence shows that the local vegetation outside the Euphrates Valley had changed from moist woodland-steppe (with areas of oak-dominated park-woodland not far away) to much drier equivalents, with park-woodland now well out-of-reach, and with dry, treeless steppe probably now encroaching from the southeast. The timing of this desiccation sequence appears to have coincided (approximately) with the sharp forest retreat indicated by the pollen diagram from Hula (Baruch and Bottema 1991) which, in turn, coincided with the point when, elsewhere, the Younger Dryas began to "bite."

In addition, the data now suggest that some of the occupants of Abu Hureyra responded to this sequence of drastic reductions in caloric staples by starting to cultivate at least two of the cereals and, a few centuries later, maybe some large-seeded legumes as well. The evidence is as follows:

1. From around 11,000 B.P., seeds of weeds characteristic of arid-zone rain-fed arable cultivation suddenly start to increase dramatically from the relatively low levels they had maintained for the last six centuries.
2. Wild wheats and ryes continue to be used and make a regular (albeit now much smaller) input into the charred remains (this despite the dry conditions of the Younger Dryas having almost certainly eliminated all wild stands from the area).
3. Despite finding the remains of only wild-type cereals (in considerable abundance) throughout the deposits dating to the four centuries of phase 1, soon after 11,000 B.P., we encounter the first rye grains of an extreme domestic type, the earliest of which gave an AMS date of 10,930 ± 120 B.P. (OxA-6685).
4. Two or three centuries later, lentils and other large-seeded legumes suddenly reappear and start to increase (this despite the fact that they had been absent for some centuries and that the arid conditions which caused their disappearance still prevailed and would have prevented any wild stands becoming re-established in the area). A millennium later, we continue to encounter rye of similar domestic morphology together with lentils. By then, however, they were accompanied by other crops such as domestic einkorn.

On their own, any one of these lines of evidence would probably be unconvincing. But the fact that all four lines seem to point in the same direction has led us to suspect cultivation, and thus, reluctantly, to overturn our earlier arguments for the occupation of Abu Hureyra having remained exclusively hunter-gatherer throughout the Epipaleolithic. The fact that the start of cultivation around 11,000 B.P. seems to have coincided with the loss of local stands of wild wheats and ryes suggests that it was perhaps this particular loss that triggered the decision to start cultivating them.

The evidence also indicates that the transition from seemingly exclusive dependence on wild, gathered foods and hunting to exclusive reliance on cultigens to provide caloric staples appears to have taken 2,500 years. During the final thousand years of this process, however, the use of starch staples gathered from the wild was seemingly piecemeal and marginal (de Moulins 1997; de Moulins in Moore et al. 2000). In addition, our evidence suggests that not all families at Abu Hureyra adopted cultivation immediately. Indeed, it was probably only a small minority that did so. Thereafter, the numbers of cultivators steadily increased until the entire population came to rely on cultivation to provide almost all its energy needs.

What, then, of the evidence which had previously persuaded us (Hillman et al. 1989a) that the "weed" seeds from the classic segetals had originated from plants growing on the native steppe rather than from weeds of cultivated crops? The new data reaffirm that, for the first few centuries of occupation at Abu Hureyra, all such seeds had, indeed, come from plants growing in the uncultivated steppe. Their regular recovery in the context of domestic fires suggests that several of them were probably gathered as supplementary foods, in much the same way as closely related plants occurring in equivalent resource environments are exploited by recent hunter-gatherers. However, their abrupt and synchronous increase at ca. 11,000 B.P. indicates a radical change in the situation.

We had previously considered that what had initially appeared to be a piecemeal increase in the remains of small-seeded legumes and grasses at Abu Hureyra around 11,000 B.P. represented the intensified gathering of low-quality foods to compensate for the loss of the preferred caloric staples (Hillman et al. 1989a). However, this interpretation cannot explain the parallel increase in the remains of inedible stony-seeded gromwells, the steepness of the increases seen in the new data, or the way that the increases steadily continue for centuries. The fact that all three groups of plants represent the classic weeds of rain-fed cereal cultivation on gypsiferous soils in the arid steppe of central Syria suggests a more convincing explanation for their synchronous and abrupt increase: namely, that they had invaded new areas of cultivation near Abu Hureyra and arrived on-site as weeds of the harvested crops. Certainly, in the area around Abu Hureyra, increasing aridity at this time should have caused these particular plants to decline in abundance. Their sudden increase is explicable only by some form of disturbance. Domestic animals were not the source of this disturbance because sheep and goats, the first domestic animals at Abu Hureyra, did not appear until ca. 9400 B.P., and domestic cattle and pigs later still (Legge 1996; Legge and Rowley-Conwy in Moore et al. 2000). Similarly, the disturbance cannot be attributed to the side effects of increased sedentism, such as more trampling and bigger middens, because some of the Abu Hureyra population had already been sedentary for the preceding 400 years. Tillage associated with cultivation in areas of enhanced water availability (from which competing vegetation had been cleared) seems to offer the only obvious explanation. The steady increases in numbers of weed seeds during the ensuing centuries probably reflects regular additions to the patches of land taken into cultivation.

But what of the other major line of evidence that had previously argued against cultivation: namely, that the grain of potentially cultivable wild-type annual cereals was associated with grain from perennial, wild mountain-rye (*Secale montanum*), which is never found growing as a weed of cultivation? In 1992, some years after advancing that argument, further ecological fieldwork took me across the high plateau of Uzun Yayla in central Anatolia. There, I found fields of wheat infested not only by the usual annual weed-rye, but also by large numbers of tussocks of an exceptionally tall and robustly-culmed perennial rye (growing right through the crop stands).[14] Intriguingly, the grains of this robust rye proved to be morphologically identical to those of the normal perennial wild mountain rye (*Secale montanum*). It is therefore possible that the charred grains from Epipalaeolithic Abu Hureyra that I had identified as the "*S. montanum* type" could theoretically have come, not from the normal form of *S. montanum* (which does not grow as an arable weed), but rather from plants of this other (as yet unnamed) type, which positively thrives as an arable weed.[15] Clearly, therefore, the fact that, in the Abu Hureyra samples, grains of a perennial-type rye accompanied every find of grains of wild einkorn and wild annual rye, could no longer be argued to exclude any possibility that these two annual wild-type cereals were, indeed, under cultivation.

The other two lines of evidence explored by Hillman, Colledge, and Harris

(1989a) had proved inconclusive from the outset. Thus, we were now left without any unequivocal evidence to support our earlier arguments, which excluded the possibility of cultivation during the Epipaleolithic. Instead, we now have four lines of evidence suggesting that cultivation at Abu Hureyra was already under way by ca. 11,000 b.p. (ca. 13,000 B.P. cal.; Stuiver et al. 1998). Furthermore, this finding is now paralleled by new evidence for possible pre-domestication cultivation from early PPNA occupations at sites such as Jerf al-Ahmar, Netif Hagdud, and Mureybit, where the evidence again extends back into the terminal Epipaleolithic (albeit to ca. 700 years after the apparent inception of cereal cultivation at Abu Hureyra; Colledge 1994, 1999; Hillman et al. in press; Kislev 1997; Willcox 1996, 1999).

Ecological Field Studies: Some Thoughts

It has taken us 27 years of research to begin to understand ancient subsistence at Abu Hureyra. Our modeling of the resource base available to the hunter-gatherers, who first settled Abu Hureyra, and of the environmental pressures on specific resources that seemingly drove some of them to start cultivating, has been heavily dependent on our ability to model the composition and distribution of ancient plant communities at a level of detail never attempted before in Southwestern Asian archaeology. And this second tier of modeling has, in turn, relied largely on our own ecological field studies conducted in the region during these same 27 years.[16]

Had we known that, to make sense of the plant-based components of subsistence at Abu Hureyra would have required so many years of laboratory analyses of the plant remains and even more years of field studies of the relevant food plants and their source communities in locations scattered over Southwest and Central Asia, we might well have lowered our sights from the outset. Our retrospective surprise at the amount of work that has been involved perhaps reflects the fact that, for too many years, we in archaeology have imagined that we could somehow research events of considerable ecological complexity—such as those surrounding the inception of cultivation—without ever needing to come to grips with the ecological detail. All too often, the result of this reluctance to confront ecological realities has been the publication of explanatory models couched in generalities and bereft of any of the specifics that would allow them ever to be tested against the sort of data that are today becoming available from the growing number of detailed analyses of bulk-recovered biological remains from early sites.[17]

Use of palate-knife generalizations of the sort expressed in coinages such as the seasonal use of resources from broad-spectrum ecosystems are clearly appropriate at certain levels of debate on agricultural origins. However, we quickly reach a point where, for productive debate to continue, it becomes necessary to press for the resource species to be named, for the types of community, habitat, and ecosystem to be defined, and for specific patterns to be cited for the seasonal scheduling of the named resources.[18]

It is surely unrealistic to hope to understand the decision-making of the hunter-gatherers who first chose to start cultivating their caloric staples—without first acquiring at least some of that knowledge of plant and animal ecology which they themselves must have mastered so comprehensively to have subsisted successfully for centuries in those environments and which must then have played a central role in their decision to radically change that subsistence. And we cannot seriously expect to achieve this without first reconstructing their resource environment in some detail. The fact that other factors doubtless also played a role in the decision to start cultivating does not preempt the need to explore the ecology. After all, past ecological factors are amenable to investigation, which is sadly not true of some of the others. Secondly, as our work at Abu Hureyra has taught us, very few of those sites that might have witnessed on-the-spot transitions from foraging to farming have yet to be subjected to detailed analysis of their ecology. And more well-researched examples are badly needed. Thirdly, at Abu Hureyra where the requisite ecological studies have been undertaken, it has proved to have been ecological factors that triggered the initial decision to start cultivating.

Grappling with ecological complexities is time-consuming. And to post-processualists, being reminded of the universality of the impact of clearly defined ecological processes is doubtless anathema. But whatever the effects of conscious thought on the conformation of energy and matter (Bohm 1995), merely averting our gaze from ecological processes will not uncreate them. Nor will denying their existence.

A challenge to the adequate study of ecological factors involved in the origins of agriculture also comes in a more prosaic form. It remains to be seen whether the necessary long-term ecological fieldwork can ever again be adequately funded from the overstretched budgets of the key funding agencies. It is equally uncertain whether many overstretched academics can today find the time required for projects lasting as long as the one at Abu Hureyra.

REFERENCES

El Azm, A.M. 1986. "An Ethno-Agricultural Study of Certain Sieving Systems at the Village of El Findara in the Alawite Mountains." M.Sc. thesis, University College London.

———. 1992. "Crop Storage in Ancient Syria: A Functional Analysis Using Ethnographic Modelling." Ph.D. diss., University College London.

Baruch, U., and S. Bottema. 1991. "Palynological Evidence For Climatic Changes in the Levant ca. 17,000–9,000 B.P." In *The Natufian Culture in the Levant*, edited by O. Bar-Yosef and F.R. Valla, 11–20. Ann Arbor.

Bohm, D. 1995. *Wholeness and the Implicate Order*. London.

Butler, A. 1989. "Cryptic Anatomical Characters As Evidence of Early Cultivation in the Grain Legumes (Pulses)." In *Foraging and Farming: The Evolution of Plant Exploitation,* edited by D.R. Harris and G.C. Hillman, 390–407. London.

———. 1990. "Legumes in Antiquity: A Micromorphological Investigation of Seeds of the Vicieae." Ph.D. diss., University College London.

———. 1991 "The Vicieae: Problems of Identification." In *New Light on Early Farming: Recent Developments in Palaeoethnobotany*, edited by J.M. Renfrew, 61–73. Edinburgh.

Cave, M. 1989. "Chemical Criteria for Identifying Charred Remains of Wild Barleys from Epipalaeolithic Sites in SW Asia." B.Sc. thesis, University College London.

———. 2000. "The Role of Chemical Criteria in Identifying Charred Remains of Grasses and Wild Cereals from Epipalaeolithic Sites in SW Asia, with Particular Reference to the Use of Chemometrics." Ph.D. diss., University College London.

Colledge, S.M. 1988. "Scanning-Electron Microscope Studies of the Pericarp Layers of Some Wild Wheats and Ryes. Methods and Problems." In *Scanning-Electron Microscopy in Archaeology*, edited by S.L. Olsen, 225–36. *BAR-IS* 452. Oxford.

———. 1994. "Plant Exploitation on Epipalaeolithic and Early Neolithic Sites in the Levant." Ph.D. diss., University of Sheffield.

———. 1999. "Identifying Pre-Domestication Cultivation Using Multivariate Analysis." In *The Origins of Agriculture and Crop Domestication,* edited by A.B. Damania, J. Valkoun, G. Willcox, and C.O. Qualset, 121–31. Aleppo.

Dennell, R.W. 1972. "The Interpretation of Plant Remains: Bulgaria." In *Papers in Economic Prehistory*, edited by E.S. Higgs, 149–59. Cambridge.

———. 1974. "Botanical Evidence for Prehistoric Crop Processing Activities." *JAS* 1:275–84.

French, D.H. 1971. "An Experiment in Water-Sieving." *Anatolian Studies* 21:59–64.

———. 1972. "Recent Archaeological Research in Turkey: Aşvan 1971." *Anatolian Studies* 22:11–22.

———. 1973. "Aşvan Project." *Anatolian Studies* 23:191–6.

Grant, V. 1958. "The Regulation of Recombination in Plants." *Cold Spring Harbor Symposium of Quantitative Biology* 23:337–63.

———. 1963. *The Origin of Adaptations.* New York.

Harris, D.R. 1984. "Ethnohistorical Evidence for Exploitation of Wild Grasses and Forbs." In *Plants and Ancient Man: Studies in Palaeoethnobotany*, edited by W. van Zeist and W.C. Casparie, 63–9. Rotterdam.

———. 1989. "An Evolutionary Continuum of People-Plant Interaction." In *Foraging and Farming: The Evolution of Plant Exploitation,* edited by D.R. Harris and G.C. Hillman, 11–26. London.

———. 1990. *Settling Down and Breaking Ground: Rethinking the Neolithic Revolution.* Stichting Nederlands Museum voor Anthropologie en Praehistorie, Twaalfde Kroon-Voordracht. Amsterdam.

Hillman, G.C. 1973. "Crop Husbandry and Food Production: Modern Models for the Interpretation of Plant Remains." *Anatolian Studies* 23:241–4.

———. 1978. "On the Origins of Domestic Rye—Secale Cereale: The Finds from Aceramic Can Hasan III in Turkey." *Anatolian Studies* 28:157–74.

———. 1981. "Reconstructing Crop Husbandry Practices from Charred Remains of Crops and Weeds." In *Farming Practice in British Prehistory*, edited by R. Mercer, 123–62. Edinburgh.

———. 1983. "Criteria for Identifying the Rachis Remains of Free-Threshing Wheats." Paper presented at the Sixth Symposium of the International Workshop for Palaeoethnobotany: *Plants and Ancient Man: Studies in Palaeoethnobotany.* Groningen.

———. 1984. "Interpretation of Archaeological Plant Remains: The Application of Ethnographic Models from Turkey." In *Plants and Ancient Man: Studies in Palaeoethnobotany*, edited by W. van Zeist and W.C. Casparie, 1–42. Rotterdam.

———. 1996. "Late Pleistocene Changes in the Wild Plant Foods Available to Hunter-Gatherers of the Northern Fertile Crescent: Possible Preludes to Cereal Cultivation." In *The Origins and Spread of Agriculture and Pastoralism in Eurasia*, edited by D.R. Harris, 159–203. London.

———. Forthcoming. "Percival, Archaeobotany, and Problems of Identifying Wheat Remains." In *Wheat—Yesterday, Today and Tomorrow: A Celebration of the Life of John Percival (1863–1949).* Reading.

Hillman, G.C., S.M. Colledge, and D.R. Harris. 1989a. "Plant Food Economy during the Epipalaeolithic Period at Tell Abu Hureyra, Syria: Dietary Diversity, Seasonality, and Modes of Exploitation." In *Foraging and Farming: The Evolution of Plant Exploitation,* edited by D.R. Harris and G.C. Hillman, 240–68. London.

Hillman, G.C., and M.S. Davies. 1990. "Measured Domestication Rates in Wild Wheats and Barley under Primitive Cultivation, and Their Archaeological Implications." *Journal of World Prehistory* 4:157–222.

———. 1992. "Domestication Rates in Wild Wheats and Barley under Primitive Cultivation: Preliminary Results and Archaeological Implications of Field Measurements of Selection Coefficient." In *Préhistoire de l'agriculture: Nouvelle approches expérimentales et ethnographiques,* edited by P.C. Anderson, 113–58. Paris.

Hillman, G.C., R. Hedges, A.M.T Moore, S.M. Colledge, and P. Pettitt. Forthcoming. "New Evidence of Late Glacial Cereal Cultivation at Abu Hureyra on the Euphrates." *The Holocene* 11 (4).

Hillman, G.C., C.E.R. Jones, and G.V. Robins. 1982. "The Potential of Examining Archaeological Materials with Pyrolysis-Mass Spectrometry: An Exploratory Exercise." Paper presented at the 5th International Symposium on Analytical Pyrolysis, Vail, Colorado.

Hillman, G.C., E. Madeyska, and J.G. Hather. 1989. "Wild Plant Foods and Diet at Late Palaeolithic Wadi Kubbaniya (Upper Egypt): Evidence from Charred Remains." In *The Prehistory of Wadi Kubbaniya.* Vol. 2, *Stratigraphy, Subsistence and Environment,* edited by F. Wendorf, R. Schild, and A. Close, 162–242. Dallas.

Hillman, G.C., S. Wales, F.S. McLaren, J. Evans, and A.E. Butler. 1993. "Identifying Problematic Remains of Ancient Plant Foods: A Comparison of the Role of Chemical, Histological and Morphological Criteria." *World Archaeology* 25:94–121.

Hopf, M. 1983. "Jericho Plant Remains." In *Excavations at Jericho.* Vol. 5, edited by K.M. Kenyon and T.A. Holland, 576–621. London.

Jones, G.E.M. 1984. "Interpretation of Archaeological Plant Remains: Ethnographic Models from Greece." In *Plants and Ancient Man: Studies in Palaeoethnobotany,* edited by W. van Zeist and W.C. Casparie, 43–61. Rotterdam.

———. 1987. "A Statistical Approach to the Archaeological Identification of Crop Processing." *Journal of Archaeological Science* 14:311–23.

Kislev, M.E. 1997. "Early Agriculture and Paleoecology of Netiv Hagdud." In *An Early Neolithic Village in the Jordan Valley.* Pt. 1, *The Archaeology of Netiv Hagdud,* edited by O. Bar-Yosef and A.Q. Gopher, 209–36. Cambridge.

Kobyljanskij, V.D. 1989. *Flora of Cultivated Plants of the USSR.* Vol. 2, pt. 1, *Rye.* Leningrad.

Legge, A.J. 1996. "The Beginning of Caprine Domestication in Southwest Asia." In *The Origins and Spread of Agriculture and Pastoralism in Eurasia,* edited by D.R. Harris, 238–62. London.

Mason, S.L.R., and G.C. Hillman. Forthcoming. *Handbook of Wheat Identification.* London.

McLaren, F.S. Forthcoming. "The Role of Chemical Criteria in the Identification of Ancient Old World Cereals." Ph.D. diss., University College London.

McLaren, F.S., J. Evans, and G.C. Hillman. 1990. "Identification of Charred Seeds from Epipalaeolithic Sites in SW Asia from Chemical Criteria." In *Archaeometry* 90: *Proceedings of the 15th International Archaeometry Symposium, Heidelberg,* 797–806. Basel.

Moore, A.M.T. 1979. "A Pre-Neolithic Farmers' Village on the Euphrates." *Scientific American* 241(2):50–8.

Moore, A.M.T., G.C. Hillman, and A.J. Legge, 1975. "The Excavation of Tell Abu Hureyra in Syria: A Preliminary Report." *Proceedings of the Prehistoric Society* 41:50–77.

———. 2000. *Village on the Euphrates: From Foraging to Farming at Abu Hureyra.* New York.

Moulins, M.D. de. 1997. *Agricultural Changes at Euphrates and Steppe Sites in the Mid-8th to the 6th Millennium B.C. BAR-IS* 683. Oxford.

Peña-Chocarro, L. 1990. "Early Agriculture in Spain: Evidence from Plant Remains and Ethnographic Studies of Traditional Crop Husbandry." Ph.D. diss., University College London.

———. 1996. "*In situ* Conservation of Hulled Wheat Species: The Case of Spain." In *Hulled Wheats: Proceedings of the First International Workshop on Hulled Wheats; 21–22 July 1995; Castelveccio Pascoli, Tuscany, Italy,* edited by S. Padulosi, K. Hammer, and J. Heller, 126–46. Rome.

Riley, R. 1965. "The Evolution of Wheat." In *Essays in Crop Plant Evolution,* edited by J. Hutchinson, 103–22. Cambridge.

Rowley-Conwy, P. 1998. "The Origins and Spread of Agriculture and Pastoralism: Are the Grey Horses Dead?" *International Journal of Osteoarchaeology* 8:218–27.

Sikkink, L.L. 1988a. "Ethnoarchaeology of Harvest and Crop-Processing in Traditional Households in the Montara Valley, Central Andes of Peru." M.A. thesis, University of Minnesota, Minneapolis.

———. 1988b. "Traditional Crop Processing in Central Andean Households: An Ethnoarchaeological Perspective." In *Multidisciplinary Studies in Andean Anthropology,* edited by V. Vitzhum. Ann Arbor.

Stebbins, G.L. 1950. *Variation and Evolution in Plants.* New York.

———. 1974. *Flowering Plants: Evolution above the Species Level.* London.

Stocks, D. 1979. "Testa Micromorphology and Dormancy Behaviour in Species of Vicia." B.Sc. thesis, University College Cardiff, Wales.

Stuiver, M., P.J. Reimer, E. Bard, J.W. Beck, G.S. Burr, K.A. Hughen, B. Kromer, G. McCormac, J. van der Plicht, and M. Spurk. 1998. "INTCAL98 Radiocarbon Age Calibration, 24,000-0 cal. B.P." *Radiocarbon* 40:1041–83.

Willcox, G. 1996. "Evidence for Plant Exploitation and Vegetation History from Three Early Neolithic Pre-Pottery Sites on the Euphrates (Syria)." *Vegetation History and Archaeobotany* 5:143–52.

———. 1999. "Archaeobotanical Evidence for the Beginnings of Agriculture in Southwest Asia." In *The Origins of Agriculture and Crop Domestication,* edited by A.B. Damania, J. Valkoun, G. Willcox, and C.O. Qualset, 25–38. Aleppo.

Zeist, W. van, and J.A.H. Bakker-Heeres. 1979. "Some Economic and Ecological Aspects of the Plant Husbandry at Tell Aswad." *Paleorient* 5:161–9.

———. 1986. "Archaeobotanical Studies in the Levant. 3: Late Palaeolithic Mureybit." *Palaeohistoria* 26:171–99.

Zeist, W. van, and S. Bottema. 1968. "Wild Einkorn Wheat and Barley from Tell Mureybit in Northern Syria." *Acta Botanica Neerlandica* 17:44–53.

NOTES

[1] The BIAA Director, David French, had a clear vision of the potential role of the sciences (especially the natural sciences) in archaeology, and, for its time, his Aşvan Project was revolutionary in incorporating an exceptionally broad spectrum of bioarchaeological, geoarchaeological, pedological, and ethnographic studies.

[2] Equivalent studies that followed elsewhere include: El Azm 1986,1992; Jones 1984, 1987; Peña-Chocarro 1990, 1996; Sikkink 1988a, 1988b.

[3] With some of the key specimens of wheat, it was necessary to germinate a few grains so I could make chromosome counts from root-tip squashes and double-check ploidy level. Although some of the results from the search for new criteria were subsequently presented at conferences (e.g., Hillman 1983), made available as student handouts, or, in a very few cases, published (e.g., in Hillman et al. 1993; Hillman forthcoming; McLaren et al. 1990), work on the bulk of them still continues (e.g., Mason and Hillman forthcoming; McLaren forthcoming).

[4] Our studies of the use of internal anatomy of seeds and fruits are currently available as unpublished notes.

[5] Of these field experiments on harvesting, bulk-processing, and preparing wild plants as food, and the laboratory analyses of their nutrient status and toxicity, the only ones published in any detail so far are those concerning food plants from the Late Paleolithic hunter-gatherer site of Wadi Kubbaniya in Egypt (Hillman et al. 1989b). Nevertheless, a few of the experiments that targeted the food plants used at Abu Hureyra are outlined briefly in Moore et al. (2000, 349–65). Work on this subject area continues apace, in collaboration with Tony Leeds of the Department of Nutrition and Dietetics at Kings College London, Ray Mears (the international authority on bushcraft), Michelle Wollstonecroft (who works on root foods from aquatics), and Sarah Mason, an expert on the ethnobotany of acorns and several other wild foods.

[6] I chose to measure selection pressure (or "selection coefficient") because, of the factors determining domestication rate, this was the principal unknown.

[7] The rationale is summarized in Colledge 1999; Hillman et al. 1989a, 253–5; and Hillman and Davies 1990, 207–8; 1992, 146.

[8] These traditional systems of husbandry and crop processing were eventually published in Hillman (1981, 1984).

[9] These examples of pristine steppe were located over 50 km from the nearest areas of arable farming, and so the presence of "weed" species could not be attributed to invasions of weeds from local fields.

[10] Unpublished field notes, 1970–1992.

[11] This and similar research breaks were made possible through the generous hospitality of Andrew and Barbara Moore at their beautiful woodland home then near Yale, where I was well out of reach of my students in London.

[12] Using the map of potential present-day vegetation as the starting point for generating a map of local vegetation distribution at the start of the Epipaleolithic occupation would have been even more complex (given that woodland-steppe plants in this area were at the time still spreading rapidly from their erstwhile Pleistocene *refugia* in the Levant). Fortunately, our models of the spread of these plants indicated that, by this time, they had already been established in the Abu Hureyra area for several centuries, even though further to the east the same plants were still migrating rapidly southeastward (Hillman 1996; Moore et al. 2000, fig. 3.18, maps a–d).

[13] Nevertheless, like many hunter-gatherers ancient and modern, they had probably sown or planted the occasional patch (or individual plants) of non-staples that fulfilled a role that was unlikely ever to lead to the systematic cultivation of caloric staples. Indeed, such practices had doubtless gone on already for many millennia (Harris 1989, 1990).

[14] Although I initially identified this robust perennial weed-rye as an aberrant form of *Secale montanum* Guss. subsp. *anatolicum* (Boiss.) Tsvelev, infrared spectra from propanol extractions made by Frances McLaren (pers. comm. 1993, forthcoming) surprisingly indicated strong affinities with the annual *S. cereale* L. subsp. *ancestrale* Zhuk, but suggested no close affinity with the normal *S. montanum*. Its identity therefore remains unclear. It does not match any of the species described in Kobyljanskij (1989), which is by far the most reliable taxonomic treatment of rye yet published.

[15] It should be mentioned that, growing around the rocky margins of these same rye-infested fields, just beyond the limits of any plowing, was a solid swathe of the normal form of mountain rye (*S. montanum* Guss.), the grains of which yielded infrared spectra that Frances McLaren (pers. comm. 1993, forthcoming) identified as entirely typical of *S. montanum*, and entirely different from those she obtained from the new, and as yet unnamed, species of perennial rye that successfully infested the wheat crop.

[16] Some of the core ecological field studies involved in the Abu Hureyra Project have been mentioned above. They are discussed in greater detail in Hillman (1996), and in chs. 3 and 12 of Moore et al. (2000). However, to provide here a clearer picture of the diversity of the sort of field studies necessarily entailed in a project of this type, it is perhaps worth itemizing some of the more narrowly targeted of the supplementary ecological investigations that we undertook in order to provide specific additional information that was required (often urgently) for interpreting data from the Abu Hureyra remains. Eight examples should suffice: (a) the composition of weed floras in rye crops under hoe cultivation in areas of partially cleared oak-scrub woodland in mountain areas of central Anatolia in 1974; (b) composition of one of the last areas of pristine riverine forest surviving in Southwest Asia: in the Kelkit Çay Valley in northern Turkey in 1976; (c) El Koum, Syria, 1986: the composition of "weed" floras in small refugia of terebinth *(Pistacia atlantica)* woodland-steppe on Jebel Bishri, the composition of various forms of crucifer-dominated steppe around the El Koum Basin, and the distribution of traditional streambed cultivation in and around the basin; (d) ecology of the feather-grasses and other food plants in the desert-steppe ecotone of eastern Jordan in 1987; (e) the seasonality of food-seed production by winter-flowering shrub chenopods in the Azraq Desert Basin (Jordan) in 1988 and a follow-up study with Chris Stele in 1990, complete with experiments on methods of harvesting and processing the seeds as food; (f) the composition of relatively pristine pistachio *(Pistacia vera)* woodland-steppe in the Batkhyz area of Turkmenistan in 1991; (g) the composition of degraded *Populus euphratica–Tamarix* sp. riverine forest surviving in the upper Tedzhen Valley (Turkmenistan) in 1991; and (h) the ecology and biotaxonomy of wild annual ryes in one of their last major refugia on Mount Ararat in 1992.

[17] This point is discussed more cogently by David Harris in ch. 3 of this volume and Peter Rowley-Conwy (1998).

[18] Any suggested patterns of resource scheduling should ideally be supported (for each of the relevant plant or animal foods) by relevant ecological models of seasonal patterns of availability, accessibility, and variations in nutrient status and toxicity (Hillman et al. 1989b). If it is also possible to cite ethnographic parallels for the suggested patterns of scheduling, so much the better. This is especially the case where the parallels involve well-defined models based on cross-cultural, common denominators isolated from a range of different, culturally independent groups that utilize equivalent resource spectra from habitats analogous to those proposed for the site.

— 6 —

Early Domestic Animals in Europe: Imported or Locally Domesticated?

Peter Rowley-Conwy

This chapter considers the evidence for early domestic animals in Europe, specifically the evidence for the local domestication of native cattle and pig, and for the pre-Neolithic introduction of domesticates.[1] I argue that there is little good evidence for either of these. In most cases the domestic animals were apparently imported as a group at the start of the Neolithic, although there are exceptions. The belief that some Mesolithic peoples domesticated animals derives from the assumption that these societies were going through a unilinear development identical to that of the Near East—only a few thousand years "later." I see no justification for this assumption. In addition, the idea that the Mesolithic developed to the point where it began domesticating animals just before the Near Eastern package arrived seems a great coincidence. Such vitalist or progressivist perspectives are thus open to question. Although I argue for a fairly abrupt introduction of domestic animals in any particular area, it should be noted that this does not have any bearing on whether the people were immigrants or not.

Sheep And Goats

It is now virtually universally agreed that there were no native sheep or goats in Holocene Europe (Poplin 1979; Vigne 1988, 174–89). This has been demonstrated metrically for sheep in the western Mediterranean (Uerpmann 1987). Near Eastern sites are characterized by many small animals (inter-

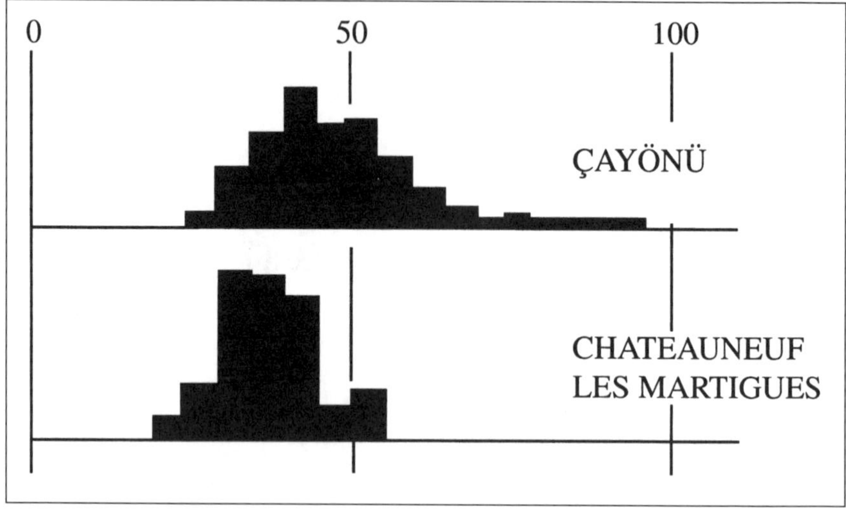

Fig. 6.1. Distribution of sheep measurements at Çayönü (Turkey) and Chateauneuf-les-Martigues (southern France). (Redrawn from Uerpmann 1987, fig. 1)

preted as domestic) and a few large animals (interpreted as hunted wild individuals). In figure 6.1, Çayönü has a typical distribution of this kind with a "tail" of large individuals projecting to the right; it does not form a normal distribution as would be expected if a single biological population was represented. Individuals from two genetically distinct populations must be present, and these must be wild and domestic—even though not every specimen can be diagnosed. Chateauneuf-les-Martigues differs in having no large wild individuals, only the smaller domestic ones forming a more or less normal distribution. The absence of hunted wild individuals indicates that there were no Holocene wild populations in this area.

Further east, claims for both wild and domestic sheep in the Mesolithic at La Adam in Romania are not paralleled at other sites in the region, and are likely to result from stratigraphic problems (Bökönyi 1977). It currently looks as if neither wild sheep nor wild goats were present in early Holocene Europe.

Claims have been advanced that domestic sheep/goat were introduced into the final Mesolithic ahead of the appearance of the formal Neolithic. But these claims remain problematic. For example, the site of Deby in Poland yielded 29 fragments identified as terminal Mesolithic sheep/goat. But there is radiocarbon evidence of later contamination (Lasota-Moskalewska 1998); the bones themselves have not been directly dated. The western Mediterranean is the area where terminal Mesolithic sheep/goat are most often claimed (e.g., Ducos 1977; Geddes 1985). Some doubt has recently been cast on these claims for two main reasons. First, caves are notoriously complex archaeological sites: deposits frequently are disturbed by burrowing animals and later human activity. And there is considerable evidence of this at the relevant sites (Zilhão 1993). Renewed excavations at Chateauneuf failed to find sheep in the Mesolithic layers (Courtin et al. 1985). Secondly, the presence of wild chamois and ibex in the region are a complicating factor: their bones can be very similar to those of domestic caprines, particularly if fragmented and/or juvenile (Uerpmann 1987). In the recent publication of the important site of Dourgne, Guilaine considers both problems and is very cautious about possible pre-Neolithic caprines (Guilaine 1993, 452–8). At Arene Candide in Liguria, sheep were present from the start of the Neolithic; goats were apparently absent until the start of the middle Neolithic (Rowley-Conwy 1997, 1998, 2000a). As the western Mediterranean Neolithic appears to have spread along the coast, this indicates that even early Neolithic claims of domestic goats further to the west may have to be reexamined.

Cattle

Research on cattle is more complex because of the presence of wild aurochs (*Bos primigenius*) throughout Europe. The possibility that animals were locally domesticated therefore exists, and the proposal has been advanced in several regions. Some of these regions are in areas of little zooarchaeological research and few comparative samples, such as southeastern Italy (Cipolloni-Sampò 1987). This review will therefore concentrate on two claims in better-researched areas of Europe: Hungary and southern Scandinavia. In Hungary, the claims involve Neolithic farmers seeking to increase their holdings of domestic cattle by domesticating more from the wild. In southern Scandinavia, the claimed domesticates appear in late Mesolithic contexts.

The Hungarian case will be considered first. Bökönyi argued for local domestication of aurochs. He believed that the first domestic cattle were introduced from elsewhere in the early Neolithic, but that these were supplemented by a much larger number of local domesticates in the middle and later Neolithic. He advances four proofs of local domestication: first, the presence of both wild and domestic forms on archaeological sites; second, the presence of forms transitional between wild and domestic; third, a change in the age and sex of the wild animals killed; and fourth, implements or buildings designed to capture the wild form (Bökönyi 1974, 111).

Of these four, the first would also be the result if domesticates were introduced and wild animals continued to be hunted, while the fourth may be discounted as no such evidence would ever be unambiguous. In support of the third, Bökönyi (1974, 112) states that at the site of Berettyószentmárton, "a typical settlement of the domestication fever of the late Neolithic where the domestication of cattle was one of the most important parts of animal husbandry," most of the wild cattle were adult males. In order to domesticate the young animals, these protective males would first have to be killed; their presence is thus evidence that domestication was taking place. This reasoning is, however, unlikely to be correct, since adult male aurochs would probably have lived apart from the females for much of the year, joining them for the mating season. We cannot be sure how aurochs would have organized their social lives, but the pattern of some males living apart for part of the year recurs in their closest surviving relatives. It is reported among semi-feral cattle (Ewer 1968, 89), bison in both Europe and North America (Fuller 1960; Jaczewski 1958), and to an extent African buffalo (Sinclair 1974). Protective females would be more likely to pose a threat to anyone seeking to interfere with their calves, so the bones of wild adult males on a settlement can hardly be taken as evidence that young were being domesticated.

Bökönyi's second criterion, animals transitional between wild and domestic, is therefore the crucial one—yet it is questionable whether Bökönyi's methods allowed him to distinguish even between wild and domestic, male and female, far less between these and "transitional" animals. The only example he gives is one that compares proximal metacarpals from Seeberg Burgäschisee-Sud (originally published by Boessneck et al. 1963) with those from Berettyószentmárton (Bökönyi 1974, table 1). The measurements are plotted in figure 6.2. At Seeberg, wild and domestic animals were identified (Boessneck et al. 1963), falling into clearly separate size groups. At Berettyószentmárton, however, the two categories run together, which Bökönyi regards as evidence for transitional animals.

There are two problems with this conclusion. First, the proximal metacarpal is a bad bone to use. The epiphysis is fused at birth so there is no direct indication of age if the bone is broken and the distal end (which does fuse) is missing. The presence of subadults may thus be a complicating

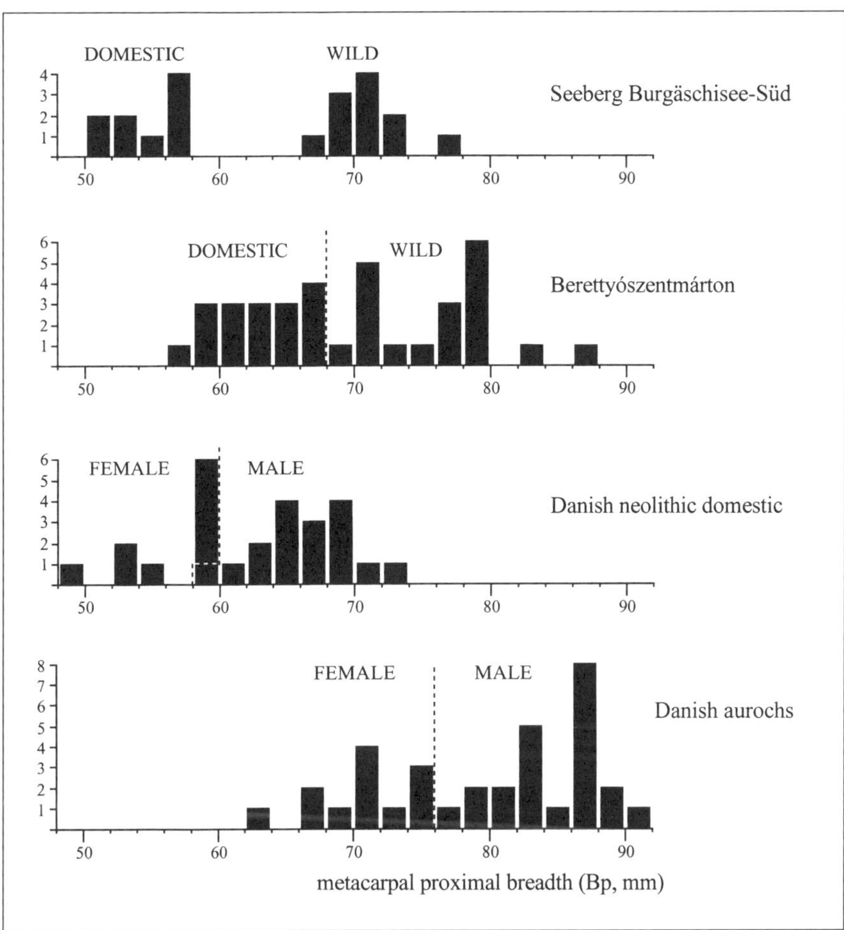

Fig. 6.2. Breadth of cattle proximal metacarpals (measurement Bp as defined by von den Driesch 1976). Seeberg Burgäschisee-Süd from Boessneck et al. (1963, 183–5); Berettyószentmárton from Bökönyi (1974, table 1); the Danish Neolithic domestic sample comprises those listed by Degerbøl and Fredskild (1970, table 11) and the complete bones from Troldebjerg listed by Higham and Message (1968, table C); Danish aurochs from Degerbøl and Fredskild (1970, table 11).

factor. Secondly, Bökönyi takes no account of sexual dimorphism. Work on naturally occurring aurochs skeletons from Denmark has demonstrated that males are larger than females (Degerbøl and Fredskild 1970); the sex of these specimens can be determined from their associated horn cores, which is not possible with fragmentary and dispersed material from archaeological settlements. The aurochs measurements are plotted in figure 6.2. Also plotted are the measurements from Neolithic animals of known sex; these are either from partially complete skeletons that may be sexed by their horns (Degerbøl and Fredskild 1970) or are unbroken bones from the site of Troldebjerg that may be ascribed to sex on the basis of their length—males being longer than females (Higham and Message 1968). The sexual divisions for both Danish aurochs and Neolithic domestic cattle in figure 6.2 are thus likely to be reliable and are not based just on drawing a line through the approximate center of the distributions. It is clear from figure 6.2 that the distributions of both Danish samples are very similar to that from Berettyószentmárton, which therefore probably comprises just a single population—presumably domestic, though one or two hunted aurochs cannot be excluded. Little justification can be found for dividing the Hungarian sample into wild and domestic, and none for assuming the pattern indicates local domestication. It is not clear why the Seeberg pattern is so dichotomous; this assemblage was published before Degerbøl and Higham demonstrated the degree of sexual dimorphism to be found in cattle. It should perhaps be reexamined in the light of more recent findings.

The distal end of the metacarpal is altogether more useful, because all fused specimens must come from animals older than about two years of age. Bökönyi curiously does not give the measurements from Berettyószentmárton, but lists a few Neolithic specimens from other sites (Bökönyi 1974, 461). The sample is, however, too small and scattered to offer much support for any argument. Figure 6.3 plots these and various other samples. Those from Seeberg fall into two distinct groups, once again identified as wild and domestic (Boessneck et al. 1963). The Danish aurochs in figure 6.3 are also bimodal, as are the Neolithic specimens. This definitely results from sexual dimorphism; once again this raises doubts about the Seeberg sample. In the absence of sufficient published data, the Hungarian situation cannot be discussed further.

The conclusion for Hungary is thus that there is no support for the hypothesis of local domestication. This is by no means the first time that objections have been raised (e.g., Bogucki 1989). It seems to be most unlikely that middle and late Neolithic farmers requiring more domestic cattle would choose the domestication of more wild individuals as the means to achieve this. Why not simply breed from the existing domestic stock? It is even more implausible that such new domesticates would then be kept genetically separate from the preexisting domesticates—which they would have to be for Bökönyi to be able to recognize them. First generation domes-

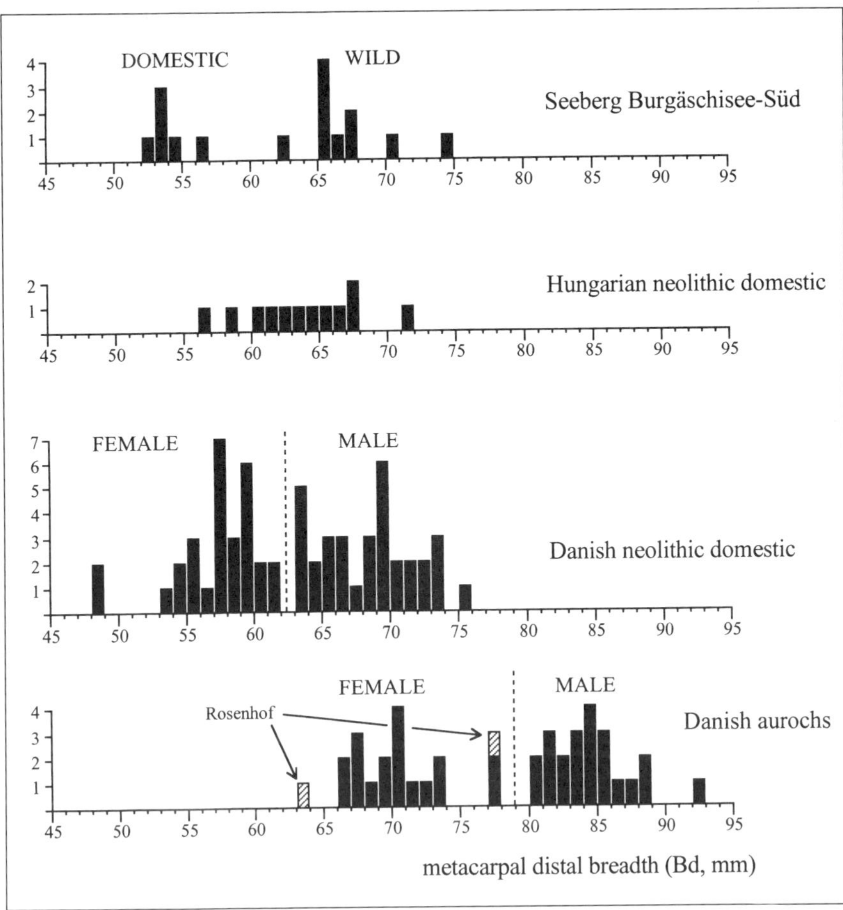

Fig. 6.3. Breadth of cattle distal metacarpals (measurement Bd as defined by von den Driesch 1976). Seeberg Burgäschisee-Süd from Boessneck et al. (1963, 183–5); Hungarian sample from Bökönyi (1974, 461); the Danish Neolithic domestic sample comprises those listed by Degerbøl and Fredskild (1970, table 11) plus all the distal ends from Troldebjerg plotted by Higham and Message (1968, fig. 43); Danish aurochs from Degerbøl and Fredskild (1970, table 11).

ticates would not of course show any size change, so this separation would have to be maintained for many generations for the effects to become visible. Bökönyi's scenario is therefore most implausible, and not supported by the data he presents.

In Denmark and northern Germany, a few domestic cattle have been claimed in late Mesolithic contexts. Farming was present not far to the south, so such cattle (or their meat) could certainly have been imported from the farmers. The claims are, however, based on metrical evidence, which is in no case unequivocal. The naturally occurring aurochs skeletons mentioned above have led most to reject the claimed domesticates in Denmark.

Four lower third molars from Dyrholm I were initially thought to be so small that they must be domestic (Degerbøl 1942, 1963). More finds of skeletons, however, extended the aurochs' size range downward, and better dating indicated that many of the smaller specimens fell later in time, contemporary with the late Mesolithic (Ertebølle) in the late Atlantic period. This led Degerbøl to reconsider his position; in 1970 he concluded that all four Dyrholm I specimens were probably wild (Degerbøl and Fredskild 1970), a conclusion that has gone largely unchallenged within Denmark. One tooth from Rosenhof in northern Germany has more recently been claimed as domestic by Nobis (1975, 1983; Heinrich 1993, 84; Persson 1999, 47–8). But it falls very close to the four from Dyrholm I, and Degerbøl's argument applies equally well: the Rosenhof tooth is insufficiently far removed from the aurochs' size range to be accepted as definitely domestic (see Rowley-Conwy 1995).

For other bones in the skeleton, Degerbøl demonstrates that domestic males and wild females are similar in size (figs. 6.2–6.3). A bone sample containing both wild and domestic animals should thus contain three size groups: (1) the very large wild males, (2) the small domestic females, and (3) the intermediate group of both wild females and domestic males. Degerbøl's conclusion is straightforward: since wild males are commonly found on late Mesolithic settlements, but domestic females never are, it is logical to conclude that the intermediate size group represents only wild females and not domestic males. Given the considerable size difference between wild and domestic cattle, it is most likely that the domestic ones were introduced (Degerbøl and Fredskild 1970, 134). Two distal metacarpals from Rosenhof in northern Germany are plotted in figure 6.3; the smaller has been claimed as definitely domestic, the larger as falling in the "wild-domestic-transitional-field" (*Ur-Hausrind-Übergangsfeld*; Nobis 1975). Degerbøl's argument continues to apply, however, and it is likely that both specimens are from wild females (Rowley-Conwy 1995). The unlikelihood of anyone keeping very small numbers of wild and transitional cattle in two separate genetic stocks is reiterated.

There is therefore no good evidence for domestic cattle in the Danish late Mesolithic, though direct dates on some specimens are needed. The earliest published bones from Denmark likely to come from domestic cattle are those from Åkonge, where 16 fragments appear alongside very large numbers of red deer and wild boar. The site is ^{14}C dated to ca. 3000 B.C. (uncalibrated)—right at the transition to the Neolithic (Gotfredsen 1998). It is important that the bones themselves be directly dated, but their domestic status is supported both metrically (Gotfredsen 1998, fig. 3) and by the fact that Åkonge lies on the island of Zealand, on which aurochs had long been extinct (Aaris-Sørensen 1980). The cattle must therefore have been introduced. In northern Germany, two late Mesolithic scapulae are claimed to come from domestic animals, one from Rosenhof and one from Bregendtwedt-Förstermoor at Satrup (Nobis 1962, 1975). This is, however, based on collum length (the width of the "neck," measurement SLC according to von den Driesch 1976), an even less reliable element than the proximal metacarpals from Hungary. It is highly age dependent: for example, in female red deer, it increases by 50% after fusion (Legge and Rowley-Conwy 1988).

Pigs

As with cattle, wild pigs are present throughout Europe. This has often caused problems for separating wild from domestic. But the arguments have been more complex than for cattle: some authorities, having failed to demonstrate whether the animals were wild or domestic, have gone on to question whether the distinction between wild and domestic is at all meaningful. In other words, osteological uncertainty is assumed to mean behavioral intermediacy—not necessarily a valid extrapolation.

A good example comes from the work of Jarman (1976, 528), considering the pigs from the north Italian Neolithic site of Molino Casarotto. He states (528) that the Molino Casarotto pigs "bridge the accepted size ranges of wild pigs and Neolithic domestic pigs from sites such as Seeberg Burgäschisee-Süd. Furthermore, there is no indication that we are dealing with two separate populations of pigs as regards size, as no strongly bi-modal tendency is apparent in the size distribution of the bones." The difficulty of classifying as clearly "wild" or "domestic" such practices as Medieval pannage (driving pigs into woodland to feed on acorns) or some New Guinea pig husbandry is then often invoked in support of the "intermediate" behavioral interpretation (Jarman and Wilkinson 1972).

This argumentation is questionable. First, the comparison between Seeberg Burgäschisee-Süd and Molino Casarotto is dubious. Lower M3 length (the only measurement actually listed by Jarman for Molino Casarotto) is plotted in figure 6.4. Many specimens at Molino Casarotto are indeed smaller than the wild boar at Seeberg—but Molino Casarotto is to the south of the Alps, while Seeberg is to the north. The mountains separate two wild boar populations occupying different climatic and vegetational zones. Like many other animals, wild boars diminish in size as temperatures increase (for archaeological examples, see Davis 1981; Rowley-Conwy 1995). Animals at the two sites would therefore be expected to differ in size. Seeberg is therefore not an appropriate comparison.

But does the absence of visible bimodality at Molino Casarotto indicate that there was a single pig population living in some "intermediate" manner? This is doubtful; one may ask, how does an "intermediate" pig actually live? Most exploitation practices can in fact be categorized as "wild" or "domestic" fairly easily. Medieval pannage was carefully regulated, and the pigs were in no sense "feral" or "semi-wild." In medieval Welsh laws, "Pannage . . . was reserved for the animals of authorized persons during a defined season in autumn and early winter. . . . These woods were guarded, and the entry of

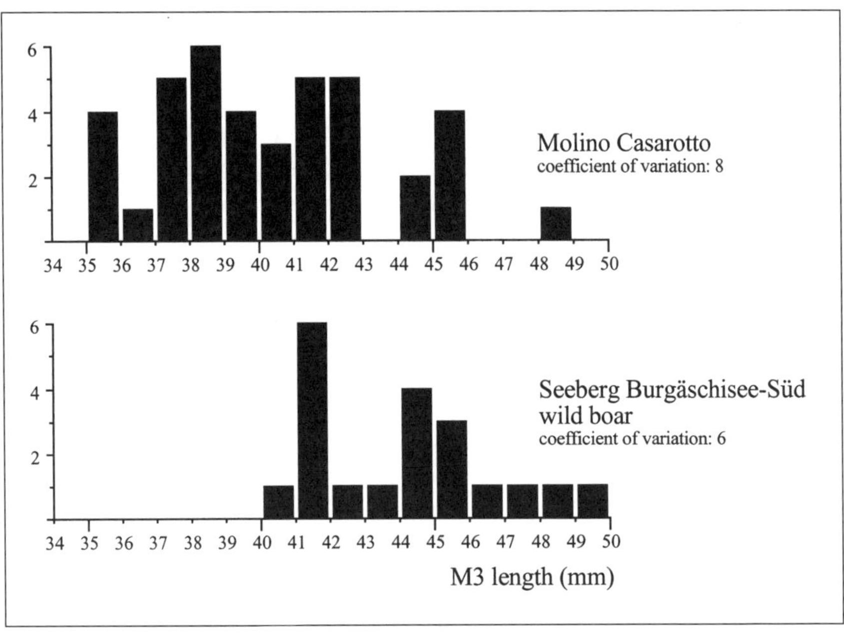

Fig. 6.4. Lengths of lower third molars of pigs. Molino Casarotto from Jarman (1976, table 4); Seeberg Burgäschisee-Süd from Boessneck et al. (1963, table 8).

unauthorized swine during this close season was an act of trespass to be compensated for or punished as such" (Linnard 1982, 16). Pigs at pannage were accompanied by swineherds (Edlin 1970, 97); the Medieval *Luttrell Psalter* contains an illustration of a swineherd up an oak tree shaking acorns onto the ground for his pigs to consume (the relevant illustration is folio 59, verso). There is no difficulty in regarding such pigs as fully domestic, even though they did not live in a field like sheep. Most practices in New Guinea, such as those listed by Rosman and Rubel (1989), are also classifiable. The "intermediate" situations involve female pigs roaming freely in and around a village, and breeding with feral boars encountered outside the village. Among the Etoro, litters of piglets are brought into the longhouses. "The piglets are subsequently fed and fondled for three to six months so that they will develop a permanent attachment to their owners. . . . Each piglet is individually named, its ears are clipped to make it readily distinguishable from wild pigs (which are hunted), and the males are castrated." As they get older they wander more widely, but are "frequently encountered in the course of daily activities. On these occasions a pig is invariably called by name, stroked, scratched, and fed bits of food in order to renew its familiarity with it. Any pig, whether recently reared or mature, will be sought out by its owner to receive such ministrations if it has not been sighted by some member of the community for more than a week" (Kelly 1988, 115–6). Why should pigs like this not be considered fully domestic? The crucial point is that by breeding with wild males, there is no genetic distinction between wild and domestic. A bone assemblage from an Etoro village would thus have the characteristics of a single population, although it comprises both wild and domestic individuals.

In prehistoric Europe, any pig assemblage that has the characteristics of a single population could thus be fully domestic, fully wild, or represent an Etoro-like situation where domestic females breed with wild males. A single-population assemblage is thus quite difficult to interpret. For Molino Casarotto, Jarman gives no reason for choosing any one of these in preference to another. But has Jarman demonstrated that only one population is present at Molino Casarotto?

A significant development in the study of pig populations has been the work of Payne and Bull (1988), in which the metrical parameters of a single population are established. For any given measurement (e.g., lower M3 length), the mean and standard deviation are calculated. Pearson's coefficient of variation (the standard deviation as a percentage of the mean) is a measure

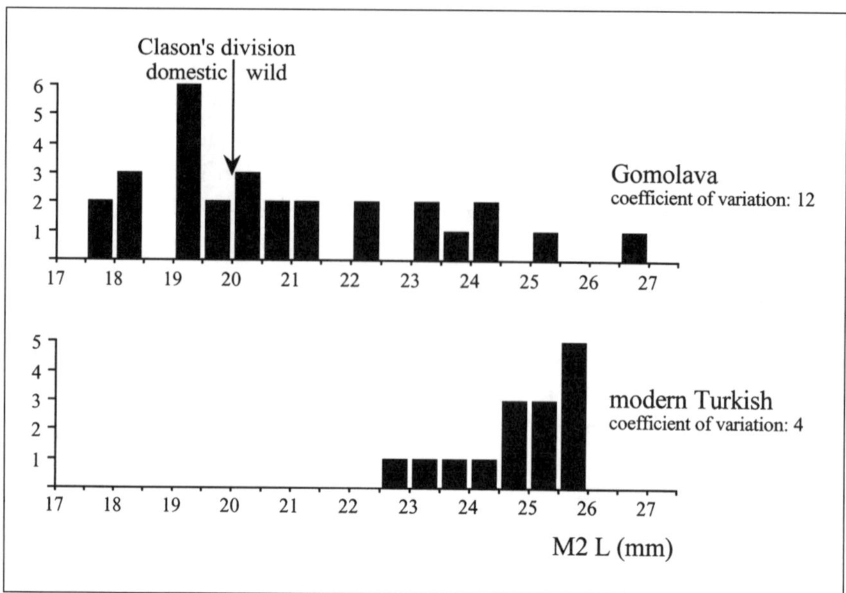

Fig. 6.5. Lengths of lower second molars of pigs. Gomolava from Clason (1979, table 6); modern Turkish from Payne and Bull (1988, table 1a).

of the spread of the measurements independent of the absolute size of the individuals. Figure 6.5 presents an example, based on M2 because insufficient M3s were available to Payne and Bull. The modern Turkish distribution is visibly tight and unimodal, and it has a coefficient of variation of 4. The Neolithic site of Gomolava in Yugoslavia was examined by Clason (1979), who believed that both wild and domestic pigs were present (fig. 6.5). The Gomolava spread is wide; Payne and Bull applied their method to these measurements, and a coefficient of variation of 12 was obtained. This must mean that two genetically separate populations of different sizes are represented. If all pigs were part of a single gene pool, the spread would be similar to the modern Turkish sample. The only way that two separate populations could exist is if one were wild, the other domestic. The domestic animals must have been under close human control. If any significant number of illicit liaisons in the undergrowth had occurred between wild boar and domestic females, the size difference between the two populations would disappear. This conclusion of Payne and Bull's thus supports Clason's contention that both wild and domestic pigs were present—even though the Gomolava sample is not bimodal, and whether or not Clason's actual point of division is correct.

Payne and Bull's method is a useful tool for examining assemblages. On occasion, interpretation becomes difficult if only few wild individuals are hunted because the coefficient of variation does not depart very far from that expected in one population (Rowley-Conwy 2000b). This may be the case at Molino Casarotto, where the coefficient for M3 length is 8—only a little larger than expected for a single population (cf. the value of 6 for Seeberg). More data might resolve this. But the Molino Casarotto sample in figure 6.4 shares one typical feature with many other assemblages: a major peak of smaller individuals and a rightward-projecting tail of fewer larger ones. This is the classic distribution expected if just a few large wild animals were hunted to supplement the main domestic kill (cf. also Uerpmann's sheep from Çayönü in fig. 6.1). Such a pattern is typical for most sites in Neolithic Europe; three examples from different regions are given in figure 6.6. The widespread nature of this two-population pattern is revealed by its occurrence at Peschany 1, a first-millennium B.C. site near Vladivostok on the Sea of Japan (Rowley-Conwy 2000; Rowley-Conwy and Vostretsov 1997).

When the two-population pattern is present, there is little doubt that one population must comprise fully domestic pigs under close control, and there is no need to invoke any intermediate status. Single-population distributions are rare in Neolithic Europe, but when they do occur the problem is more difficult. This is the case for two areas: the island of Gotland in the Baltic and parts of the western Mediterranean.

Middle Neolithic assemblages from Gotland provide single-population patterns, and opinion is divided as to whether the pigs were domestic or wild (Lindquist and Possnert 1997 argue for domestic status; Rowley-Conwy and Storå 1997 for wild). The single population pattern is in contrast to that put forward by Benecke (1993), who argued that local domestication was occurring. But there is no evidence for this in the results obtained by Rowley-Conwy and Storå (1997). The status of Neolithic pigs in the east of the Baltic is little known; sometimes small numbers of domestic animals are claimed (e.g., Dolukhanov 1979) but few metrical data are available. Until more data are published, claims for later Mesolithic "intensification" put forward by Zvelebil (1995) are hard to sustain. If late Mesolithic human populations were larger, for example, because they were coastal, then increased hunting pressure might take place, but evidence for a trend toward local domestication is lacking (Rowley-Conwy and Storå 1997).

The situation in the western Mediterranean is also complex. At Arene

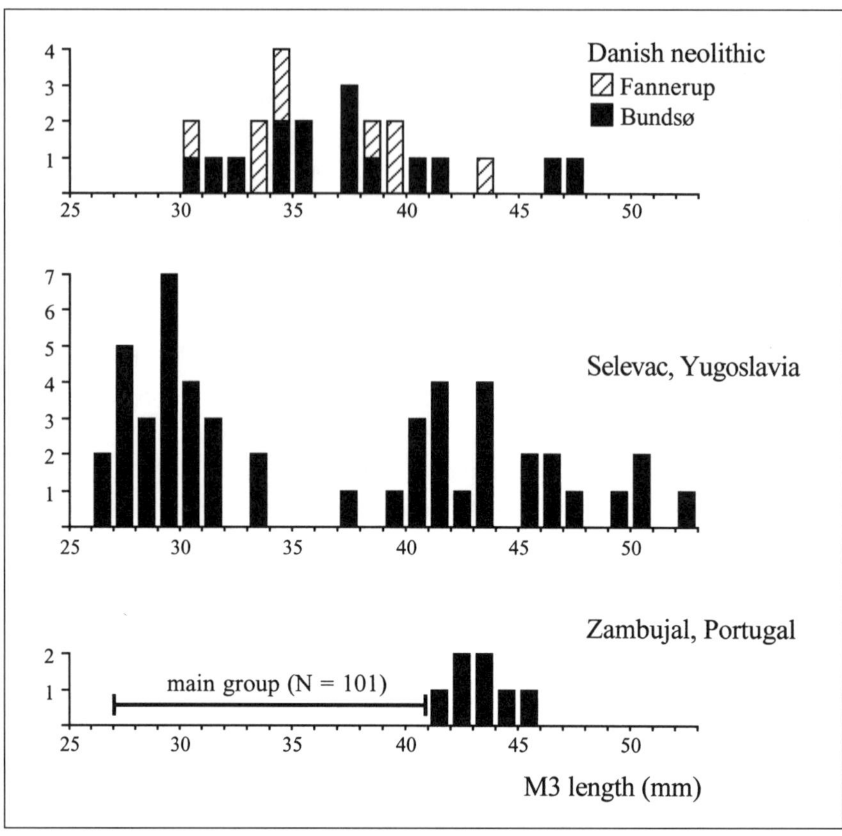

Fig. 6.6. Lengths of lower third molars of pigs. Bundsø from Degerbøl (1939, table 9); Fannerup from Rowley-Conwy (1984, app. 2a); Selevac from Legge (1990, app. 5.1); and Zambujal from von den Driesch and Boessneck (1976, table 22).

Candide in Liguria, pigs are believed to have been wild for most of the Neolithic, based on the coefficients of variation of most (though not of all) elements and on the distribution of naturally shed milk teeth. Such teeth are shed during normal chewing and testify that live (and therefore domestic) animals were penned within the cave. They are found only in the terminal Neolithic and later, while those of cattle and caprines occur throughout the Neolithic sequence (Rowley-Conwy 1997). Shed caprine teeth are commonly found in Neolithic caves (Helmer 1984). The situation in southern France is unclear. Small numbers of Neolithic domestic pigs are claimed, for example, at Gazel (Geddes 1980). Samples are often small and highly fragmented, as at Dourgne (Geddes 1993), making interpretation difficult, but domestic pigs may be absent during the earliest Neolithic (Geddes 1980; Helmer 1987). A further complicating factor has been the claim for early Neolithic and even terminal Mesolithic domestic pigs at Sarsa, Nerja, and Parralejo in southeastern Spain (Boessneck and von den Driesch 1980). None of these caves is free of stratigraphic disturbances at the critical point, so that the bones in question cannot be ascribed to a cultural horizon with certainty (Zilhão 1993). Even if the bones are correctly dated, however, there is no reason to assume that they must derive from domestic animals; metrically they could equally well derive from wild boar (Rowley-Conwy 1995).

Conclusion

I have tried to argue that in most cases the appearance of domestic animals may have been a tidier process than is sometimes envisaged. Evidence for local domestication of cattle and pigs, whether in advance of or after the arrival of the conventional Neolithic, is very thin. In most areas, the evidence is most simply interpreted to support the hypothesis that all the major animal species were introduced from elsewhere—at about the same time. This was usually at the start of the conventional Neolithic, although not always. The western Mediterranean pig problem has been mentioned; and it is possible that in Norway and Ireland some Neolithic cultural items spread ahead of the domestic animals (Prescott 1996; Burenhult 1984, pers. comm.).

New zooarchaeological methods and ongoing zooarchaeological work have been crucial. We are now better equipped than before to understand measurement distributions, although there is clearly a long way to go. A second crucial feature has been the radiocarbon accelerator, which allows the dating of individual animal bones to test whether they are contemporary with the layers in which they are found or whether they are intrusive. This combination of new zooarchaeological and chronometric methods means that it is worth reanalyzing many bone assemblages excavated perhaps many years ago. Rather than dismissing them as useless, we should rework such assemblages in the full knowledge that we may not be able to extract from them as much information as we would hope. At the same time, they will provide some information—and the questions they raise will direct our research when new assemblages appear.

REFERENCES

Aaris-Sørensen, K. 1980. "Depauperation of the Mammalian Fauna of the Island of Zealand during the Atlantic Period." *Videnskabelig Meddelelser fra Dansk Naturhistorisk Forening* 142:131–8.

Benecke, N. 1993. "The Exploitation of Sus scrofa (Linné, 1758) on the Crimean Peninsula and in Southern Scandinavia in the Early and Middle Holocene: Two Regions, Two Strategies." In *Exploitation des Animaux Sauvages a Travers le Temps,* 233–45. Juan-les-Pins.

Boessneck, J., and A. von den Driesch. 1980. "Tierknochenfunde aus vier Südspanischen Höhlen." In *Studien über frühe Tierknochenfunde von der Iberischen Halbinsel* 7, edited by J. Boessneck and A. von den Driesch, 1–83. Munich.

Boessneck, J., J.P. Jéquier, and H.R. Stampfli. 1963. *Seeberg Burgäschisee-Süd: Teil 3, Der Tierreste.* Bern.

Bogucki, P. 1989. "The Exploitation of Domestic Animals in Neolithic Central Europe." In *Early Animal Domestication and its Cultural Context,* edited by P.J. Crabtree, D. Campana, and K. Ryan, 118–34. Philadelphia.

Bökönyi, S. 1974. *History of Domestic Mammals in Central and Eastern Europe.* Budapest.

———. 1977. "The Introduction of Sheep-Breeding to Europe." *Ethnozootechnie* 21:65–70.

Burenhult, G. 1984. *The Archaeology of Carrowmore.* University of Stockholm, Institute of Archaeology, Stockholm.

Cipolloni-Sampò, M. 1987. "Problèmes des débuts de l'economie de production en Italie sud-orientale." In *Premières Communautés Paysannes en Méditerranée Occidentale,* edited by J. Guilaine, J. Courtin, J.-L. Roudil, and J.-L. Vernet, 181–8. Paris.

Clason, A.T. 1979. "The Farmers of Gomolava in the Vinca and La Tène period." *Palaeohistoria* 21:41–81.

Courtin, J., J. Évin, and Y. Thommeret. 1985. "Révision de la stratigraphie et de la chronologie absolue du site de Châteauneaf-du-Martigues (Bouches-du-Rhône)." *L'Anthropologie* 89:543–56.

Davis, S.J.M. 1981. "The Effects of Temperature Change and Domestication on the Body Size of Late Plaistocene to Holocene Mammals of Israel." *Paleobiology* 7:101–14.

Degerbøl, M. 1939. "Dyreknogler." *Aarbøger for Nordisk Oldkyndighed og Historie* 1939:85–198.

———. 1942. "Et knoglemateriale fra Dyrholm-bopladsen, en ældre stenalder køkkenmødding." In *Dyrholm: En Stenalderboplads på Djursland,* edited by T. Mathiassen, M. Degerbøl, and J. Troels-Smith, 77–136. Copenhagen.

———. 1963. "Prehistoric Cattle in Denmark and Adjacent Areas." In *Man and Cattle,* edited by A.E. Mourant and F.E. Zeuner, 68–79. London.

Degerbøl, M., and B. Fredskild. 1970. *The Urus (Bos primigenius Bojanus) and Neolithic Domesticated Cattle (Bos taurus domesticus Linné) in Denmark.* Copenhagen.

Dolukhanov, P. 1979. *Ecology and Economy in Neolithic Eastern Europe.* London.

Driesch, A.E. von den. 1976. *A Guide to the Measurement of Animal Bones from Archaeological Sites.* Cambridge, Mass.

Driesch, A.E. von den, and J. Boessneck. 1976. "Die Fauna vom Castro do Zambujal." In

Studien über frühe Tierknochenfunde von der Iberischen Halbinsel 5, edited by A. von den Driesch and J. Boessneck, 4–129. Munich.

Ducos, P. 1977. "Le mouton de Châteauneuf-lez-Martiques." In *L'Élevage en Méditerrannée Occidentale*, edited by J.-L. Miege, 77–85. Paris.

Edlin, H.L. 1970. *Trees, Woods and Man*. 3rd ed. London.

Ewer, R.F. 1968. *Ethnology of Mammals*. London.

Fuller, W.A. 1960. "Behaviour and Social Organisation of the Wild Wood Bison of Wood Buffalo National Park, Canada." *Arctic* 13:3–19.

Geddes, D. 1980. *De la Chasse au Troupeau en Méditerranée Occidentale: Les Débuts de l'Elevage dans le Bassin de l'Aude*. Toulouse.

———. 1985. "Mesolithic Domestic Sheep in West Mediterranean Europe." *Journal of Archaeological Science* 12:25–48.

———. 1993. "La faune de l'abri de Dourgne: Paléontologie et paléoéconomie." In *Dourgne: Derniers Chasseurs-Collecteurs et Premiers Éleveurs de la Haute-Vallée de l'Aude*, edited by J. Guilaine, 365–97. Toulouse.

Gotfredsen, A.B. 1998. "En rekonstruktion af palæomiljøet omkring tre senmesolitiske bopladser i Store Åmose, Vestsjælland-baseret på pattedyr-og fugleknogler." *Geologisk Tidsskrift* 1998:92–103.

Guilaine, J. 1993. *Dourgne: Derniers Chasseurs-Collecteurs et Premiers Éleveurs de la Haute-Vallée de l'Aude*. Toulouse.

Heinrich, D. 1993. "Die Wirbeltierreste vom ellerbekzeitlichen Siedlungsplatz Schlamersdorf LA 5, Kreis Stormarn." *Zeitschrift für Archäeologie* 27:67–89.

Helmer, D. 1984. "Le parcage des moutons et des chèvres au neolithique ancien et moyen dans le sud de France." In *Animal and Archaeology*. Vol. 3, *Herders and Their Flocks*, edited by J. Clutton-Brock and C. Grigson, 39–45. *BAR-IS* 202. Oxford.

———. 1987. "Les suidés du Cardial: sangliers ou cochons?" In *Premières Communautés Paysannes en Méditerranée Occidentale*, edited by J. Guilaine, J. Courtin, J.-L. Roudil, and J.-L. Vernet, 215–20. Paris.

Higham, C.F.W., and M. Message. 1968. "An Assessment of a Society's Attitude towards Bovine Husbandry." In *Science in Archaeology*, edited by D. Brothwell and E.S. Higgs, 315–30. London.

Jaczewski, Z. 1958. "Reproduction of the European Bison, Bison bonasus (L.), in Reserves." *Acta Theriologica* 1:333–71.

Jarman, M.R. 1976. "Prehistoric Economic Development in Sub-Alpine Italy." In *Problems in Economic and Social Archaeology*, edited by G. de G. Sieveking, I.H. Longworth, and K.E. Wilson, 375–99. London.

Jarman, M.R., and P.F. Wilkinson. 1972. "Criteria of Animal Domestication." In *Papers in Economic Prehistory*, edited by E.S. Higgs, 83–96. Cambridge.

Kelly, R.C. 1988. "Etoro suidology: A Reassessment of the Pig's Role in the Prehistory and Comparative Ethology of New Guinea." In *Mountain Papuans*, edited by J.F. Weiner, 111–86. Ann Arbor.

Lasota-Moskalewska, A. 1998. "Archaeological Reconstruction of the Animal Husbandry at Deby in Kujavia, Poland." In *Harvesting the Sea, Farming the Forest*, edited by M. Zvelebil, R. Dennell, and L. Domanska, 135–6. Sheffield.

Legge, A.J. 1990. "Animals, Economy and Environment." In *Selevac: A Neolithic Village in Yugoslavia,* edited by R. Tringham, 216–42. Los Angeles.

Legge, A.J., and P.A. Rowley-Conwy. 1988. *Star Carr Revisited: A Re-Analysis of the Large Mammals.* London.

Linnard, W. 1982. *Welsh Woods and Forests: History and Utilization.* Cardiff.

Lindqvist, C., and G. Possnert. 1997. "The Subsistence Economy and Diet at Jakobs/Ajvide, Eksta Parish and Other Prehistoric Dwelling and Burial Sites on Gotland in Long-Term Perspective." In *Remote Sensing.* Vol. 1, edited by G. Burenhult, 29–90. Institute of Archaeology, University of Stockholm, Stockholm.

Nobis, G. 1962. "Die Tierreste prähistorischer Siedlungen aus dem Satrupholmer Moor (Schleswig-Holstein)." *Zeitschrift für Tierzüchtung und Zuchtungsbiologie* 77:16–30.

———. 1975. " Zur fauna des ellerbekzeitlichen Wohnplatzes Rosenhof in Ostholstein I." *Schriften des Naturwissenschaftlichen Vereins für Schleswig-Holstein* 45:5–30.

———. 1983. "Wild- und Haustierknochen des frühneolithischen Fundplatzes Siggeneben-Süd." In *Siggeneben-Süd. Ein Fundplatz der frühen Trichterbecherkultur an der holsteinischen Ostseeküste,* edited by J. Meurers-Balke, 115–8. Neumunster.

Payne, S., and G. Bull. 1988. "Components of Variation in Measurements of Pig Bones and Teeth, and the Use of Measurements to Distinguish Wild from Domestic Pig Remains." *ArchæoZoologia* 2:27–65.

Persson, P. 1999. *Neolitikums Början.* Department of Archaeology, University of Gothenburg, Gothenburg.

Poplin, F. 1979. "Origine du Mouflon de Corse dans une nouvelle perspective paléontologique: Par marronage." *Annales Génétique et Sélection Animale* 11:133–43.

Prescott, C. 1996. "Was There *Really* a Neolithic in Norway?" *Antiquity* 70:77–87.

Rosman, A., and P.G. Rubel. 1989. "Stalking the Wild Pig: Hunting and Horticulture in Papua New Guinea. In *Farmers as Hunters,* edited by S. Kent, 27–36. Cambridge.

Rowley-Conwy, P. 1984. "Mellemneolitisk økonomi i Danmark og Sydengland: Knoglefundene fra Fannerup." *Kuml* 1984:77–111.

———. 1995. "Wild or Domestic? On the Evidence for the Earliest Domestic Cattle and Pigs in South Scandinavia and Iberia." *International Journal of Osteoarchaeology* 5:115–26.

———. 1997. "The Animal Bones from Arene Candide: Final Report." In *Arene Candide: Functional and Environmental Assessment of the Holocene Sequence*, edited by R. Maggi, 153–277. Rome.

———. 1998. "Improved Separation of Neolithic Metapodials of Sheep (*Ovis*) and Goats (*Capra*) from Arene Candide Cave, Liguria, Italy." *Journal of Archaeological Science* 25:251–8.

———. 2000a. "Milking Caprines, Hunting Pigs: The Neolithic Economy of Arene Candide in Its West Mediterranean Context." In *Animal Bones, Human Societies*, edited by P. Rowley-Conwy, 124–32. Oxford.

———. 2000b. "East is East and West is West but Pigs Go On Forever: Domestication from the Baltic to the Sea of Japan." In *Current and Recent Research in Osteoarchaeology* 2, edited by S. Anderson, 35–40. Oxford.

Rowley-Conwy, P., and J. Storå. 1997. "Pitted Ware Seals and Pigs from Ajvide, Gotland:

Methods of Study and First Results." In *Remote Sensing,* vol. 1, edited by G. Burenhult, 113–27. Institute of Archaeology, University of Stockholm. Stockholm.

Rowley-Conwy, P., and Y.I. Vostretsov. 1997. "Animal Bones from Peschany 1: Peter the Great Bay, Russian Far East." *EAANnouncements* 23:4–6.

Sinclair, A.R.E. 1974. "The Social Organisation of the East African Buffalo (*Syncerus caffer* Sparrman)." In *The Behaviour of Ungulates and its Relation to Management,* vol. 2, edited by V. Geist and F. Walther, 676–89. Morges.

Uerpmann, H.-P. 1987. "The Origins and the Relations of Neolithic Sheep and Goats in the Western Mediterranean." In *Premières Communautés Paysannes en Méditerranée Occidentale*, edited by J. Guilaine, J. Courtin, J.-L. Roudil, and J.-L. Vernet, 175–9. Paris.

Vigne, J.-D. 1988. *Les Mammifères Post-Glaciaires de Corse. Étude Archéozoologique.* Paris.

Zilhão, J. 1993. "The Spread of Agro-Pastoral Economies across Mediterranean Europe: A View from the Far West." *Journal of Mediterranean Archaeology* 6:5–63.

Zvelebil, M. 1995. "Hunting, Gathering, or Husbandry? Management of Food Resources by the Late Mesolithic Communities of Temperate Europe." *MASCA Research Papers in Science and Archaeology* Suppl. 12:79–104.

NOTE

[1] I would like to thank Albert Ammerman and the University of Venice for organizing the conference at which an earlier version of this paper was delivered, and to all present for useful comments.

PART III

The Transition in Southern Europe

— 7 —

The Origins of the Greek Neolithic: A Personal View

Curtis Runnels

The presence of farmers employing a fully agricultural economy is documented in Greece at the beginning of the seventh millennium B.C. in calendar years. The abrupt appearance of this farming-village culture on Greek soil marks in traditional terms the beginning of the Neolithic period in Greece, which is contemporary with or earlier than all other Neolithic cultures in Europe.[1] Speculation concerning its origins has been ongoing for more than a century (Finlay 1869; Weinberg 1970; Theocharis 1973), and it is perhaps not overstating the case to say that theoretical interpretations have been conditioned by the limitations of the data and by the theoretical and social environments of archaeologists working on the problem. In the period from Oscar Montelius (1880s) to V. Gordon Childe (1920s) a hypothesis of diffusion, only rarely discussed as a formal process, was developed as the chief explanation for the appearance of Neolithic farmers in southeastern Europe, and this diffusionary hypothesis was the basis for mainstream thinking for many decades. It is surely not a coincidence that this hypothesis was articulated during a period that saw the spread of Europeans and European culture in the Americas, Australia, Africa, and Asia. Nor is it surprising that interest in the diffusionary hypothesis would fade at the end of the modern age of mass migration. Beginning in the 1960s, with a peak of interest in the late 1970s and early 1980s, a new hypothesis of agricultural origins and dispersals was introduced that shifted the explanatory focus away from diffusion. A new model was introduced that posited local experimentation with agricultural domesticates by indigenous peoples who selectively adopted elements (e.g., animal species, plant species, and pottery) from neighboring agricultural economies. As Albert Ammerman (1989) has argued, this "indigenist" hypothesis coincides with a period of nationalism and emerging ethnic self-consciousness among European scholars in particular. After the introduction of the indigenist hypothesis, the role of demic diffusion was downplayed or abandoned as the chief explanation for the origins of agriculture, although cultural, or stimulus, diffusion was invoked occasionally to explain the apparent spread of certain limited features of the Near Eastern agricultural economy to western Europe (Dennell 1983, 152–89). In the most fully developed expressions of the indigenist hypothesis (Dennell 1983; Gimbutas et al. 1989; Theocharis 1973; Zvelebil 1986), diffusion was rejected and the hypothesis of multiple independent episodes of agricultural origins, which occurred at different times in different geographic regions, was promulgated.

Cavalli-Sforza and Albert Ammerman challenged the received wisdom of the indigenists in 1971 by introducing the wave of advance model with an hypothesis of demic diffusion to account for the dispersal of agriculture from the Near East to Europe. Their new approach received its most sophisticated expression in the publication of their 1984 book on the Neolithic transition (Ammerman and Cavalli-Sforza 1984), which served to revive interest in the process of agricultural dispersals as a diffusionary process. It was also in the 1970s that archaeologists working in Greece and other Mediterranean lands began to make greater use of regional intensive surveys with a multidisciplinary structure involving archaeologists, geoarchaeologists, and other specialists. The new methods of fieldwork were better suited to the gathering of data useful for the study of agricultural origins (Runnels forthcoming) and their introduction was at least in part responsible for the rapidly increasing archaeological data for the Neolithic and pre-Neolithic periods that became available in the 1970s and 1980s. Regional surveys made it possible for the first time to evaluate hypotheses of agricultural origins, and by the 1990s accumulating evidence suggested that agriculture was introduced into Greece as the result of some form of diffusion and not as the result of independent development of agriculture by indigenous peoples.[2] The Neolithic economy of agriculture and village life was introduced to the Greek mainland in what appears to be an episode of demic diffusion, which most likely originated in Turkey, perhaps on the southern coast in the region of Antalya where there are numerous sites belonging to late Pleistocene and early Holocene foragers. Geography is a critical factor here. Greece is the part of Europe that lies closest to Turkey, and it is increasingly clear that the precocious pre-pottery and early pottery Neolithic cultures on the Anatolian plateau and in southeast Turkey are to be regarded as part of the larger core area where agricultural originated, and as such are the likely jumping off point for the early migrants to the Aegean.[3]

The recolonization of Greece after the apparent abandonment of the region at the end of the Pleistocene was probably carried out in a two-stage process similar to that proposed for northern Europe by Rupert Housley and his colleagues (1997). Early Holocene migrants to Greece brought with them a completely new cultural package of domesticated plants, animals, and material culture, and they established themselves more or less simultaneously in central and southern Greece and on some of the larger Greek islands, for instance Crete, Corfu, and perhaps some of the Cyclades.[4] The wave of advance model and hypothesis of demic diffusion have thus far proven to be robust methods for the study of the origins of the Greek Neolithic.

Investigating the Mesolithic-Neolithic Transition in Greece

My involvement and interest in these questions arose in the context of a seminar on the Franchthi Cave given by Thomas Jacobsen in 1973 at Indiana University, where I was a first-year graduate student. Having participated in the excavations at Franchthi, I was keenly aware that the work at this significant site was likely to contribute to the understanding of the transition from the Paleolithic to the Mesolithic and the Neolithic in southeastern Europe because of its long stratigraphic sequence covering the periods in question (Farrand 1993, 2000). In this seminar, the recent paper by Ammerman and Cavalli-Sforza (1971) was discussed and debated with some enthusiasm, and this debate continued to occupy our attention in later seminars at Indiana University, where the wave of advance model and demic diffusion hypothesis were compared with the work of archaeologists advocating independent agricultural origins (e.g., Gimbutas et al. 1989; Dennell 1983; Theocharis 1973). The majority opinion, as it was often expressed by my colleagues, was that Ammerman and Cavalli-Sforza were wrong, and most of the available evidence from Greece, including the Franchthi Cave sequence, could be interpreted as the result of local, independent experimentation with plant domestication which led to the independent evolution of the village farming way of life. The fluidity of opinions on this matter at the time is reflected in the work of one Aegean scholar, Colin Renfrew (1973a, 84–119), who stressed the internal dynamics of indigenous societies but who also drew attention to the strong evidence for demic diffusion in Greece before the Bronze Age (Renfrew 1973b, 1987, 145–59).

For my own part, I found Cavalli-Sforza and Ammerman convincing for a number of reasons. One was the abrupt appearance of a new and completely alien cultural pattern at Franchthi Cave[5] and other Greek sites at the beginning of the Neolithic period. Another reason, even more compelling, was the similarity of the structure, or composition, of the Neolithic package of cultural traits with the structure of contemporary Neolithic cultures at early sites as far afield as Anatolia, the Near East, Turkmenistan, and Pakistan. The Neolithic suite, as I chose to call the highly structured "package," includes village plans based on rectilinear buildings with stone foundations and mudbrick superstructures, often having internal buttresses; the use of painted pottery; the production of ceramic anthropomorphic figurines, typically of robust females either seated or standing with arms across the chest or abdomen; blade-based lithic technologies that made extensive use of raw materials imported from distant sources; polished stone axes; and mortuary practices (infant interments in domestic areas, secondary interments of adults, and cremation).

The core of the Neolithic suite over the vast geographic area encompassing the Aegean was an economy of food production based on domesticated forms of cattle, pigs, sheep, goats, barley, wheat, peas, and lentils, which dominate the new village economies, often at the expense of indigenous species of plants and animals that had formed the basis of Paleolithic and Mesolithic subsistence. It seemed to me then, as it does now, that the indigenist model could not account as economically as the diffusionary hypothesis for the fact that this suite has the same content and structure, with very little modification of the Near Eastern-Anatolian Neolithic economy. The wave of advance model and hypothesis of demic diffusion is the most economical way to explain the dispersal of the Neolithic suite that we observe in the Aegean.

In the 1970s, the chief obstacle to the evaluation of the indigenist and demic diffusion hypotheses was the lack of adequate information about both

Mesolithic and Early Neolithic sites and settlement patterns in Greece. Claims of substantial indigenous populations in the Mesolithic period (Dennell 1983, 152–68; Theocharis 1973, 17–58) immediately preceding the advent of the Neolithic could neither be supported nor rejected as long as Franchthi Cave and one other site (Sidari in Corfu) remained as the only excavated Mesolithic sites in all of Greece. Thirty years ago, the principal areas of dense Neolithic settlement such as Thessaly and Macedonia had not been investigated for Mesolithic sites in any systematic manner. My own involvement with early survey work around Franchthi in the mid 1970s convinced me that systematic regional surveys were essential for investigating the Mesolithic to Neolithic transition. An opportunity to put a plan into action presented itself when I was invited by the co-directors of the Stanford University Archaeological and Environmental Survey of the Southern Argolid, Michael Jameson and Tjeerd van Andel, to design and direct the archaeological field survey. One of the principal objectives of this survey was the investigation of the region around the Franchthi Cave for Paleolithic through Neolithic sites (Jameson et al. 1994; Runnels 1985; Runnels et al. 1995; Runnels and van Andel 1987; van Andel and Runnels 1987; van Andel et al. 1986). The success of this survey convinced me to carry out similar survey work elsewhere, often in close collaboration with Tjeerd van Andel, who contributed his geological and geomorphological expertise and insight. The purpose of these surveys was in part to evaluate the evidence for pre-Neolithic indigenous populations, and we selected regions for survey that had long been regarded as crucial for understanding the origins of the Greek Neolithic. Referring only to those surveys that have been completed and published, the areas covered include the northern Argolid, Thessaly, and Epirus.[6]

One important result of these surveys is definitive support for the diffusionary hypothesis of the origin of the Greek Neolithic. It should be noted in this context that these surveys were specially designed to detect late Paleolithic and Mesolithic sites. Intensive search methods were employed, such as tightly-spaced lines of fieldwalking, within a research design that made great use of geomorphological data in assessing the relative probability of finding sites in different parts of the landscape and in evaluating the effects of erosion and alluviation in destroying or obscuring late Pleistocene and early Holocene sites. The publication of the lithics from Sidari by A. Sordinas (1970) and Franchthi by C. Perlès (1990) provided the necessary information about the lithic industries we expected to encounter in a prospection for surface sites. As a consequence of these advantages, the results of the surveys can be taken as reasonably accurate assessments of the distribution and density of Mesolithic and other pre-Neolithic sites in Greece.

The results of the survey in Thessaly, which is manifestly at the center of developed Neolithic culture, are particularly significant. No Mesolithic sites were discovered in eastern Thessaly even though large areas were carefully searched. This finding is in accord with those from some other areas of the Balkans.[7] The discovery by a Greek team of the Directorate of Speleology and Palaeoanthropology of a Mesolithic site at Theopetra in the foothills of the Pindos Mountains on the western edge of the Thessalian region does not alter the conclusion that Mesolithic sites are manifestly rare in northern Greece (Runnels 1995). Mesolithic sites were discovered, however, in both the Argolid (Runnels 1996) and in Epirus (Runnels 1995; Runnels et al. forthcoming). In these regions the lithic industries found on Mesolithic sites are clearly similar to those from Sidari and Franchthi, and the discovery of sites in the same general regions as the only two sites already known to archaeology is not a wholly unexpected result. The implications of these findings are clear. Places with numerous early Neolithic sites, such as Thessaly were established in areas where there is no evidence for substantial native populations, a conclusion which strongly supports the hypothesis that the Neolithic village economy was introduced to these regions by immigrants (as it was on Crete; Broodbank and Strasser 1991). In Epirus and the Argolid, where Mesolithic populations are known to exist, the picture is somewhat more complicated. Later Neolithic sites are located in different parts of the landscape and there is very little evidence of cultural continuity between the Mesolithic and Neolithic sites (Jameson et al. 1994, 325–48), but Catherine Perlès has noted some continuity of Mesolithic lithic technology at Franchthi in the later Neolithic levels at the same site (Perlè 1987, 1990). This finding of limited continuity is not supported by other materials from the site that have been so far published (e.g., Hansen 1991), but Perlès has nevertheless used the lithics at Franchthi to suggest that there were two patterns of Greek Neolithization. In her view, regions without native populations (e.g., Thessaly and Macedonia) were occupied by immigrant farmers, while regions such as the Argolid and Epirus with small indigenous populations of Mesolithic foragers made the transition to farming through a process that may have involved both demic and cultural diffusion (in the form of contact between natives and immigrants), perhaps a form of the "mutualism" discussed by Ammerman and Cavalli-Sforza.[8]

Conclusion

The hypothesis of demic diffusion as an explanation of the origins of the Neolithic in Greece has been amply supported by data collected in the last 25 years. The wave of advance model adequately describes the process of agricultural dispersals and, when combined with the demic diffusion hypothesis, such dispersals can be viewed as a function of local population dynamics on frontiers. It is a sound mechanism that explains the movement of agricultural populations on both a local and a continental scale. It is nevertheless not so much an explanation of the phenomenon of agricultural dispersal as it is a quantitative description of the process, and the specific explanation of the movement of farming peoples from Anatolia to the Aegean must take into account a range of factors that includes, but is not limited to, a consideration of individual decision making and human volition, the consideration of topographic features such as mountains and bodies of water that may have affected local rates of movement, and the different scales of agricultural dispersals in regions of small size, perhaps consisting of no more than a single island or drainage, or small aggregates of such areas. Some of these factors are difficult to include in any theory of agricultural dispersals at this time, although some attempts have been made to consider topographic features and the element of human volition when investigating demic diffusion in individual regional case studies (e.g., Runnels 1989; Runnels and van Andel 1988).

There is good reason to think that the demic diffusion hypothesis can be sharpened further by dividing the process into two stages, similar to those proposed for the recolonization of northern Europe after the last deglaciation (Housley et al. 1997). For Housley and his colleagues the colonization of uninhabited territory consisted of two different but interrelated stages of one process, with the first stage of exploration designated as the initial "pioneer" phase of colonization followed by a second "residential" phase of more substantial logistical land use, which may or may not have been based on permanent settlements (Housley et al. 1997, 44–7). This attempt to model demic diffusion on a regional scale is applicable also to the Greek Mesolithic-Neolithic transition, where any attempt to explain the demic diffusion that brought Neolithic farmers to Greece can be seen as benefiting from significant information about regional conditions and geography from earlier visitors (Broodbank and Strasser 1991) and based squarely on the driving force of human volition pulled forward by identifiable incentives such as preferred habitats with perennial water sources.[9]

In my view, the regional model of the Neolithization of Greece is analogous to the recolonization of recently deglaciated parts of northern Europe. As a working hypothesis, we can assume there was no substantial late Pleistocene population in mainland Greece, where most Paleolithic sites appear to be unoccupied after about 13,000 B.P. (Runnels 1995), and it is at least worth considering the possibility that Greece was essentially uninhabited at the beginning of the Holocene. If this proves to be the case, it may be worth reconsidering the view put forth 20 years ago by Grahame Clark (1980, 59–100) that the Mediterranean Mesolithic was the result of the spread of seafaring foragers around the Mediterranean basin in the early Holocene in response to major environmental changes that followed the end of the Pleistocene.

On the Greek mainland, the movement of peoples in the Mesolithic was both small and very localized, and all but one of the known sites is located directly on the coast in the vicinity of areas with abundant freshwater and marine resources (Runnels 1995). In this model, the resettlement of Greece in the Mesolithic period may be viewed as the "pioneer" phase of a process of demic diffusion that later would bring Neolithic farmers from Anatolia to Greece in a "residential" phase of the same process (Runnels 1989; van Andel and Runnels 1995). In this reformulation, the wave of advance model remains at the center of any discussion of agricultural dispersals, with future progress expected by focusing on the leading edge of the wave of advance where it intersects with specific, often small, regions where the two-phase process of colonization of new territory may one day be observed in detail.

REFERENCES

Ammerman, A.J. 1989. "On the Neolithic Transition in Europe: A Comment on Zvelebil and Zvelebil (1988)." *Antiquity* 63:162–5.

Ammerman, A.J., and L.L. Cavalli-Sforza. 1971. "Measuring the Rate of Spread of Early Farming in Europe." *Man* 6:674–88.

———. 1984. *The Neolithic Transition and the Genetics of Populations in Europe*. Princeton.

Andreou, S., M. Fotiadis, and K. Kotsakis. 1996. "The Neolithic and Bronze Age of Northern Greece." *American Journal of Archaeology* 100:537–97.

Bailey, G.N., E. Adam, E. Panagopoulou, C. Perlès, and K. Zachos, eds. 1999. *The Palaeolithic Archaeology of Greece and Adjacent Areas: Proceedings of the ICOPAG Conference*. Athens.

Broodbank, C., and T.F. Strasser. 1991. "Migrant Farmers and the Neolithic Colonization of Crete." *Antiquity* 65:233–45.

Chapman, J.C., R.S. Shiel, and S. Batovi. 1987. "Settlement Patterns and Land Use in Neothermal Dalmatia, Yugoslavia: 1983–1984 Seasons." *Journal of Field Archaeology* 14:123–46.

Clark, G. 1980. *Mesolithic Prelude: The Palaeolithic-Neolithic Transition in Old World Prehistory*. Edinburgh.

Demoule, J.-P., and C. Perlès. 1993. "The Greek Neolithic: A New Review." *Journal of World Prehistory* 7:55–416.

Dennell, R. 1983. *European Economic Prehistory: A New Approach*. London.

Farrand, W.R. 1993. "Discontinuity in the Stratigraphic Record: Snapshots from Franchthi Cave." In *Formation Processes in Archaeological Context*, edited by P. Goldberg, D.T. Nash, and M.D. Petraglia, 85–96. Madison.

———. 2000. *Depositional History of Franchthi Cave: Sediments, Stratigraphy, and Chronology. Franchthi* 12. Bloomington.

Finlay, George. 1869. *Παϱατηϱησεις επιτης ευ Ελβετι και Ελλαδι Πϱοιστοϱικης Αϱχαιολογι ας* [*Remarks on the Prehistoric Archaeology of Switzerland and Greece*]. Athens.

Gimbutas, M., S. Winn, and D. Shimabuku. 1989. *Achilleion: A Neolithic Settlement in Thessaly, Greece, 6400–5600 BC*. Los Angeles.

Hansen, J.M. 1988. "Agriculture in the Prehistoric Aegean: Data Versus Speculation." *American Journal of Archaeology* 92:39–52.

———. 1991. *The Palaeoethnobotany of Franchthi Cave. Franchthi* 7. Bloomington.

Honea, K. 1975. "Prehistoric Remains on the Island of Kythnos." *American Journal of Archaeology* 79:277–9.

Housley, R.A., C.S. Gamble, M. Street, and P. Pettit. 1997. "Radiocarbon Evidence for the Lateglacial Human Recolonisation of Northern Europe." *Proceedings of the Prehistoric Society* 63:25–54.

Jameson, M.H., C.N. Runnels, and Tj.H. van Andel. 1994. *A Greek Countryside: The Southern Argolid from Prehistory to the Present Day*. Stanford.

Kyparissi-Apostolika, N. 1994. "Prehistoric Inhabitation in Theopetra Cave, Thessaly." In *La Thessalie*, 103–8. Athens.

Özdoğan, M., and N. Basgelen, eds. 1999. *Neolithic in Turkey: The Cradle of Civilization*. Istanbul.

Papathanassopoulos, G.A., ed. 1996. *Neolithic Culture in Greece*. Athens.

Perlès, C. 1987. "Les industries du néolithique 'préceramique' de Grèce: Nouvelles études, nouvelles interpretations." In *Chipped Stone Industries of the Early Farming Cultures in Europe*, 19–39. Warsaw.

———. 1988. "New Ways with an Old Problem: Chipped Stone Assemblages as an Index of

Cultural Discontinuity in Early Greek Prehistory." In *Problems in Greek Prehistory*, edited by E.B. French and K.A. Wardle, 477–87. Bristol.

———. 1990. *Les Industries Lithiques taillées de Franchthi (Argolide, Grèce).* Tome II, *Les Industries du Mésolithique et du Néolithique Initial. Franchthi* 5. Bloomington.

———. 1995. "La transition Pleistocène/Holocène et le problème du Mésolithique en Grèce." In *Los Ultimos Cazadores: Transformaciones Culturales y Economicas durante el Tardiglaciar y el Inicio del Holocene en el Ambito Mediterraneo*, edited by V.V. Bonilla, 179–206. Alicante.

Renfrew, C. 1973a. *Before Civilization: The Radiocarbon Revolution and Prehistoric Europe.* New York.

———. 1973b. "Problems in the General Correlation of Archaeological and Linguistic Strata in Prehistoric Greece: The Model of Autochonous Origin." In *Bronze Age Migration in the Aegean*, edited by R.A. Crossland and A. Birchall, 263–76. London.

———. 1987. *Archaeology and Language: The Puzzle of Indo-European Origins.* New York.

Runnels, C. 1985. "Trade and the Demand for Millstones in Southern Greece in the Neolithic and the Early Bronze Age." In *Prehistoric Production and Exchange. The Aegean and Eastern Mediterranean*, edited by B. Knapp and T. Stech, 30–43. Los Angeles.

———. 1988. "A Prehistoric Survey of Thessaly: New Light on the Greek Middle Palaeolithic." *Journal of Field Archaeology* 15:277–90.

———. 1989. "Trade Models in the Study of Agricultural Origins and Dispersals." *Journal of Mediterranean Archaeology* 2:149–56.

———. 1995. "The Stone Age of the Aegean from the Palaeolithic to the Advent of the Neolithic." *American Journal of Archaeology* 99:699–728.

———. 1996. "The Palaeolithic and Mesolithic Remains from the Berbati-Limnes Survey." In *The Berbati-Limnes Archaeological Survey 1988–1990*, edited by B. Wells and C. Runnels, 23–34. Stockholm.

———. Forthcoming. "Anthropogenic Soil Erosion in Prehistoric Greece: The Contribution of Regional Surveys to the Archaeology of Environmental Disruptions and Human Response." In *Environmental Disruptions and the Archaeology of Human Response*, edited by G. Bawden. Albuquerque.

Runnels, C., D.J. Pullen, and S. Langdon, eds. 1995. *Artifact and Assemblage: The Finds from a Regional Survey of the Southern Argolid.* Vol. 1, *The Prehistoric Pottery and the Lithic Artifacts.* Stanford.

Runnels, C., and T. van Andel. 1987. "The Evolution of Settlement in the Southern Argolid, Greece: An Economic Explanation." *Hesperia* 56:303–34.

———. 1988. "Trade and the Origins of Agriculture in the Eastern Mediterranean." *Journal of Mediterranean Archaeology* 1:83–109.

———. 1993a. "The Lower and Middle Paleolithic of Thessaly, Greece." *Journal of Field Archaeology* 20:299–317.

———. 1993b. "A Handaxe from Kokkinopilos, Epirus, and Its Implications for the Paleolithic of Greece." *Journal of Field Archaeology* 20:191–203.

Runnels, C., and T.H. van Andel. Forthcoming. "The Early Stone Age Prehistory of the Nome of Preveza (Greece): A Paleoenvironmental and Archaeological Study of Landscape and Settlement." In *Landscape Archaeology in Southern Epirus, Greece,* edited by J.R. Wiseman and K. Zachos. Princeton.

Runnels, C., T.H. van Andel, K. Zachos, and P. Paschos. 1999. "Human Settlement and Landscape in the Preveza Region, Epiros, in the Pleistocene and Early Holocene." In *The Palaeolithic Archaeology of Greece and Adjacent Areas. Proceedings of the ICOPAG Conference, Ioannina,* edited by G. Bailey, 120–9. Athens.

Sordinas, A. 1970. *Stone Implements from Northwestern Corfu, Greece.* Memphis.

Theocharis, D. 1973. *Neolithic Greece.* Athens.

van Andel, T.H., and C. Runnels. 1987. *Beyond the Acropolis: A Rural Greek Past.* Stanford.

———. 1995. "The Earliest Farmers in Europe." *Antiquity* 69:481–500.

van Andel, T.H., C. Runnels, and K. Pope. 1986. "Five Thousand Years of Land Use and Abuse in the Southern Argolid." *Hesperia* 55:103–28.

Weinberg, S.S. 1970. "The Stone Age in the Aegean." *Cambridge Ancient History* I.1:557–672.

Wells, B., and C. Runnels, eds. 1996. *The Berbati-Limnes Archaeological Survey 1988–1990.* Stockholm.

Wells, B., C. Runnels, and E. Zangger. 1990. "The Berbati-Limnes Valley Project: The 1988 Season." *Opuscula atheniensia* 18:207–38.

Zvelebil, M. 1986. "Mesolithic Societies and the Transition to Farming: Problems of Time, Scale and Organisation." In *Hunters in Transition,* edited by M. Zvelebil, 167–88. Cambridge.

NOTES

[1] For a review of the Greek Neolithic, see Andreou et al. (1996), Demoule and Perlès (1993), and the excellent collection of articles in Papathanassopoulos (1996); Runnels (1995) reviews the pre-Neolithic cultures in Greece.

[2] The evidence for the indigenous development or autochthonous origins of the Greek Neolithic is summarized by Demoule and Perlès (1993), where a modified model of demic diffusion allowing for limited interaction between incoming farmers and local people in some areas is followed. Perlès (1995) elaborates on these views. Hansen (1988, 1991) discusses the paleobotanical evidence for early agriculture. Demic diffusion is discussed specifically by Broodbank and Strasser (1991), Runnels (1989, 1995), and van Andel and Runnels (1995). A perusal of the papers in Papathanassopoulos (1996) indicates that the demic diffusion hypothesis now has more adherents than the indigenist hypothesis.

[3] For the new picture of the Turkish Neolithic as part of the Near Eastern core area for agricultural origins, see Özdogan and Basgelen (1999).

[4] For Knossos on Crete, see Broodbank and Strasser (1991); for Sidari on Corfu, see Sordinas (1970); Maroula on Kythnos first reported by Honea (1975) may be Mesolithic, but awaits systematic investigation.

[5] Not all authors accept this (e.g., Demoule and Perlès 1993, 365).

[6] Argolid: Wells and Runnels (1996) and Wells et al. (1990); Thessaly: Runnels (1988) and

Runnels and van Andel (1993a); Epirus: Runnels and van Andel (1993b, forthcoming) and Runnels et al. (forthcoming).

[7] For the excavations at the Theopetra Cave, see Kyparissi-Apostolika (1994), Papathanassopoulos (1996, 67–8), and the reports on Theopetra in Bailey et al. (1999). For the paucity of the Mesolithic in the Balkans, see Chapman et al. (1987).

[8] Perlès's views are laid out in a series of publications (Perlès 1987, 1988, 1995); for mutualism, see Ammerman and Cavalli-Sforza (1984, 62, passim).

[9] In the case of Thessaly, for example, the early Neolithic agriculturalists were drawn to the large, unpopulated floodplains of the Peneios River (van Andel and Runnels 1995).

— 8 —

A Review of the Late Mesolithic in Italy and Its Implications for the Neolithic Transition

Paolo Biagi

In Italy, the question of the Mesolithic-Neolithic transition began to arise when thick stratigraphic sequences with evidence of both Neolithic and lower-lying, pre-Neolithic assemblages were discovered. In the 1940s, the excavations carried out by L. Cardini and L. Bernabò Brea at the Arene Candide Cave in Liguria (Bernabò Brea 1946; Maggi et al. 1997) produced the first evidence for occupation levels and human burials which Cardini (1980) thought were Mesolithic. Radiocarbon dates subsequently showed they were older in age, dating them to the so-called final Epigravettian culture just before the Holocene (Maggi 1997). At the end of the 1960s, there was the recovery of long Mesolithic sequences at several sites in the Adige Valley of the Alps near Trent. At the opposite end of Italy, the excavations carried out a few years later at the Uzzo Cave in Sicily (Tagliacozzo 1993) led to the discovery of another Mesolithic sequence, which involved an approach to subsistence that was different from the one observed in the Adige Valley.

In the Trentino, the Mesolithic assemblages looked similar to those already known from other south European sites and also to those found at a few caves near Trieste in northeastern Italy. The lower parts of the early Holocene sequences in the Trentino are characterized by Sauveterrian assemblages, whose internal development can be traced over the course of the Preboreal and the Boreal climatic stages. The upper parts have Castelnovian industries—previously called Tardenoisian (Broglio 1971)—that date to the beginning of the Atlantic climatic stage. In the case of the Uzzo Cave in western Sicily, the

Mesolithic levels produced lithic assemblages (whose typological characteristics are still not well defined; Costantini et al. 1987) that are different from those found in northern Italy. At the time of their discovery, the sequences in both places were held to be "continuous" in nature: they were considered to have no significant gaps or breaks in them (at least when it came to the study of the Mesolithic–Neolithic transition). Now things are seen in a different light in both cases. More recently, the radiocarbon dating of the levels at the rock shelters in the Adige Valley and the Uzzo Cave, the study of their assemblages, and the sedimentary analysis of stratigraphic units at the sites all point toward the incompleteness of the respective sequences (Biagi and Spataro 2001).

In the 1980s, as part of the growing interest in the question of the Neolithic transition, some scholars working elsewhere in Europe began to turn their attention to the Italian sites essentially for the first time (Lewthwaite 1986). Previously, the Italian sites, including the ones mentioned above, had been largely ignored in the Anglo-American literature. There were, in fact, few scholars outside of the country who had firsthand knowledge of the Mesolithic sites in Italy. The debate at the time centered essentially on a tug-of-war between diffusionist and indigenist models of the Neolithic transition (Zvelebil 1986b). The former followed in some ways the ideas already put forward by V.G. Childe (1925). But this position was reformulated by Ammerman and Cavalli-Sforza (1973) in terms of demic diffusion and the wave of advance model. The latter position was advocated by M. Zvelebil (1986a), whose ideas were based initially on what he observed in the Baltic region of northern Europe (Zvelebil and Rowley-Conwy 1984). The purpose of this chapter is to review the evidence for late Mesolithic occupation in Italy as its study has developed over the course of the last 50 years and to discuss some of the implications of the Italian evidence for the current debate over the Neolithic transition in Europe. Here one of the aims will be to clear up some of the misconceptions about the situation in Italy that have accumulated in the secondary literature in recent years.

The Mesolithic Evidence

The sites of Vatte Zambana, Romagnano III, and Pradestel in the Adige Valley were all discovered at the end of the 1960s to have early Holocene sequences. Since then, prehistorians in Italy have been divided into two main schools of thought. The first simply denies that the term Mesolithic has an independent cultural meaning and uses the term Epipaleolithic with reference to all of the north Italian complexes of early Holocene age (Broglio 1971). Such a point of view is closely linked with the characteristic chronostratigraphic perspective of G. Laplace (1964), which is based almost exclusively on the typological study of the chipped stone assemblages in order to trace their internal patterns of development over time. The second group of researchers led by A.M. Radmilli (1960) uses the term Mesolithic essentially in an economic sense. Radmilli viewed the Mesolithic as a transitional epoch of food shortage between the end of the Paleolithic and the start of the Neolithic—a time when foraging and the gathering of marine resources made a growing contribution to subsistence (Radmilli and Tongiorgi 1958). M. Taschini proposed a third position when she published the radiocarbon dates from the rock shelter called Riparo Blanc in central Italy (Taschini 1968). She emphasized the autonomy of the term as a cultural phenomenon and held that the Mesolithic was not characterized by a subsistence economy that was poorer than that of the late Paleolithic. Instead, she considered the period to involve, in a more positive sense, a broader range of choices in the exploitation of environmental resources. On a separate front, some order among the various lithic industries was established by A. Bietti (1981), who used the method of cluster analysis to work out the patterns of association between the assemblages from different Mesolithic sites in Italy.

At the present time, northern Italy is one of the parts of Europe where the last Holocene hunter-gatherers have been studied in some detail (figs. 8.1–8.3). The areas where Mesolithic sites are best represented are the eastern and central Alpine arc (from Trieste in the east to Lombardy in the west) and the Apennine territories of eastern Liguria, Tuscany, and Emilia. In contrast, very few sites are known on the plains of the Po Valley (Biagi 2001). All of the Mesolithic sites near Trieste are located in karst caves. Commonly, they have yielded lithic assemblages (based for the most part on local raw materials), which span the time from the Preboreal through the early Atlantic. Following the re-examination of the lithic assemblages from one of most important sequences, that found at the Azzurra Cave (Ciccone 1993, 42), it is now clear that the Castelnovian, the last part of the Mesolithic sequence, is poorly represented at the site. It is abruptly interrupted just after the beginning of the Atlantic, when trapezes and notched blades make their first appearance in a late Sauveterrian-type assemblage. This is also observed in the excavations carried out at the Grotta della Tartaruga (Cremonesi 1984), where the Castelnovian sequence is restricted

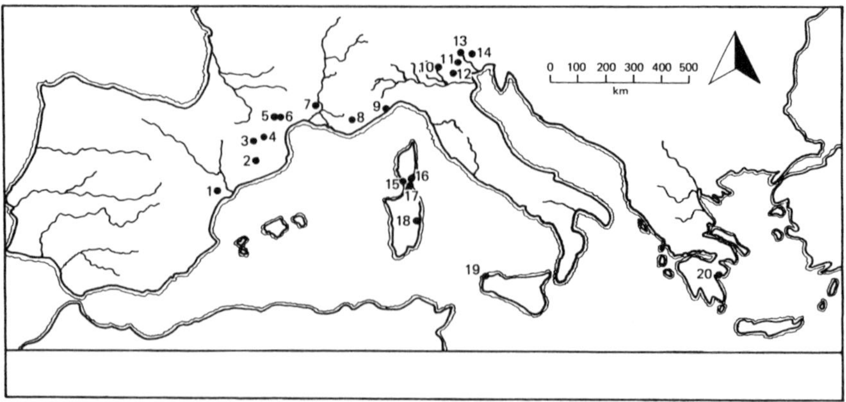

Fig. 8.1. Map of key sites of the Mesolithic and early Neolithic in the Mediterranean (After Lewthwaite 1986, 56). *1*, Cova Fosca (Castellón); *2*, Balma del Gai (Moià); *3*, Balma Margineda (Andorra); *4*, Abri du Roc de Dourgne (Fontanès-de-Sault); *5*, Grotte Gazel (Sallèles-Cabardès); *6*, Balma de l'Abeurador (Félines-Minervois); *7*, Baume de Montclus (Montclus); *8*, Baume de Fontbrégoua ((salernes); *9*, Arene Candide (Finale Ligure); *10*, Riparo Valtenesi (Manerba); *11*, Colbricon; *12*, Riparo Tagliente (Grezzana); *13*, Riparo Battaglia (Asiago); *14*, Piancavallo; *15*, Basi (Serra-di-Ferro); *16*, Curacchiaghiu (Levie); *17*, Araguina-Sennola (Bonifacio); *18*, Corbeddu Cave (Oliena); *19*, Grotta dell'Uzzo (Trapani); *20*, Franchthi Cave (Koilada).

to the uppermost artificial cuts. Much the same is shown by the Mesolithic sequence at the Grotta Benussi (Broglio 1971, 214), where the uppermost Castelnovian layers have been carbon-dated to the time between the middle and the end of the eighth millennium B.P. (Alessio et al. 1983).

Moving to the west, one finds in the Adige Valley three of the most important Mesolithic sequences of southern Europe—those at the sites of Vatte di Zambana, Romagnano III, and Pradestel mentioned before (Broglio 1971; Clark 2000). The excavations carried out at these rock shelters have led to the definition of the chronocultural and typological sequences of the Mesolithic industries of northern Italy (Broglio and Kozlowski 1983): the attribution of the Sauveterrian culture to the Preboreal and Boreal and the Castelnovian to the early Atlantic. On the basis of the geometric hypermicroliths and microliths from these sites, it has been established that, broadly speaking, the Preboreal assemblages are characterized by isosceles triangles and backed points and the Boreal ones by elongated scalene triangles and abrupt-retouched, bipolar long points. The tool kit changes with the start of the Atlantic climatic stage when isosceles and scalene trapezoidal arrowheads and notched bladelets make their appearance. Other high-altitude territories

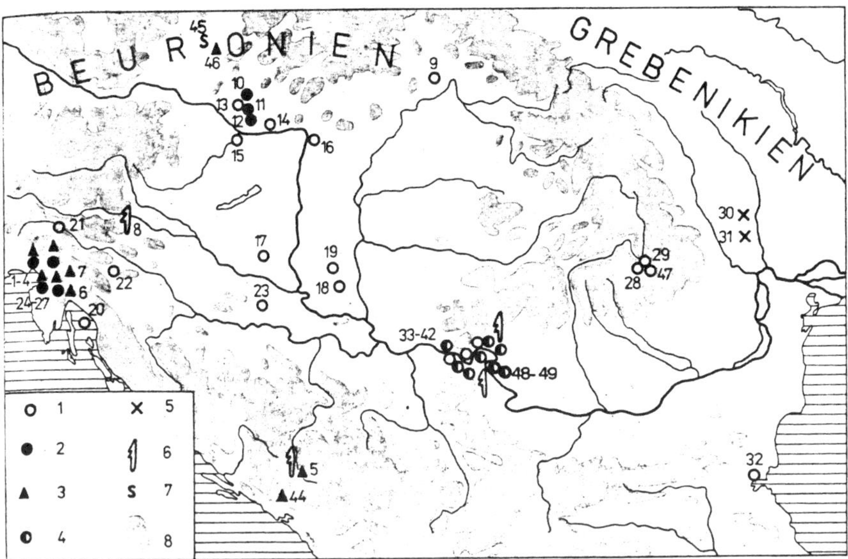

Fig. 8.2. Map showing the cultures of Mesolithic age to the east of the Alps (Kozlowski and Kozlowski 1983, 55). Key of cultures: *1*, Tardigravettian; *2*, Sauveterrien; *3*, Castelnovien; *4*, Culture de Lepenski Vir; *5*, type Maluşteni IV; *6*, type Šacvice; *8*, montagnes. Key of sites: *1*, Grotta Azzurra; *2*, Grotta Benussi; *3*, Grotta della Tartaruga; *4*, Grotta Zingari; *5*, Odmut; *6*, Pod Črmukljom; *7*, Dedkov Trebez; *8*, Spehovka; *9*, Strda n/Bodrogom; *10*, Sered I; *11*, Mostova; *12*, Tomašikovo; *13*, Dolna Streda; *14*, Hurbanovo; *15*, Györ; *16*, Szödliget; *17*, Pecs; *18*, Bacska Polanka; *19*, Hajdukovo; *20*, Kopacina Pećina; *21*, Polijšiška Cerekev; *22*, Vindja; *23*, Ruševo; *24–27*, see 1–4; *28*, Gilma Sita Buzeului; *29*, Cremenea Sita Buzgului; *30*, Maluşteni IV; *31*, Bereşti Dealui Taberei; *32*, Dekilitaž; *33*, Veterani-Terasa; *34*, Icoana; *35*, Razvrata; *36*, Ostovul Banului; *37*, Schela Cladovei; *38*, Padina; *39*, Lepenski Vir; *40*, Vlasac; *41*, Hajdučka Vodenica; *42*, Alibeg; *44*, Črvena Stijena; *45*, Šakvice; *46*, Mikulčice; *47*, Costanda Ladauti; *48*, Ostrovul Corbului; *49*, Ostrovul Mare.

where Mesolithic encampments are documented are those of the Valtrompia-Valcamonica watershed in the Alps of Lombardy. On the other hand, little evidence for Mesolithic settlement has been documented in the case of the Piedmontese Alpine arc even as far as western Liguria, where Pian del Re (at 850 m in elevation) in the interior behind San Remo is the only known site (Biagi et al. 1985).

With specific regard to the late Mesolithic period, it must be stressed that while more than 110 Castelnovian sites are currently known in northern Italy, only 25 have been excavated so far (Biagi 2001). In addition, all of the Trieste sites take the form of caves and late Mesolithic occupation is not really well represented at any of them. In most cases, they seem to have been frequented only at the beginning of the Atlantic (Ciccone 1993). Thus, continuous

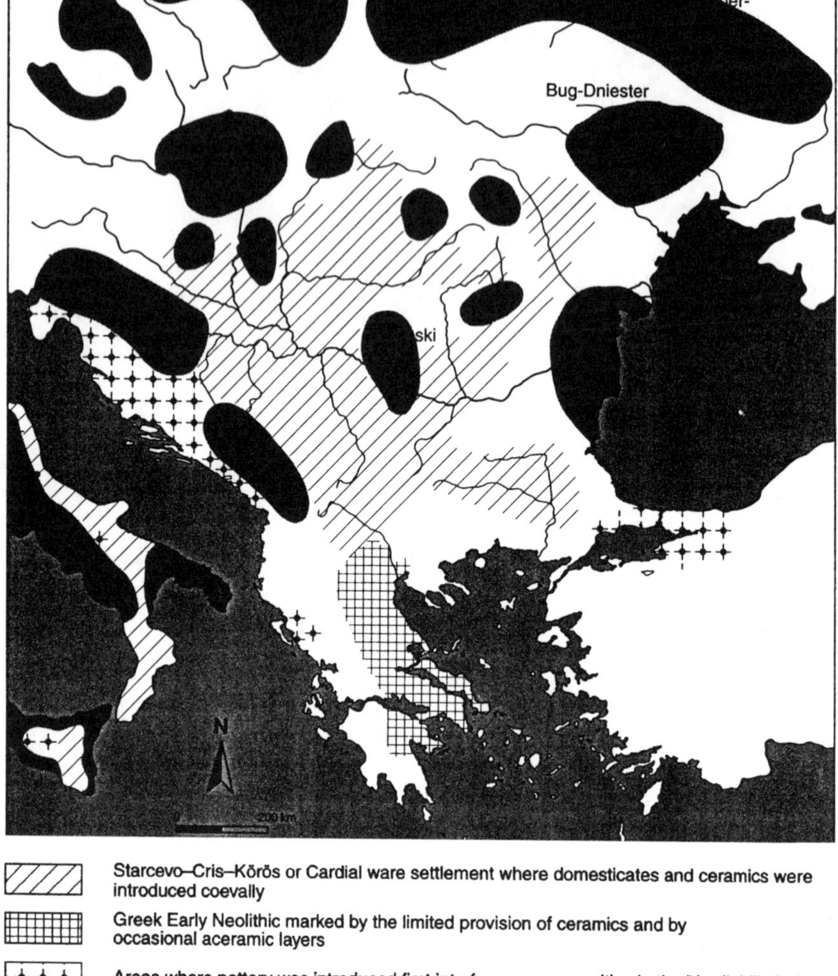

Fig. 8.3. Map showing areas of concentrated hunter-gatherer settlement. (After Zvelebil and Lillie 2000, 71)

Castelnovian stratigraphies, spanning the entire early Atlantic climatic stage, are not attested at the sites investigated near Trieste.

The situation along the North Apennine chain is different. No Castelnovian deposits have been found in situ in Liguria so far. And just a few sites dating to this time have been excavated along the Tusco-Emilian watershed and in its related valleys flowing northward. Among them, the Mesolithic sequence of the north Tuscan, valley-bottom site of Isola Santa is abruptly interrupted at the beginning of the Atlantic. In Emilia, two high-altitude sites have been carbon-dated to Castelnovian times: Passo della Comunella (at 1,619 m in elevation) to 6960 ± 130 B.P. (Birm-830) and Lama Lite (at 1764 m) to 6620 ± 80 B.P. (R-1394). The middle-altitude Tuscan site of Piazzana (820 m) has a date of 7330 ± 85 B.P. (R-400; Castelletti et al. 1994).

Martini and Tozzi (1996) summarized the evidence for Mesolithic occupation in southern Italy. They draw attention to the very low number of Castelnovian sites in the entire region, including Apulia, where just a few surface scatters have been recovered so far (Biagi 1996, 10). The scarcity of late Mesolithic sites along the Adriatic coast and in southern Italy in general (Grifoni Cremonesi 1998, 59) has been interpreted by some as possibly resulting from the precocious Neolithic adaptation of the area (Grifoni Cremonesi 1996, 70). Nevertheless, the data currently available (see fig. 8.4) indicate that Mesolithic sites of Preboreal and Boreal ages (and not just late Mesolithic ones) are rare along the whole Italian Adriatic coastline as far north as the Venetian Lagoon. Today it is difficult to maintain that our limited knowledge of the distribution of the Mesolithic encampments in this part of Italy may simply be blamed on a lack of systematic regional surveys. Enough fieldwork has been carried out over the last two decades to lead to the discovery of Mesolithic sites, if they actually existed there. It has to be remembered that the same situation (the absence of Mesolithic sites) also occurs in most parts of the Balkan Peninsula as well as in Greece. The indication then is that late Mesolithic sites are extremely rare—perhaps even absent from whole areas of study (Pluciennik 2000; Biagi and Spataro 2002)—for unknown reasons. The arguments put forward by Zvelebil and Lillie (2000, 69) to explain such a situation are inconsistent as demonstrated for Greece by C. Runnels (1995, 725) and C. Perlès (2001, 24). The Mesolithic along the North Adriatic coast and its interior is better known. Here evidence is provided by the excavations carried out in the Trieste Karst (Biagi et al. 1993) and from other sites recently excavated in Istria (Miracle and Forenbaher 2000). Long, discontinuous sequences are known in both cases: for example, the Edera Cave (Biagi and Spataro 2001) and Pupićina Peć

(Miracle 1997). The analysis of the archaeological material from these sites now in progress should help to clarify the processes of Neolithization at the head of the Adriatic.

The Early Neolithic Evidence

Recently, scholars such as Grifoni Cremonesi and Tozzi (1996), Guilaine and Cremonesi (1996), Grifoni Cremonesi (1998), and Vartanian et al. (2000) have written detailed accounts of the problems concerning the Neolithic transition in Apulia in southeastern Italy. These are mainly based on the results of the fieldwork conducted at the Impressed Ware sites of Torre Sabea, Ripa Tetta, and Trasano. Another important monograph on this subject has recently been edited by Biancofiore and Coppola (1997), who published the Impressed Ware site of Scamuso near Bari. It is mainly thanks to the work of these authors that it is now possible to say that the Neolithic transition in southeastern Italy began during the last three centuries of the eighth millennium B.P. According to the current evidence, the subsistence economy of the earliest Impressed Ware villages of Apulia was based primarily on sheep and cattle breeding as well as on the cultivation of several species of domesticated cereals (Cremonesi and Guilaine 1987, 382; Costantini and Tozzi 1987; Coppola and Costantini 1987). These data, together with those provided by the study of the lithic and ceramic assemblages, indicate that a complete Neolithic package had reached the region by around 7300 b.p. (ca. 6,000 B.C. cal.).

In this part of Italy, the number of early Neolithic settlements is extremely high, with thick concentrations in places such as the coasts of the Salentina Peninsula, the Tavoliere Plain, the Ofanto Valley, and the Materano area of Basilicata. Although only a few of these settlements have been excavated and radiocarbon dated, and the sites have been attributed to different stages in the development of the early Neolithic of southeastern Italy, it must be stressed that their number—estimated to be a few hundred villages—contrasts sharply with that of the Mesolithic sites—limited to eight in all with only five of them dating to the late Mesolithic (see fig. 8.4), including the Cave of Latronico in the nearby region of Basilicata. This cave represents the only south Italian sequence from which a good set of Mesolithic radiocarbon dates has been obtained. The dates span the interval between 7800 ± 90 B.P. (R-449) and 7420 ± 90 B.P. (respectively R-449 and R-445; Cipolloni

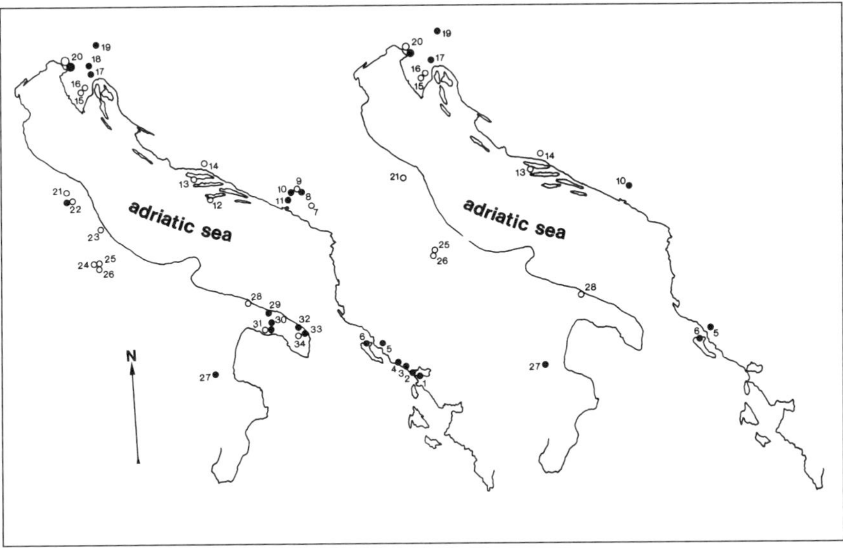

Fig. 8.4. Two maps of Mesolithic sites in the Adriatic basin (Biagi and Spataro 2002). (A) Distribution map of early Mesolithic sites, Preboreal and Boreal (circle), and late Mesolithic sites, Atlantic (dots): *1*, Tourkovouni; *2*, Preveza; *3*, Loutsa; *4*, Ammoudia; *5*, Konispol Cave; *6*, Sidari; *7*, Trebački Krš; *8*, Medena Stijena; *9*, Malisina Stijena; *10*, Odmut Cave; *11*, Crvena Stijena; *12*, Vela Spilja; *13*, Kopačina Špilja; *14*, Gopodska Pećina; *15*, Pupićina Peć; *16*, Sebrn Abri and Klanjceva Peć; *17*, Podosojna; *18*, Pod Črmukljo and Dedkov Trebec; *19*, Breg and Ljubljana marsh sites; *20*, Breg and Slovene Karst caves; *21*, Grotta del Prete; *22*, Pievetorina; *23*, Ripoli; *24*, Ortucchio; *25*, Grotta Continenza; *26*, Grotta di Pozzo; *27*, Latronico; *28*, Grotta delle Mura; *29*, Torre Testa; *30*, Oria; *31*, Terragne; *32*, S. Foca; *33*, Alimini Lakes; *34*, Grotta Marisa. (B) Distribution map showing the radiocarbon dated sites in the same region.

Sampò et al. 1999); the latter slightly antedates the oldest date so far available for the beginning of the Neolithic in the region, which comes from the coastal site of Scamuso (7290 ± 110 B.P.; Gif-6339).

At the opposite end of the peninsula, western Liguria was the first region of northern Italy to be settled by early Neolithic Impressed Ware people around the beginning of the seventh millennium B.P. (Maggi and Chella 1999). On the other hand, the excavations in the Trieste Karst and in the Friuli Plain indicate that the earliest Neolithic in northeastern Italy began only around the middle of the same millennium (Impronta and Pessina 1998; Biagi and Spataro 2002), probably representing the northernmost extension of the Danilo culture of Dalmatia. In the central Po Valley of northern Italy, the early Neolithic sites of the Vhò and Fiorano cultures, with very few exceptions, fall close to the end of the seventh millennium B.P.

Similar or slightly more recent dates have been obtained for the Gaban group in the Trentino and the Isolino group of northwestern Lombardy and the Ticino Canton of Switzerland (Bagolini and Biagi 1990; Donati and Carazzetti 1987). On the basis of this chronological evidence (together with the limited distribution of late Mesolithic sites on the Po Plain discussed below), it is now difficult to accept the hypothesis of Bagolini and Barfield (1991, 287) according to which "the formation of the early Neolithic of this region would appear to be the result of an intrusive Impressed Ware Culture from both the Adriatic and Ligurian sides of the peninsula with the subsequent acculturation of the local Castelnovian Mesolithic to form several regional Neolithic Groups." The striking thing to note is the comparatively late start of the Neolithic in the central part of northern Italy.

Late Mesolithic and Early Neolithic Site Distributions

In northern Italy, the location of the Mesolithic camps does not coincide at all with that of the early Neolithic villages. The difference between the late Mesolithic and early Neolithic settlement patterns is even clearer when we take into consideration the vertical distribution of the sites (Biagi 2001, fig. 5). The current sample of 106 late Mesolithic and 91 early Neolithic sites from the region shows a contrasting pattern in terms of site altitude. While just over half of the Castelnovian sites (54) are located at elevations between 500 and 2,000 m above sea level, only two of the early Neolithic sites occur above 500 m. Furthermore, apart from the large number of Castelnovian "sites," which are known only from light scatters of surface finds, the available archaeological evidence for the late Mesolithic is insufficient for several reasons (seven are listed below for northern Italy).

1. Only a limited number of sites have been excavated. They are mainly located at high altitudes both in the Alps and in the Apennines, and they are usually considered to be seasonal, short-visit hunting camps.
2. Little is known about the valley-bottom base camps; those located in the Trent basin have been excavated over an area of a few square meters, and they have not yielded any evidence for structural remains.
3. The stratigraphic sequences at the (presumed) base camps of

early Atlantic age in the Trent basin are always incomplete; they are found to have sedimentary breaks in them for reasons (possibly climatic) that are still not well understood. This is also the case for the cave sequences of the Trieste Karst (Ciccone 1993) as well as for the north Apennine site of Isola Santa (Biagi et al. 1979). Most likely, a similar situation also occurs in the late Mesolithic part of the sedimentary sequence at the Uzzo Cave in Sicily (Biagi and Spataro 2001).

4. Almost nothing is known of the subsistence economy of the last hunter-foragers, mainly because almost all of the sites are high-altitude camps where faunal remains are not preserved because of the acidity of the mountain soils. The only exception is that of the rock shelter called Mondeval de Sora located at 2,150 m in the Dolomites (Alciati et al. 1992).
5. The sources of raw material used for making stone tools by the Castelnovian hunter-foragers are commonly local ones. For example, this is the case of the sites near Trieste where dark varieties of flint occur as layers in local limestone outcrops and small pebbles are found in the alluvial terraces of the Timavo River (Turk pers. comm. 2000). Most of the sites that occur on the moraines of the Alpine lakes are located close to sources of flint pebbles. And sources of good-quality grey and red flint—exploited mainly during the Mesolithic period—are known in the central-eastern Alps. Biagi et al. (1979, 29) list the flint outcrops of the northern Apennines; Cipriani et al. (2000–2001) list those of the Tusco-Emilian Apennines; and Biagi et al. (1985) indicate those for Liguria.
6. On the landscape, the activities of the Castelnovian hunter-foragers do not seem to have left evidence of disturbance in the vegetation record. As reported for the northern Apennines (Lowe et al. 1994a, 184), "There is no evidence for any sustained human action during the Mesolithic in any of the sites investigated so far." In fact, the "earliest significant effects of human populations on the regional vegetation . . . suggests an age around 6300 B.P." (Lowe et al. 1994b, 157). This means that the earliest significant interference can be attributed to the first Neolithic pastoralists.

7. According to several authors (Cannarella and Cremonesi 1967; Bortolami et al. 1977; Shackleton and Van Andel 1985; Ciabatti et al. 1987), sea-level rise in the early Holocene may have submerged some of the more suitable areas for habitation along both the Ligurian coast and the upper Adriatic coast. This is certainly true for the Sauveterrian sites of the Preboreal and Boreal periods. It is not necessarily the case for the sites of early Atlantic age, however, since the coastline had already reached more or less its present position by that time. Along the north Adriatic coast, this is demonstrated by the study of the marine shellfish remains recovered from the Azzurra Cave and the Edera Cave in the Trieste Karst (Cannarella and Cremonesi 1967; Girod pers. comm. 2001). Much the same would appear to hold for the shores of the Ionian Sea as shown by the discovery of late Mesolithic sites along the sandy shoreline of northern Epirus in Greece (Runnels 1995; Biagi and Spataro 2002). In northern Italy, most of the early Neolithic settlements are distributed in the plains of the Po River and of Friuli (or along their bordering valleys).

Although both late Mesolithic and early Neolithic sites often occur within a given region, their distributions are invariably different at a more refined level of spatial resolution. Thus, with the exception of cave sites, there is seldom a direct overlap between the two distributions in a given local area in Italy. In fact, early Neolithic sites are for the most part open-air settlements that are distributed on the landscape in different ways than the late Mesolithic sites. In southeastern Italy, the large number of early Neolithic settlements stands in stark contrast with the scarcity of late Mesolithic sites. In particular, this contrast is well documented for the Tavoliere Plain near Foggia, which has been examined by a number of field surveys since the 1970s. It is also observed in the case of the Acconia Survey in Calabria (mentioned in ch. 1 of this volume). In contrast to the conclusion put forward by Zvelebil and Zvelebil (1988, 578) that "in most areas (of Europe), there is a continuity of settlement, continuity of some aspects of material culture (such as lithics) . . . across the Mesolithic-Neolithic transition," in most parts of Italy, it is difficult to make a good case for continuity in patterns of settlement over the Neolithic transition.

The Lithic Assemblages

The problems concerning changes in the chipped stone assemblages between the late Mesolithic and the early Neolithic of northern Italy have already been widely discussed (see Bagolini and Biagi 1987; Biagi 1991, 51; and Biagi et al. 1993). With regard to the old problem of the similarities or "derivation" of the early Neolithic flint assemblages from the Castelnovian ones, still stressed by Bisi et al. (1987, 421) and Bagolini et al. (1989, 100), the strong differences between the latter and the Castelnovian—both in terms of the raw materials used and in terms of the lithic technology—have been pointed out by Bagolini and Biagi (1987), Biagi (1991, 51), Biagi et al. (1993), and Starnini and Voytek (1997). The supposed similarities are restricted to the occurrence of sub-conical cores and geometrical instruments (mainly trapezoidal) in both periods, although they often show different utilization patterns (Biagi 1996). Against this, new types of blade tools arrive with the Neolithic in northern Italy: burins (often of side-notch type), abrupt retouched, straight borers, bladelets with a sinuous profile, rhomboids obtained with the microburin technique, and sickle blades. The situation in southeastern Italy is parallel in many respects to the better known one in northern Italy. Again new types make their first appearance with the early Neolithic Impressed Ware Culture. Among them are straight borers and sickle blades, although the trapezes (often of isosceles shape) obtained with the microburin technique are still present (Cremonesi and Guilaine 1987). The blade index is lower, and macrolithic bifacial tools are common at some sites (Giampietri 1996).

Beyond the Generic Mesolithic

Most of the theoretical papers devoted to the Neolithic transition have focused on different possible explanations, which have recently been listed by M. Zvelebil (2001, 58). The major weakness of most of these publications is the high level of generalization assigned by the authors to the phenomenon of the early Holocene hunter-gatherers (which, in one way or another, may have been involved in the Neolithic transition). The questions that really need to be raised are the following. Which hunter-gatherers: those in general or the last ones in a given area? Of which cultural aspect? Of which climatic period? These are, in effect, the important questions that need to be clarified.

With specific regard to Italy, incongruities have at times marked previous attempts to address the problem from a theoretical point of view. An example is provided by a paper by J. Lewthwaite (1986, 56) in which a distribution map of 19 "key sites of the Mesolithic and early Neolithic in the Mediterranean" includes a number of Pleistocene, late Paleolithic, and final Epigravettian sites. As shown in figure 8.1, four of the six "Mesolithic" sites indicated for northern Italy (Arene Candide, Riparo Tagliente, Riparo Battaglia and Piancavallo) have assemblages and radiocarbon dates that belong to the final Epigravettian culture, which comes before the Mesolithic period. Another recent contribution presents a picture of hunter-gatherers sites in a general perspective—without proper attention to chronology, cultural attribution, and site location—that is no less misleading. For example, it is of interest to compare figures 8.2 and 8.3 and draw attention to the significant differences between the two maps. Kozlowski and Kozlowski (1983, 55) give a distribution map of final Paleolithic and Mesolithic sites (fig. 8.2) with a clear indication of the specific cultures and time periods that are involved. In contrast, Zvelebil and Lillie (2000, 71) consider much the same area but their generic treatment of "hunter-gatherers" takes no account of different time periods (fig. 8.3). If we consider their "areas of concentrated hunter-gatherer settlement" and look specifically at the eastern Adriatic region of Dalmatia and its interior, we note strong discrepancies between their map and the preceding one. If what is shown in figure 8.3 is supposed to represent the late Mesolithic—that is, those hunter-gatherers directly involved in the transition—the question becomes where are the data drawn from. Clearly, as we have seen above, there is at this time a lack of good evidence for such concentrations in the area. On the other hand, if figure 8.3 is meant to represent an amalgamation of several different time periods, then it does not provide a sound basis on which to make arguments about the Mesolithic-Neolithic transition. There is a fundamental misconception here: what might be called the use and abuse of the generic Mesolithic. If we turn to southern Italy and Sicily, "concentrations" of hunter-gatherer sites are shown there as well (fig. 8.3). Here the distribution map is even more misleading, however. As mentioned above and shown in figure 8.4, there are, in fact, few Mesolithic sites that are known today in Apulia and Basilicata. If we focus specifically on the late Mesolithic in these two regions of southern Italy (that is, the sites identified by solid dots in fig. 8.4), the problem is even worse.

There is the further question of the validity of the claim that figure 8.3 makes with regard to "areas where pottery was introduced first into forager communities in the Availability phase" (notably the Dalmatian coast, Epirus, central-western Italy, and central Sicily). Much the same claim (for pottery occurring in Mesolithic contexts) was put forward in a previous distribution map of the "pottery using hunter-gatherers in Europe and temperate Asia, 6,000–1,000 b.c." (Zvelebil 1986, 171), where almost all of the Italian peninsula apparently witnessed this phenomenon (see fig. 8.5). The problem here is the paucity of well documented cases for this in Italy. Contrary to Zvelebil's claims, it must be stressed that the only Italian site known so far where pottery is found in association with a late Castelnovian complex is that of layer 3a at the Edera Cave near Trieste. There a well-defined, circular hearth yielded a poor but specialized Castelnovian lithic assemblage, including many microburins, just a few trapezoidal arrowheads, and one notched bladelet (Biagi et al. 1993, 49). This feature also produced a few

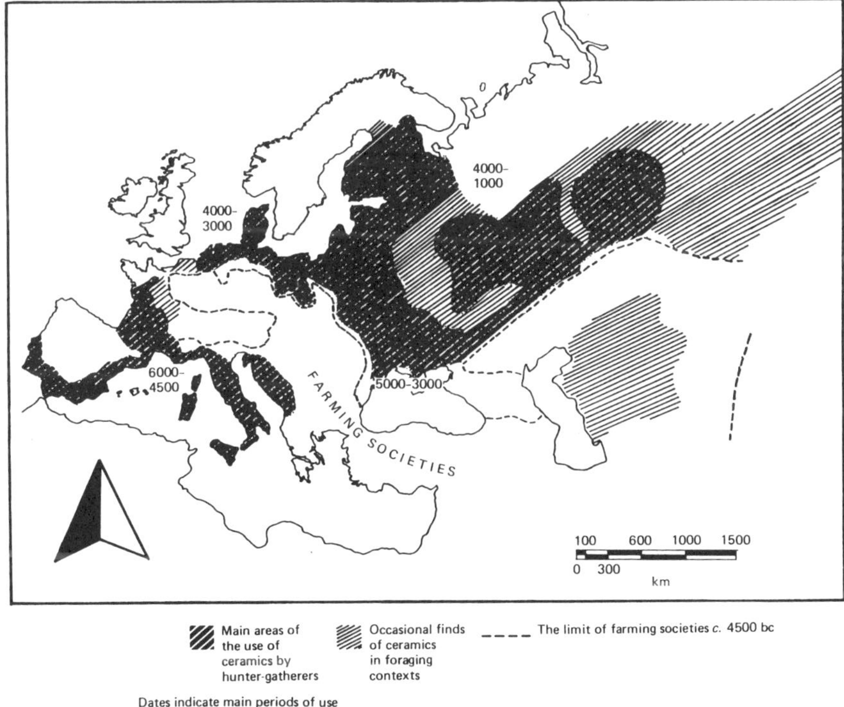

Fig. 8.5. Map of Europe showing areas of the use of ceramics by hunter-gatherers. (After Zvelebil 1986b, 71)

small potsherds, among which is a necked flask fragment of non-local production (Spataro 2001). Two radiocarbon dates are now available for this hearth: 6700 ± 130 B.P. from charcoal (GX-19569) and 6480 ± B.P. from *Patella caerulea* shellfish (GrN-25474). What needs to be emphasized here is that these dates do not support the notion of the early appearance of pottery among hunter-gatherers in northern Italy. In fact, they show just the opposite: the first evidence for pottery at the Edera Cave falls in the range of the radiocarbon dates obtained from the overlying early Neolithic levels at the same site, which contain pottery associated with the Vlaska (Danilo) group (Biagi and Spataro 2002, fig. 3). In other words, the late Castelnovian hunter-gatherers who frequented the Edera Cave first came in possession of pottery just prior to the advent of the first farmers in the Trieste area. The question of the possible interactions between the last Castelnovian hunter-gatherers at the site and the first Neolithic communities in the region at the head of the Adriatic (dating to around the middle of the seventh millennium B.P.) is one that I have considered elsewhere (Biagi 2002).

To sum up, the problem of the transition from hunting and gathering to food production in Italy should be addressed in a different way. The available archaeological evidence clearly demonstrates that the only hunter-gatherers that might have been involved in the Neolithic transition are the last hunter-gatherers: that is, the Castelnovian Mesolithic populations whose chronology covers an interval of some 1,500 radiocarbon years from roughly 8000 to 6500 B.P. As Binder (2000, 121) has pointed out: "the current evidence therefore suggests that the Sauveterrian complex is not involved in the transition to agriculture." Thus, it is difficult to understand why such generic hunter-gatherer distribution maps have been drawn in the first place. They do not provide a sound point of departure for the study of the transition in Italy. Furthermore, according to some authors, the term hunter-gatherers should be applied with caution to the Sauveterrian: "The Sauveterrian people are better described as hunters . . . because gathering is not seriously evidenced" (Binder 2000, 120).

Conclusion

There are today three main competing hypotheses for the explanation of the Neolithic transition in Europe: indigenism, demic diffusion, and leapfrog colonization. In the case of northeastern Italy, it is possible to conclude that the process of Neolithization took place according to demic diffusion—most likely in a way similar to that suggested by Kozlowski and Kozlowski (1986, 106) for central Europe. It is important to recall that the last hunter-gatherers had already disappeared from northern Italy well before the end of the seventh millennium B.P., as indicated by the available radiocarbon evidence (Biagi 1991; Pessina and Rottoli 1996; Impronta and Pessina 1999).

In opposition to what has been claimed by Donahue (1992, 77), whose argument (without references) has received perhaps undue attention in the Anglo-American literature, there is not good evidence for of the presence of pottery "in sites with distinctly Mesolithic stone artefacts," whose "technology, whether Castelnovian, Tardenoisian, Romanellian (sic), or another regional variant, continued into the early Neolithic." This assertion was based essentially on the archaeological material (for the most part still unpublished) recovered during the excavations at the site of Petriolo III South in Tuscany (Donahue et al. 1992). However, many Italian prehistorians have serious doubts about the reliability of this excavation (Gambassini pers. comm. 1999). Two radiocarbon dates from the excavation have been published (Skeates 1994, 236): 7610 ± 60 B.P. (OxA-1353) and 7480 ± 100 B.P. (OxA-2630), the first of which has been attributed to an unspecified Epigravettian context and the second to a ceramic Mesolithic complex. In addition, the claim made for the "appearance of domestic sheep and ceramic prior to the advent of sites with fully Neolithic economy," which is held to occur in "virtually all regions of Mediterranean Europe" (Donahue et al. 1992, 77), has recently been shown to be incorrect. This is one of the results of the reanalysis of the archaeozoological assemblages (including those from the Arene Candide Cave; Rowley-Conwy 1997, 2000) and the careful re-excavation of a few major sites where early sheep remains were previously thought to occur (Rowley-Conwy 1995, 347; see also ch. 6 of this volume).

In the light of what is currently known in Italy, the explanation of the transition that would seem to make the most sense is that of a demic expansion along the lines proposed by Ammerman and Cavalli-Sforza (1973, 1984), which has been reinforced for Europe as a whole by a number of new scientific results (Cavalli-Sforza et al. 1994, 297). Thus, even though the "wave of advance" model might not be accepted *in toto* in the "mathematical" way it was originally formulated (Zilhao 2001, 14184), it is difficult, in view of the current evidence, to demonstrate much in the way of cultural or subsistence linkage between the last hunter-foragers and the first farmers in

Italy. This is so for a reason that most authors have never taken into proper account in their treatment of the problem: the shortage or even absence of late hunter-gatherers in many areas of Italy. Thus, I would like to encourage other scholars to think more carefully about the following question: Where are the late Castelnovian hunters and gatherers? In a given area under investigation, how many late (Castelnovian) Mesolithic foragers were there at the start of the Atlantic climatic stage? One possibility that we may have to consider in certain parts of Italy is that of a decline in population numbers at some time in the interval between the end of the Boreal and the local onset of the transition itself. This may be one way to explain some of the gaps that are observed in the archaeological record during the time just before the transition. Is it possible that such a decline was related to the circulation of new diseases and epidemics like those that took their toll on some of the north American Indian tribes before the arrival of the "more civilized" newcomers (Diamond 1997)?

REFERENCES

Alciati, G., L. Cattani, F. Fontana, M. Gerhardinger, A. Guerreschi, S. Milliken, P. Mozzi, and P. Rowley-Conwy. 1992. "Mondeval de Sora: A High Altitude Mesolithic Campsite in the Italian Dolomites." *Preistoria Alpina* 28:355–66.

Alessio, M., L. Allegri, F. Bella, A. Broglio, G. Calderon, C. Cortesi, S. Improta, M. Preite Martinez, V. Petrone, and B. Turi. 1983. "14-C Datings of Three Mesolithic Series of Trento Basin in the Adige Valley (Vatte di Zambana, Pradestel, Romagnano) and Comparisons with Mesolithic Series of Other Regions." *Preistoria Alpina* 19:245–54.

Ammerman, A.J., and L.L. Cavalli-Sforza. 1973. "A Population Model for the Diffusion of Early Farming." In *The Explanation of Culture Change*, edited by C. Renfrew, 343–57. London.

———. 1984. *The Neolithic Transition and the Genetics of Populations in Europe*. Princeton.

Bagolini, B., and L.H. Barfield. 1991. "The European Context of Northern Italy during the Third Millennium." In *Die Kupferzeit als historische Epoche*, edited by J. Lichardus, 287–97. Bonn.

Bagolini, B., and P. Biagi. 1987. "The First Neolithic Chipped Stone Assemblages of Northern Italy." In *Chipped Stone Industries of the Early Farming Cultures in Europe*, edited by J.K. Kozłowski and S.K. Kozłowski, 423–49. Warsaw.

———. 1990. "The Radiocarbon Chronology of the Neolithic and Copper Age of Northern Italy." *Oxford Journal of Archaeology* 9:1–24.

Bagolini, B., O. Delucca, A. Ferrari, A. Pessina, and B. Wilkens. 1989. "Insediamenti neolitici ed eneolitici di Miramare (Rimini)." *Preistoria Alpina* 25:53–120.

Bernabò Brea, L. 1946. *Gli scavi nella Caverna delle Arene Candide: Parte I Gli strati con ceramiche*. Bordighera.

Biagi, P. 1991. "The Prehistory of the Early Atlantic Period along the Ligurian and Adriatic Coasts of Northern Italy in a Mediterranean Perspective." *Rivista di Archeologia* 15:45–54.

———, ed. 1995. *L'insediamento neolitico di Ostiano-Dugali Alti (Cremona) nel suo contesto ambientale ed economico*. Monografie di Natura Bresciana 22. Brescia.

———. 2001. "Some Aspects of the Late Mesolithic and Early Neolithic Periods in Northern Italy." In *From the Mesolithic to the Neolithic*, edited by R. Kertész and J. Makkay. *Archaeolingua* 11:71–88. Budapest.

———. 2002. "New Data on the Early Neolithic of the Upper Adriatic Basin. In *La Cèramique Imprimée dans le Néolithique Méditerranéen*, edited by D. Binder, I. Caneva, and M. Özdogan. Proceedings of the 2000 Sophia Antipolis Seminar.

Biagi, P., L. Castelletti, M. Cremaschi, B. Sala, and C. Tozzi. 1979. "Popolazione e territorio nell'Appennino Tosco-Emiliano e nel tratto centrale del bacino del Po tra il IX e il V millennio." *Emilia Preromana* 8:13–36.

Biagi, P., M. Cremaschi, and R. Nisbet. 1993. "Soil Exploitation and Early Agriculture in Northern Italy." *The Holocene* 3:164–8.

Biagi, P., R. Maggi, and R. Nisbet. 1985. "Liguria: 10,000–7000 B.P." In *The Mesolithic in Europe. Papers Presented at the Third International Symposium,* edited by C. Bonsall, 533–40. Glasgow.

Biagi, P., and M. Spataro. 2001. "Plotting the Evidence: Some Aspects of the Radiocarbon Chronology of the Mesolithic-Neolithic Transition in the Mediterranean Basin." *Atti della Società per la Preistoria e Protostoria della Regione Friuli-Venezia Giulia* 12:15–54.

———. 2002. "The Mesolithic/Neolithic Transition in North Eastern Italy and in the Adriatic Basin." *Saguntum* 34.

Biagi, P., E. Starnini, and B. Voytek. 1993. "The Late Mesolithic and Early Neolithic Settlement of Northern Italy: Recent Considerations." *Poročilo o raziskovanju paleolita, neolita in eneolita v Sloveniji* 21:45–67.

Biagi, P., and B.A. Voytek. 1994. "The Neolithization of the Trieste Karst in North-eastern Italy and Its Relationships with the Neighbouring Countries." *A Nyíregyházi Jósa Andras Múzeum Évkönyve* 31:63–73.

Biancofiore, F., and D. Coppola. 1997. *Scamuso: Per la storia delle comunità umane tra il VI e il III millennio nel Basso Adriatico*. Rome.

Bietti, A. 1981. "The Mesolithic Cultures in Italy: New Activities in Connection with Upper Palaeolithic Cultural Traditions." *Veröffentlichungen des Museums für Ur-und Frühgeschichte Potsdam* 14–15:33–50.

Binder, D. 2000. "Mesolithic and Neolithic Interaction in Southern France and Northern Italy: New Data and Current Hypotheses." In *Europe's First Farmers*, edited by T.D. Price, 117–43. Cambridge.

Bisi, F., A. Broglio, G. Dalmeri, M. Lanzinger, and A. Sartorelli. 1987. "Bases Mésolithiques du Néolithique ancien au sud des Alpes." In *Chipped Stone Industries of the Early Farming Communities in Europe*, edited by J.K. Kozlowski and S.K. Kozlowski, 381–422. Warsaw.

Bortolami, G.C., J.Ch. Fontes, V. Markgraf, and J.F. Saliege. 1977. "Land, Sea and Climate in the Northern Adriatic Region during the Late Pleistocene and Holocene." *Palaeogeography, Palaeoclimatology and Palaeoecology* 21:139–55.

Broglio, A. 1971. "Risultati preliminari delle ricerche sui complessi epipaleolitici della Valle dell'Adige." *Preistoria Alpina* 7:135–241.

Broglio, A., and S.K. Kozłowski. 1983. "Tipologia ed evoluzione delle industrie mesolitiche di Romagnano III." *Preistoria Alpina* 19:93–148.

Cannarella, D., and G. Cremonesi. 1967. "Gli scavi nella Grotta Azzurra di Samatorza nel Carso Triestino." *Rivista di Scienze Preistoriche* 22:281–330.

Cardini, L. 1980. *Gli scheletri Mesolitici della Caverna delle Arene Candide (Liguria).* Memorie dell'Istituto Italiano di Paleontologia Umana, N.S. 3. Rome.

Castelletti, L., A. Maspero, and C. Tozzi. 1994. "Il popolamento della Valle del Serchio (Toscana settentrionale) durante il Tardiglaciale Würmiano e l'Olocene antico." In *Highland Zone Exploitation in Southern Europe*, edited by P. Biagi and J. Nandris, 189–204. Monografie di Natura Bresciana 20. Brescia.

Cavalli-Sforza, L.L., P. Menozzi, and A. Piazza. 1994. *The History and Geography of Human Genes*. Princeton.

Childe, V.G. 1925. *The Dawn of European Civilisation*. London.

Ciabatti, M., P.V. Curzi, and F. Ricci Lucchi. 1987. "Quaternary Sedimentation in the Central Adriatic Sea." *Giornale di Geologia* 49:113–25.

Ciccone, A. 1993. "L'industria mesolitica della Grotta Azzurra di Samatorza: scavi 1982." *Atti della Società per la Preistoria e Protostoria della Regione Friuli-Venezia Giulia* 7:13–45.

Cipolloni Sampò, M., C. Tozzi, and M.L. Verola. 1999. "Le Néolithique ancien dans le sud-est de la Péninsule italienne: Caractérisation culturelle, économie, structures d'habitat." In *Le Néolithique du nord-ouest Méditerranéenne*, edited by J. Vaquer, 13–24. Paris.

Cipriani, N., M. Dini, M. Ghinassi, F. Martini, and C. Tozzi. 2000–2001. "L'approvvigionamento della materia prima in alcuni tecnocomplessi della Toscana Appenninica." *Rivista di Scienze Preistoriche* 51:337–88.

Clark, R. 2000. *The Mesolithic Hunters of the Trentino: A Case Study in Hunter-Gatherer Settlement and Subsistence from Northern Italy. BAR-IS* 832. Oxford.

Coppola, D., and L. Costantini. 1987. "Le Néolithique ancien littoral et la diffusion des céréales dans le Pouilles durant le VIe Millénaire: Les sites de Fontanelle, Torre Canne et Le Macchia." In *Premières Communautés Paysannes en Méditerranée Occidentale*, edited by J. Guilaine, J. Courtin, J-L. Roudil, and J-L. Vernet, 249–56. Paris.

Costantini, L., M. Piperno, and S. Tusa. 1987. "La néolithisation de la Sicilie occidentale d'après les résultats des fouilles a la Grotte de l'Uzzo (Trapani)." In *Premières Communautés Paysannes en Méditerranée Occidentale*, edited by J. Guilaine, J. Courtin, J-L. Roudil, and J.-L. Vernet, 397–405. Paris.

Costantini, L., and C. Tozzi. 1987. "Un Gisement à Céramique Imprimée dans le subappennin de la Daunia (Lucera, Foggia): Le village de Ripa Tetta. Economie et culture matérielle." In *Premières Communautés Paysannes en Méditerranée Occidentale*, edited by J. Guilaine, J. Courtin, J.-L. Roudil, and J.-L. Vernet 387–94. Paris.

Cremonesi, G. 1984. "I livelli mesolitici della Grotta della Tartaruga." In *Il Mesolitico sul Carso Triestino. Società per la Preistoria e Protostoria della Regione Friuli-Venezia Giulia. Quaderno* 5:65–108.

Cremonesi, G., and J. Guilaine. 1987. "L'habitat de Torre Sabea (Gallipoli, Puglia) dans le Cadre du Néolithique Ancien de l'Italie du Sud-Est." In *Premières Communautés Paysannes en Méditerranée Occidentale*, edited by J. Guilaine, J. Courtin, J.-L. Roudil, and J.-L. Vernet, 377–85. Paris.

Diamond, J. 1997. *Guns, Germs, and Steel: The Fate of Human Societies*. New York.

Donahue, R.E. 1992. "Desperately Seeking Ceres: A Critical Examination of Current Models for the Transition to Agriculture in Mediterranean Europe." In *Transition to Agriculture in Prehistory*, edited by A.B. Gebauer and T.D. Price, 73–80. Madison.

Donahue, R.E., D.B. Burroni, G.M. Coles, R.H. Colten, and C.O. Hunt. 1992. "Petriolo III South: Implications for the Transition to Agriculture in Tuscany." *Current Anthropology* 33:328–31.

Donati, P., and R. Carazzetti. 1987. "La stazione neolitica di Castel Grande a Bellinzona (Ticino, Svizzera)." *Atti della XXVI Riunione Scientifica dell'IIPP*, 467–78. Florence.

Giampietri, A. 1996. "Torre Sabea, Trasano, Ripa Tetta, Santo Stefano." In *Forme e tempi della neolitizzazione in Italia Meridionale e in Sicilia*, edited by V. Tiné, 327–9. Soneria Mannelli.

Grifoni Cremonesi, R. 1996. "La Neolitizzazione dell'Italia: Italia Centro Meridionale." In *The Neolithic in the Near East and Europe*, edited by R. Grifoni Cremonesi, J. Guilaine, and J. L'Helgouach, 69–79. XIII International Congress of Prehistoric and Protohistoric Sciences. Forlí.

———. 1998. "Il Neolitico Antico nella Fascia Peninsulare Adriatica." In *Settemila Anni fa. Il Primo Pane: Ambienti e Culture delle Società Neolitiche*, edited by A. Pessina and G. Muscio, 59–70. Udine.

Grifoni Cremonesi, R., and C. Tozzi. 1996. "Torre Sabea, Trasano, Ripa Tetta." In *Forme e Tempi della Neolitizzazione in Italia Meridionale e in Sicilia*, edited by V. Tiné, 442–8. Genoa.

Guilaine, J., and G. Cremonesi. 1996. "La Chronologie du Néolithique Ancien a Trasano (Matera, Basilicata) dans le Contexte de la Méditerranée Centrale." In *Forme e Tempi della Neolitizzazione in Italia Meridionale e in Sicilia*, edited by V. Tiné, 433–41. Genoa.

Impronta, S., and A. Pessina. 1998. "La Neolitizzazione dell'Italia Settentrionale: Il Nuovo Quadro Cronologico." In *Settemila Anni fa: il Primo Pane. Aspetti e Culture delle Società Neolitiche*, edited by A. Pessina and G. Muscio, 107–15. Udine.

Kozłowski, J.K., and S.K. Kozłowski. 1983. "Le Mésolithique à l'est des Alpes." *Preistoria Alpina* 19:37–56.

———. 1986. "Foragers of Central Europe and Their Acculturation." In *Hunters in Transition: Mesolithic Societies of Temperate Eurasia and Their Transition to Farming*, edited by M. Zvelebil, 95–108. Cambridge.

Laplace, G. 1964. *Essai de typologie systématique*. Annali dell'Università di Ferrara. Nuova Serie Paleontologia Umana e Paletnologia. Ferrara.

Lewthwaite, J. 1986. "The Transition to Food Production: A Mediterranean Perspective." In *Hunters in Transition: Mesolithic Societies of Temperate Eurasia and Their Transition to Farming*, edited by M. Zvelebil, 53–66. Cambridge.

Lowe, J.J., N. Branch, and C. Watson. 1994a. "The Chronology of Human Disturbance of the Vegetation of the Northern Apennines during the Holocene." In *Highland Zone*

Exploitation in Southern Europe, edited by P. Biagi and J. Nandris, 169–87. Monografie di Natura Bresciana 20. Brescia.

Lowe, J.J., C. Davite, D. Moreno, and R. Maggi. 1994b. "Holocene Pollen Stratigraphy and Human Interference in the Woodlands of the Northern Apennines, Italy." *The Holocene* 4:153–64.

Maggi, R. 1997. "The Radiocarbon Chronology." In *Arene Candide: A Functional and Environmental Assessment of the Holocene Sequence (Excavations Bernabò Brea-Cardini 1940–50)*, edited by R. Maggi, E. Starnini, and B. Voytek, 31–52. Memorie dell'Istituto Italiano di Paleontologia Umana, N.S. 5. Rome.

Maggi, R., and P. Chella. 1999. "Chronologie par le radiocarbone du Néolithique des Arene Candide (Fouilles Bernabò Brea)." In *Le Néolithique du nord-ouest Méditerranéen*, edited by J. Vaquer, 99–110. Paris.

Maggi, R., E. Starnini, and B. Voytek, eds. 1997. *Arene Candide: A Functional and Environmental Assessment of the Holocene Sequence (Excavations Bernabò Brea-Cardini 1940–50)*. Memorie dell'Istituto Italiano di Paleontologia Umana, N.S. 5. Rome.

Martini, F., and C. Tozzi. 1996. "Il Mesolitico in Italia Centro-Meridionale." In *The Mesolithic*, edited by S.K. Kozłowski and C. Tozzi, 47–58. XIII International Congress of Prehistoric and Protohistoric Sciences. Forlì.

Miracle, P. 1997. "Early Holocene Foragers in the Karst of Northern Italy." *Poročilo i raziskovanju paleolitika, neolitika in eneolitika v Sloveniji* 24:43–61.

Miracle, P., and S. Forenbaher. 2000. "Pupićina Cave Project: Brief Summary of the 1998 Season." *Histria Archaeologica* 29:27–48.

Perlès, C. 2001. *The Early Neolithic in Greece*. Cambridge.

Pessina, A., and M. Rottoli. 1996. "New Evidence on the Earliest Farming Cultures in Northern Italy: Archaeological and Palaeobotanical Data." *Poročilo o raziskovanju paleolitika, neolitika in eneolitika v Sloveniji* 23:77–103.

Pluciennik, M. 2000. "Reconsidering Radmilli's Mesolithic." In *Studi sul paleolitico, Mesolitico e Neolitico del bacino dell'Adriatico in Ricordo di Antonio M. Radmilli*, edited by P. Biagi, 171–83. Società per la Preistoria e la Protostoria della Regione Friuli-Venezia Giulia. Quaderno 8. Trieste.

Radmilli, A.M. 1960. "Considerazioni sul Mesolitico Italiano." *Annali dell'Università di Ferrara* 15.1:29–48.

Radmilli, A.M., and E. Tongiorgi. 1958. "Gli Scavi nella Grotta La Porta di Positano: Contributo alla Conoscenza del Mesolitico Italiano." *Rivista di Scienze Preistoriche* 13:91–109.

Rowley-Conwy, P. 1995. "Making First Farmers Younger: The West European Evidence." *Current Anthropology* 36:346–53.

———. 1997. "The Animal Bones from Arene Candide." In *Arene Candide: A Functional and Environmental Assessment of the Holocene Sequence (Excavations Bernabò Brea-Cardini 1940–1950)*, edited by R. Maggi, E. Starnini, and B. Voytek, 31–52. Memorie dell'Istituto Italiano di Paleontologia Umana, N.S. 5. Rome.

———. 2000. "Milking Caprines, Hunting Pigs: The Neolithic Economy of Arene Candide in Its Western Mediterranean Context." In *Animal Bones, Human Society*, edited by P. Rowley-Conwy, 124–32. Oxford.

Runnels, C. 1995. "Review of Aegean Prehistory IV: The Stone Age of Greece from the Palaeolithic to the Advent of the Neolithic." *American Journal of Archaeology* 99:699–728.

Shackleton, J.C., and T.H. Van Andel. 1985. "Late Palaeolithic and Mesolithic Coastlines of the Western Mediterranean." *Cahiers Ligure de Préhistoire et de Protohistoire* 2:7–19.

Skeates, R. 1994. "A Radiocarbon Date-List for Prehistoric Italy (c. 46,400 BP–2450 BP/400 Cal. BC)." In *Radiocarbon Dating and Italian Prehistory*, edited by R. Skeates and R. Whitehouse, 147–288. Rome.

Spataro, M. 2001. "An Interpretative Approach to the Prehistory of the Edera Cave in the Trieste Karst (North-Eastern Italy): The Archaeometry of the Ceramic Assemblage." *Accordia Research Papers* 8:83–99. London.

Starnini, E., and B. Voytek. 1997. "The Neolithic Chipped Stone Artifacts from the Bernabò Brea Excavation." In *Arene Candide: A Functional and Environmental Assessment of the Holocene Sequence (Excavations Bernabò Brea-Cardini 1940–50)*, edited by R. Maggi, E. Starnini, and B. Voytek, 349–426. Memorie dell'Istituto Italiano di Paleontologia Umana, N.S. 5. Rome.

Tagliacozzo, A. 1993. *Archeozoologia della Grotta dell'Uzzo, Sicilia.* Bullettino di Paletnologia Italiana Suppl. 84. Rome.

Taschini, M. 1968. "La Datatation au C14 de l'Abri Blanc (Mont Circé): Quelque observations sur le Mésolithique en Italie." *Quaternaria* 10:137–65.

Vartanian, E., P. Guibert, C. Ney, F. Bechtel, M. Schvoerer, J. Guilaine, and G. Cremonesi. 2000. "Chronologie de la Néolithisation en Italie du sud-est: Nouvelles datations grâce a la thermoluminescence (TL) sur le site de Matera-Trasano." In *Studi sul Paleolitico, Mesolitico e Neolitico del Bacino dell'Adriatico in Ricordo di Antonio M. Radmilli*, edited by P. Biagi, 245–68. Società per la Preistoria e Protostoria della Regione Friuli-Venezia Giulia, Quaderno 8. Trieste.

Zilhão, J. 2001. "Radiocarbon Evidence for Maritime Pioneer Colonization at the Origins of Farming in Western Mediterranean Europe." *Proceedings of the National Academy of Sciences USA* 98:14180–85.

Zvelebil, M. 1986a. "Mesolithic Prelude and Neolithic Revolution." In *Hunters in Transition: Mesolithic Societies of Temperate Eurasia and Their Transition to Farming*, edited by M. Zvelebil, 5–16. Cambridge.

———. 1986b. "Mesolithic Societies and the Transition to Farming: Problems of Time, Scale and Organisation." In *Hunters in Transition: Mesolithic Societies of Temperate Eurasia and Their Transition to Farming,* edited by M. Zvelebil, 167–88. Cambridge.

———. 2001. "The Social Context of the Agricultural Transition in Europe." In *Archaeogenetics: DNA and the Population Prehistory of Europe*, edited by C. Renfrew and K. Boyle, 57–79. Cambridge.

Zvelibil, M., and M. Lillie. 2000. "Transition to Agriculture in Eastern Europe." In *Europe's First Farmers*, edited by T.D. Price, 57–92. Cambridge.

Zvelebil, M., and P. Rowley-Conwy. 1984. "Transition to Farming in Northern Europe: A Hunter-Gatherer Perspective." *Norwegian Archaeological Review* 17:104–28.

Zvelebil, M., and K.V. Zvelebil. 1988. "Agricultural Transition and Indo-European Dispersal." *Antiquity* 62:574–83.

— 9 —

Radiocarbon Dating and Interpretations of the Mesolithic-Neolithic Transition in Italy

Robin Skeates

Radiocarbon dating has contributed to interpretations of the Mesolithic–Neolithic transition in Italy since the 1960s. In this chapter, I examine the development of this relationship over the last 35 years and assess the current situation before offering some suggestions for the future.

My involvement in this area of research began 15 years ago. I was first introduced to the uncalibrated radiocarbon chronology of the Mesolithic-Neolithic transition in Italy by Professor John Evans in the mid 1980s, when I was an undergraduate at the Institute of Archaeology in London. He provided me with a radiocarbon datelist for the central Mediterranean region, on which, I recall, the early Coppa Nevigata determination on shell still featured prominently. My interest in the problems of the transition then deepened in the late 1980s, when I was a postgraduate student at the University of Oxford. In 1987, Sarah Milliken and I carried out a detailed field survey of Mesolithic and Neolithic sites around the Alimini Lakes in the Salento Peninsula of southern Apulia (Milliken and Skeates 1989). Although we became convinced that we had identified at least one site that could be assigned, on typological grounds, to the Mesolithic-Neolithic transition; ultimately, this work highlighted to me the need to work with radiocarbon dates rather than artifact typologies. At about the same time, I began a critical assessment of the radiocarbon chronology of east-central Italy, as part of my doctoral thesis on the Neolithic and Copper Age of the Abruzzo-Marche region (Skeates 1992). With the encouragement of my supervisor, Dennis

Britton, I also began to compile, for comparative purposes, a radiocarbon datelist for the Neolithic and Copper Age of the whole of Italy. I calibrated this using the CALIB computer program, which had recently become available (Stuiver and Reimer 1986).

This growing interest in the radiocarbon chronology of prehistoric Italy and its associated problems prompted Ruth Whitehouse of the University of London and me to organize a conference on "Radiocarbon Dating and Italian Prehistory," which was held in 1991 at the British School at Rome, where I then held a one-year scholarship. Our aim was to publicize, to Italian archaeologists in particular, the importance of radiocarbon dating. At the end of the conference, some of the participants persuaded me to expand my thesis datelist to cover the whole of prehistoric Italy and to incorporate it into the ensuing conference proceedings (Skeates 1994b; Skeates and Whitehouse 1994). Having achieved this task, it remained clear that we still needed many more radiocarbon dates before we could talk seriously about absolute chronologies for prehistoric Italy, let alone the Mesolithic-Neolithic transition. Fortunately, more dates have been produced, and in the mid 1990s Ruth Whitehouse and I published supplementary lists of these dates (Skeates and White house 1994, 1995–1996, 1997–1998).

In the light of the Venice meeting, however, I have become even more concerned about the poor quality of the samples used to date the Mesolithic-Neolithic transition in Italy. In my opinion, this transition is a particularly complex and geographically varied social phenomenon (Skeates 1999), and it requires precise dating. We may even have to start all over again, particularly if we are to talk about the spread of agriculture, using large numbers of Accelerator Mass Spectrometer (AMS) determinations on domesticated cereal grains and sheep bones. Currently, we have just five of these—from only two sites.[1]

Review

This chapter on radiocarbon dating and the interpretation of the Mesolithic-Neolithic transition in Italy concentrates upon major published syntheses of groups of radiocarbon determinations (as opposed to detailed site excavation reports). The data discussed here are restricted to radiocarbon determinations that have produced dates in the eighth, seventh, and sixth millennia B.C. (cal.) (i.e., determinations between 8890 ± 90 b.p. and 6080 ± 50 b.p.):

that is, the millennia that appear to bracket the Mesolithic-Neolithic transition in Italy. Details, including bibliographic references, of the radiocarbon determinations mentioned below are provided in four published radiocarbon datelists for prehistoric Italy (Skeates 1994a, 162–73; Skeates and Whitehouse 1994, 1995–1996, 1997–1998), and for reasons of space, they will not be reiterated here. Uncalibrated determinations are referred to as "b.c.," whereas calibrated dates are referred to as "Cal B.C." Also, for the purposes of this paper, I have used a simple archaeological definition of the Neolithic, in which it is characterized by the presence of ceramics and/or domesticates in archaeological deposits.

The Late 1950s and Early 1960s

The first radiocarbon determinations relating to Mesolithic and early Neolithic sites in Italy were announced in 1959 by the Laboratory of Nuclear Geology of the University of Pisa (Pi) with more following in 1961. Three were published in the new international journal *Radiocarbon*, but two others were only announced by means of a mimeograph whose circulation appears to have been extremely restricted (Tongiorgi et al. 1959). These five determinations were produced from samples of either marine shell or charcoal obtained from five recently excavated sites situated in the north, center, and south of Italy.[2] They were cautiously provided with large standard deviations ranging from ±130 to ±200 years.

The Mid 1960s

Donald Brown (1964) and David Trump (1966) readily accepted the published determinations in the mid 1960s, but simply added them to the established chronological framework for Italian prehistory, which was based upon relative sequences of pottery types and stratigraphic deposits. The start of the lower or early Neolithic phase in Italy, characterized by the arrival of the first Neolithic farmers using Impressed Ware pottery, was dated to around 5000 b.c. with particular reference to the two published determinations from central Italy. David Trump (1966, 16) favored the Apulian coast as the first port of call for the immigrants, based on the style of the decorated ceramics found at Coppa Nevigata, whereas Grahame Clark (1965, 48) prioritized east-central Italy over the southeast, on the basis of the distribution of known radiocarbon determinations. There was general agreement on placing the start of a later expansion of the new economy to other parts of Italy at around 4000 b.c.

A further 11 new determinations relating to four Mesolithic and early Neolithic cave sites in Italy were published during the mid 1960s. In 1964, the Pisa Laboratory published another determination on marine shell from Arene Candide but with a large standard deviation of ±250 years. Then, between 1965 and 1967, the new Radiocarbon Dating Laboratory at the University of Rome (R) published a series of determinations for Arene Candide and three other sites.[3] In contrast to the Pisa Laboratory, their samples were predominantly of mixed fragments of either charcoal or burnt bones, which were assigned relatively optimistic standard deviations averaging ±109 years. The Rome Laboratory also expressed doubts over the consistency of radiocarbon determinations made on marine shells (Alessio et al. 1967, 357).

The Late 1960s and Early 1970s

Ruth Whitehouse (1968, 1969) attempted to refine the established chronological sequence for the early Neolithic of southern Italy during the late 1960s. However, in dating the arrival of the first farmers in this region, she was hampered by the lack of published radiocarbon determinations for Neolithic sites in Apulia, where she regarded the initial area of settlement to lie (and by a lack of knowledge of the unpublished determination for Coppa Nevigata) as well as by the apparent anomaly of the determination on charcoal from the early Neolithic deposits in Grotta del Santuario della Madonna (R-285), which stood out as particularly early by comparison with the other available determinations. She therefore chose to ignore this "anomalous" determination and placed the arrival of the first farmers at a date "shortly before around 5000 b.c.," between the dates provided by the existing determinations for Greece and central Italy, the latter which she regarded as a secondary area of colonization. For northern Italy, Lawrence Barfield (1971) also placed the appearance of the first farmers at around 5000 b.c., whereas Colin Renfrew (1970, 287) placed it at around 4500 B.C. (cal.), according to his experimental calibrated chronology. For the western Mediterranean as a whole, Jean Guilaine and A. Calvet felt that all of the available radiocarbon determinations for the early Neolithic "tallied perfectly," and that they confirmed the relative antiquity of the "Neolithization" in this region (Guilaine and Calvet 1970, 89, 91).

During the late 1960s and early 1970s, 28 more radiocarbon determinations relating to another eight Mesolithic and early Neolithic sites in Italy became available. The Rome Laboratory published a further 22 determinations relating to six sites in the north, center, and south of Italy.[4] The Geological Sciences Laboratory of Birmingham University (Birm) also published six determinations relating to Neolithic sites 3 and 4 at Molino Casarotto (Vicenza). The Rome Laboratory generally continued to assign relatively low standard deviations, averaging ±93 years, to their samples of charcoal fragments, whereas the Birmingham Laboratory assigned more cautious standard deviations, averaging ±133 years, to their samples of timber, charcoal, and peat. The Rome Laboratory also expressed concern over "inconsistencies" between their radiocarbon determinations and the relative stratigraphic and typological sequence in the Riparo Arma di Nasino (Alessio et al. 1968, 356).

Albert Ammerman and Luca Cavalli-Sforza (1971) then included five of the Italian radiocarbon determinations among their selection of 53 determinations relating to early Neolithic sites in Europe, which they used in a new approach to measure the rate of spread of early farming. The Italian examples were chosen as the most reliable determinations for the earliest probable farming sites in the south, center, and north of Italy.[5] They were used as uncalibrated radiocarbon determinations, since the available calibration curves did not extend back much before 6000 B.P. Tentative regional average diffusion rates were calculated with an increase noted for the west Mediterranean (2.08 km per year) compared to the Mediterranean as a whole (1.52 km per year) and Europe as a whole (1.08 km per year). At this broad geographical level, the analysis contributed nothing to interpretations on an Italian scale, although the authors pointed out that, "As more early Neolithic dates in Europe accumulate, it will be possible to attempt a more detailed consideration of regional rates" (Ammerman and Cavalli-Sforza 1971, 684–5).

The Mid 1970s

Only nine more radiocarbon determinations were published in the mid 1970s. They related to five new sites and one previously dated site in Italy.[6] These were published by five different laboratories.[7] Their standard deviations remained relatively cautious, with an average of ±103 years, although the Rome Laboratory, which remained somewhat overoptimistic, provided an estimate of ±50 years for their sample of charcoal from Grotta delle Soppressate (R-676).

More scholars now incorporated selections of radiocarbon determinations within their chronological frameworks for the Mesolithic and Neolithic, choosing them from the 50 or so available radiocarbon determinations for Italy (e.g., Phillips 1975; Barker 1975; Guilaine 1976; Tiné 1976). There was still no attempt, however, to calibrate these determinations, because of the lack of a calibration curve for the earliest determinations and confusion over calibrating the more recent determinations (Phillips 1975, 20). Furthermore, these determinations were still generally used to enhance traditional, regional, ceramic-based chronostylistic and cultural sequences rather than to replace them. Graeme Barker (1975, 132) was an exception, in rejecting the traditional ceramic-based "early," "middle," and "late" chronological divisions for the Neolithic in central Italy with reference to the available radiocarbon determinations. This was an optimistic stance, given that large parts of Italy were still poorly supplied with radiocarbon determinations (Phillips 1975, 20). Opinions varied as to the reliability of individual determinations. On the one hand, the Coppa Nevigata determination of around 8150 b.p. (6200 b.c.), produced but never published by the Pisa Laboratory in the late 1950s, was now widely disseminated and accommodated as the earliest date for the Neolithic in Italy, despite previous doubts about its reliability (e.g., Phillips 1975, 46; Tiné 1976, 74; Guilaine 1976, 45). On the other hand, the "anomalous" determination for the early Neolithic in Grotta del Santuario della Madonna (R-285) was now further criticized by Patricia Phillips (1975, 47), who, following Albert Ammerman's doctoral thesis, noted that the charcoal for the sample was obtained from both the last Mesolithic stratigraphic level (40) and the earliest Impressed Ware level (41), whereas Jean Guilaine (1976, 45; 1979) used it, along with the Coppa Nevigata determination and other early determinations in Corsica, France, and Spain, to continue to argue for "the precocity of the Neolithic manifestations in the western Mediterranean." Pat Phillips (1975, 32) also echoed the Rome Laboratory's concern over their determinations from Arma dello Stefanin (R-148, R-126, R-145, R-109), describing them as a "rather confusing series of dates," because of their poor correspondence with the stratigraphic sequence of the cave.

The Late 1970s and Early 1980s

During the late 1970s and early 1980s there was a significant increase in the number of available radiocarbon determinations of charcoal for the Mesolithic and early Neolithic particularly for northern Italy. The Rome Laboratory published 20 new determinations for five previously undated north Italian cave sites.[8] Another 22 determinations were published by seven more laboratories for nine sites.[9]

Both Ruth Whitehouse (1978) and Jean Guilaine (1980) maintained a partially critical approach toward these determinations. In theory, Whitehouse intended to exclude radiocarbon determinations from "uncertain cultural contexts," those that seemed "very anomalous," and those that had not been published by the laboratory. She admitted, however, that this practice was "somewhat arbitrary in this selection." For example, despite reiterating criticisms of the early determinations for the early Neolithic at Coppa Nevigata and Grotta del Santuario della Madonna, and questioning the determination for Casa San Paolo (P-1999), "since the material from the deposit in question was of middle or late Neolithic type rather then early Neolithic," she went on to argue that "we must not dismiss these dates without further consideration, since they fit well into a pattern that is emerging in the West Mediterranean, where we now have a dozen sites with dates in the 6th millennium b.c., which are associated with pottery" (Whitehouse 1978, 78). Jean Guilaine (1980, 7) was similar in his approach. On the one hand, he accepted that there were suspect determinations, particularly those made on samples of shell. On the other hand, he chose not to discard the early Coppa Nevigata determination (of a sample of shell) and instead used it to help place the whole of the southeast Italian Early Impressed Ware phase in the sixth millennium b.c., prior to the start of a Masseria La Quercia phase at the end of the sixth millennium b.c. Both scholars agreed, however, that not enough determinations were yet available to be able to weed out "anomalous" or "suspect" examples.

Despite these doubts, scholars were now able to use the available determinations to provide a coarse radiocarbon chronology for the major stylistically-defined, archaeological cultures and phases throughout Italy. For southern Italy, the first appearance of Impressed Ware was now placed somewhere in the sixth millennium b.c. (Whitehouse 1978; Guilaine 1980). For the center, Graeme Barker (1981, 67) regarded the first pottery as having come into use by the mid fifth millennium b.c., but noted that he would not be surprised if future determinations lay in the sixth millennium b.c., in line with those accumulating elsewhere in the central and western Mediterranean. And for the north, Bernardino Bagolini and Paolo Biagi placed the arrival of the first farmers on the Ligurian coast (characterized by Ligurian Impressed Ware) between the beginning and middle of the fifth-millennium b.c.; their arrival along the northern Adriatic coast toward the end of the first half of the fifth millennium b.c.; their spread inland together with the emergence of distinctive ceramic styles (characterized by the Vhò and Fiorano groups) between the middle and the end of the fifth millennium b.c.; and pottery's first appearance in the Adige Valley (characterized by the Gaban group) between the end of the fifth and the beginning of the fourth millennia b.c. Whitehouse (1978, 80) even made a first attempt at calibrating this chronology, although she pointed out that most of the determinations relating to the Mesolithic-Neolithic transition still remained too early to calibrate.

The Mid 1980s

There was another significant increase in the number of available radiocarbon determinations for the Mesolithic and early Neolithic during the mid 1980s: 57 new determinations were published for 27 sites, distributed throughout Italy.[10] A few of these sites were now provided with a series of radiocarbon determinations as opposed to single determinations. Out of the samples whose details were published, 34% were now made on human and animal bones as opposed to 66 % on charcoal. Traditional radiocarbon determinations were provided by 12 laboratories—almost exclusively on charcoal.[11] New AMS determinations were provided by a further three laboratories, using smaller samples of charcoal and also bone.[12]

The available radiocarbon determinations for the Mesolithic-Neolithic transition in Italy (which numbered 152 by 1987) were now discussed much more widely by both Anglo-American and Italian scholars. The publication of radiocarbon datelists for particular regions and periods helped in this process (e.g., Allegri et al. 1987; Cipolloni Sampò 1987). That is not to say that all Italian scholars had accepted the technique of radiocarbon dating, let alone calibration. For example, a contrast can be seen in Bernardino Bagolini and Giuliano Cremonesi's combined paper on "The Process of

Neolithization in Italy," between Bagolini's discussion of the diffusion of the Neolithic in northern Italy with reference to radiocarbon determinations, and Cremonesi's exclusive reliance upon relative stratigraphic and stylistic sequences for his own discussion of the south of Italy (Bagolino and Cremonesi 1987). However, other scholars working in the south now began to question the traditional ceramic-based, relative chronologies—especially Santo Tiné's sequence for the Tavoliere—with reference to the small but growing number of radiocarbon determinations available for the early Neolithic (e.g., Allegri et al. 1987, 69; Cipolloni Sampò 1987, 182; Lewthwaite 1987, 547).

Some confusion still surrounded the radiocarbon dating of the earliest Neolithic in the south of Italy. The status of the early Coppa Nevigata determination was again left in the balance with some rejecting it outright (e.g., Lo Porto 1987, 120), others accepting it uncritically (e.g., Cipolloni Sampò 1987; Batovic 1987, 347), and others sitting on the fence (e.g., Whitehouse 1984, 1112; Sargent 1985, 32, 25, 39). The new British Museum determination for this site did not help matters, since it was assigned a massive standard deviation of ±320 years (Whitehouse 1987). Similar conflicting opinions continued to surround the early determinations, lying in the sixth millennium b.c., for the early Neolithic sites of Casa San Paolo, Santa Tecchia, and Grotta del Santuario della Madonna (e.g., Cassano 1985, 736; Sargent 1985, 33–4, 39). Confusion also developed around a determination on charcoal from spits 13 and 14 in the Grotta dell'Uzzo in Sicily (P-2734). For a start, it was widely misquoted as 6180 ± 80 b.c. (8130 ± 80 b.p.; e.g., Lewthwaite 1986, 60; 1987; Evans 1987, 324), rather than as 5960 ± 70 b.c. (7910 ± 70 b.p.; e.g., Meulengracht et al. 1981; Tusa 1985, 67). Secondly, the deposits with which it was associated, which were characterized by an increase in the exploitation of marine resources, the import of a few fragments of obsidian, and the initial development of a blade-based lithic technology (but no pottery or domesticates), were ambiguously described by the excavator as an "aceramic," "intermediate," or "transitional" phase between the Epipaleolithic and the early Neolithic. This was taken to imply that early Neolithic ceramics and agriculture followed soon after (around the middle of the sixth millennium b.c.; e.g., Lewthwaite 1986, 60; Evans 1987, 324; Tusa 1985, 69–70) despite the lack of stratigraphic continuity between the "transitional" and early Neolithic phases. Despite these doubts concerning individual determinations, scholars broadly agreed that the start of the

Neolithic in southern Italy could be placed, fairly safely, somewhere in the first half of the sixth millennium b.c. (e.g., Cassano 1985, 734; Sargent 1985, 39; Whitehouse 1985, 1112; Lewthwaite 1986, 58; Evans 1987, 324; Follieri 1987; Lo Porto 1987, 121; Tusa 1987, 363). As for Sicily, Richard Burleigh (1984, 280) noted that, despite its size and archaeological importance, it remained very poorly covered by radiocarbon dating.

For the north of Italy, the established radiocarbon chronology was further refined by Bernardino Bagolini (1987; Bagolini and Cremonesi 1987, 21). In parts of western Liguria, the appearance of the Impressed Ware facies was placed at around 5000 b.c., whereas in the Adige Valley, the transition between the late Mesolithic Castelnovian culture and the early Neolithic Gaban group was placed at 4500 b.c., as was the start of the colonization of Romagna by Adriatic Impressed Ware groups.

Chronological differences between the north and south of Italy consequently became more evident (e.g., Ammerman and Cavalli-Sforza 1984; Sargent 1985; Lewthwaite 1986, 1987; Batovic 1987). Albert Ammerman and Luca Cavalli Sforza's revised isochron map of the spread of early farming in Europe, using 500-year time intervals, indicated its arrival in southeast Italy between 5500 and 5000 b.c., then its spread to the rest of southern and central Italy between 5000 and 4500 b.c., and its onward movement to northern Italy between 4500 and 4000 b.c. (Ammerman and Cavalli-Sforza 1984, fig. 4.5). In contrast, Andrew Sargent (1985, 38–9) noted a gap of a millennium between the inception of the Neolithic in southern Italy (in the mid sixth millennium) and its appearance in Adriatic central Italy and the Po Valley (in the mid fifth millennium b.c.) and suggested that "the first florescence of a Neolithic in the West Mediterranean bypassed the north of Italy." Lewthwaite (1985, 58) also suggested processual differences in the transition from foraging to farming between different regions with village farming replacing foraging at around 8000 b.c. in southern Italy, foraging and coastal fishing at around 5000 b.c. in Sicily, and foraging and animal husbandry at around 4200 b.c. in northern Italy.[13]

The Late 1980s and Early 1990s

A further 45 radiocarbon determinations were published in the late 1980s and early 1990s for 20 Mesolithic and Early Neolithic sites in Italy.[14] Nine laboratories produced traditional determinations.[15] A further three laboratories produced AMS determinations on charcoal, charred grain, bone, a grape pip, and a piece of daub.[16]

Donald Brown (1992) attempted another review of the radiocarbon chronology of prehistoric Italy during the early 1990s but with reference to only a small sample of the actual number of available sites with radiocarbon determinations for the Mesolithic-Neolithic transition. Bernardino Bagolini and Paolo Biagi carried out more successful analyses (1990; Biagi 1990); they continued to refine the radiocarbon chronology for the Mesolithic-Neolithic transition in northern Italy. They saw late Mesolithic bands continuing "at least until the middle of the seventh millennium b.p." (i.e., the mid fifth millennium b.c.), the settlement of the first Neolithic communities (Vhò and Fiorano) on the Po Plain "around the end of the seventh millennium b.p." (i.e., the late fifth millennium b.c.), and a gap of at least 350 radiocarbon years between the disappearance of the last hunter-gatherers and the arrival of the first farmers Biagi 1990, 49). They also noted that the earliest radiocarbon dates for Ligurian Impressed Ware (from Arene Candide) were some 400 years earlier than those for Adriatic Impressed Ware (from Maddalena di Muccia), which they related to a western (French) origin for the former, as opposed to an eastern origin for the latter (Biagi 1990, 51–5). In the south of Italy, Albert Ammerman (1990, 495) now gleefully interpreted the new British Museum determination for Coppa Nevigata, which placed the site at a much more recent date than previously imagined, as "a major setback for the indigenist school of thought, which," he held, "had pinned its hopes on an early starting date for the Neolithic in Southern Italy as a sign of independent development there."

The Mid 1990s

At least another 92 radiocarbon determinations were published for at least 32 Mesolithic and early Neolithic sites in Italy during the mid 1990s.[17] Nine laboratories produced traditional determinations.[18] A further three laboratories produced AMS determinations on samples of bone and charcoal.[19]

Paolo Biagi and his colleagues again continued to refine the uncalibrated radiocarbon chronology for the "Neolithization" of northern Italy in a series of works of synthesis (e.g., Biagi et al. 1993; Biagi 1996; Pessina and Rottoli 1996). For northeastern Italy, they now proposed that the first Neolithic settlement of the Friuli Plain and Trieste Karst took place "around the middle of the seventh millennium B.P. or slightly later," at the same time as the first agricultural settlement of the Abruzzo and Marche regions.

It was, however, following a conference on "Radiocarbon Dating and Italian Archaeology," which was held in Rome in 1991 (Skeates and Whitehouse 1994), that significant developments took place in the analysis of radiocarbon determinations relating to the Mesolithic-Neolithic transition in Italy. Rigorous critical analyses of large numbers of determinations were carried out, which considered sample types, laboratories and the comparability of determinations, calibration curves, and "radiocarbon plateaux," the quality of primary publications, and the spatial distribution of existing determinations (e.g., Skeates 1994a; Tykot 1994; Pluciennik 1994, 1997). Some controversial early Neolithic contexts on the Tavoliere Plain were also redated under the initiative of Ruth Whitehouse. The date of early Neolithic Coppa Nevigata was securely brought forward to the early fifth millennium b.c. (early sixth millennium Cal B.C.) according to a pair of AMS determinations on grain from the lowest levels of the site's ditch (OxA-1474 and OxA-1475), and the earlier Pisa and British Museum determinations were now finally rejected. And Santa Tecchia was placed closer to the mid fifth millennium b.c. (mid sixth millennium Cal B.C.) according to a British Museum determination for spit 4 of the ditch (BM-2414; Whitehouse 1994, 86–7). New calibrated regional chronologies were also established for the Mesolithic-Neolithic transition in central and southern Italy, Sicily, and Sardinia (e.g., Skeates 1994a; Tykot 1994; Pluciennik 1994, 1997; Leighton 1996; Tusa 1996). These provided new patterns with which to discuss the direction and rate of the spread of the Neolithic in Italy. In southern Italy, the earliest Neolithic was thought to have probably appeared in northern Apulia and perhaps eastern Basilicata in the first quarter of the sixth millennium Cal B.C. (i.e., much more recently than was supposed in the mid 1980s). Then, from the second quarter of the sixth millennium Cal B.C., following a "time lag" of about 300–400 calendar years, there was thought to have been a rapid spread of various traits associated with the Neolithic to other parts of the south, including southern Apulia, western Basilicata, Calabria, Sicily, and Sardinia (Pluciennik 1994, 133–8; Tykot 1994, 121). In east-central Italy, there appeared to have been a "chronological break" of about 800 calendar years between the first appearance of the "Neolithic package" in Apulia and its earliest appearance in east-central Italy, at around 5750 Cal B.C. Next, the package appears to have spread rapidly northward:

first into Molise, Abruzzo, and Marche, and then, in two slightly later stages, into Umbria and southern Emilia-Romagna (Skeates 1994a, 64). The time-lags or chronological breaks in the diffusion of the Neolithic package lent support to a "frontier model" of adoption (Skeates 1994a, 64; Pluciennik 1994, 135). These developments in the analysis of radiocarbon determinations were, however, somewhat restricted to Anglo-American scholars and their closest English-speaking Italian colleagues. As Sebastiano Tusa (1996, 45) noted, "Unfortunately not all scholars accept radiometric dating, especially the earlier dates and the calibrated chronology."

Current Patterns

In order to gain a clearer impression of where we currently stand in relation to radiocarbon dating the Mesolithic-Neolithic transition in Italy, I shall now examine the patterns that the available dates provide on a more detailed, local level than has been attempted before. I have therefore listed the single *latest* acceptable Mesolithic and single *earliest* acceptable Neolithic radiocarbon determinations for Italy (the seventh and sixth millennia Cal B.C.) on a province-by-province basis after rejecting 20 determinations (see appendix). I have also plotted the earliest dates for the pottery-using Neolithic for each province on a map, using the central date from each calibrated date-range (see fig. 9.1). My interpretations of this map, in terms of the spread of the Neolithic package, are based purely on the chronological patterns that seem to emerge from the dates (rather than upon patterns of associated material culture).

The clearest pattern emerges for eastern peninsular Italy. The earliest Neolithic dates, falling between about 6150 and 5950 Cal B.C., occur in the provinces of Bari, Foggia, Matera, and Potenza in Apulia and Basilicata. Here, the ceramics are characterized by Impressed and Painted Ware. The Neolithic then appears to have spread southwards: southeast to the provinces of Brindisi and Lecce in southern Apulia (from about 5800 Cal B.C.) in association with Impressed Ware and south to the Calabrian provinces of Cosenza and Catanzaro between about 5800 and 5750 Cal B.C. in association with Cardial and Stentinello Impressed Ware. It also spread to the northwest, after a delay of up to 400 calendar years to the provinces of Campobasso, Chieti, Pescara, and Macerata in the regions of Molise, Abruzzo, and southern Marche at around 5500 Cal B.C. Here the pottery is

Fig. 9.1. Map of the single "earliest" radiocarbon dates (Cal B.C.) for the Neolithic in the provinces of Italy

characterized by Adriatic Impressed, Incised, and Painted Ware. The Neolithic then spread again at around 5400 Cal B.C.: to the west, across the Apennines, to the Umbrian provinces of Terni and Perugia, and to the northwest into the province of Ancona at around 5250 Cal B.C. In these areas, the earliest pottery is again characterized by Impressed, Incised, and Painted Ware. On the basis of these figures, it would appear that the

Neolithic spread up the western Adriatic coastal lowlands between Bari and Ancona—over a distance of 465 km and a period of about 900 calendar years (1,000 radiocarbon years)—at an estimated rate of 0.52 km per calendar year (or 0.47 km per year using radiocarbon years). This is a slower rate than the average rate calculated by Ammerman and Cavalli Sforza (1971, 684) for the western Mediterranean (about 2 km per year) and Europe as a whole (about 1 km per year). It emphasizes the gradual and sometimes delayed nature of the diffusion of the pottery-using Neolithic in this particular region. Elsewhere, the dates form less coherent patterns.

For Sicily, we need more dates, particularly for the eastern half of the island. For the moment, however, the Neolithic appears to have arrived in the Trapani province (on the northwest coast) by around 5650 Cal B.C.; it may have reached the Agrigento province (on the southwest coast) somewhat later by around 5050 Cal B.C. Here the pottery is characterized by Impressed and Stamped Ware.

For the west coast of Italy, an interesting pattern is perhaps beginning to emerge. But it is extremely problematic, and we need more published details and more dates in order to test it. We have a group of three very early dates lying between around 6300 Cal B.C. and 6050 Cal B.C. for the provinces of Siena, Viterbo, and L'Aquila (Petriolo III South, Grotta delle Settecannelle, and Grotta Continenza). These contrast with the earliest dates of between around 5400 Cal B.C. and 5150 Cal B.C. for the provinces of Terni, Rome, and Firenze and with a complete lack of acceptable dates for Campania and southeast Lazio. According to their relevant publications, all of the samples come from *presumably* undisturbed deposits containing pottery sherds and other "Neolithic" traits: Cardial Impressed sherds at the Grotta delle Settecannelle, 15 plain sherds in unit 2 of Petriolo III South, and seven Impressed and Incised sherds in spit 22 of Grotta Continenza as well as a few ovicaprid bones (Gnesutta and Bertagnini 1993; Donahue et al. 1992; Barra et al. 1989–1990). However, all of the deposits also contain assemblages of flint artifacts and faunal remains that are suggestive of subsistence economies predominantly based upon hunting and gathering. Taking the published details at face value, what may be emerging in west-central Italy are dates for the elusive Mesolithic-Neolithic transition, as opposed to a more fully established early Neolithic, which may still be the case in southeast Italy. Leading scholars believe, however, that there are problems with the archaeological contexts of these particular samples (P. Biagi and J. Guilaine, pers. comm. at the Venice meeting; see ch. 8 of this volume). For example, not only is Petriolo III South held by many prehistorians in Italy to be a disturbed site, but the excavations have not been published in detail. And the large size of much of the debitage found there is uncharacteristic of late Mesolithic sites in Italy.

The situation in the northwest of Italy is clearer. The earliest date of around 5800 Cal B.C. occurs in the Savona province, where the pottery is characterized by Ligurian Impressed Ware. The Neolithic may then have spread out from this core: to the south to Corsica and the provinces of Sassari and Nuoro in northern Sardinia in association with Cardial Impressed Ware (at around 5650 Cal B.C.) and to the northwest to the Piacenza province in western Emilia-Romagna (by around 5450 Cal B.C.). We still need dates for southern Sardinia.

In the northeast of Italy, a more coherent pattern is now emerging. The earliest dates—between about 5650 Cal B.C. and 5600 Cal B.C.—appear around the northern Adriatic: in the easternmost parts of the Friuli-Venezia Giulia region, in the provinces of Trieste and Udine (the former in association with Vlaška style pottery), in eastern Emilia Romagna (in the provinces of Ravenna and Modena), and just across the Apennines in the Tuscan province of Lucca. The Neolithic then appears to have spread westward into the regions of Friuli, the Veneto, and the Trentino-Alto Adige between about 5500 Cal B.C. and 5400 Cal B.C., where the earliest dates occur in the provinces of Pordenone, Vicenza, Verona, and Trento (in association with Fagnigola, Fiorano, and Impressed Ware).

On the Po Plain, the Neolithic appears to have begun somewhat later: between about 5300 and 4950 Cal B.C. in the provinces of Varese, Cremona, and Reggio Emilia. Here, the pottery is characterized by the Vhò, Fiorano, and Square Mouthed Pottery styles. There is some doubt, however, over the "early" dating of deposits containing Square Mouthed Pottery, whose first phase is traditionally assigned to the Middle Neolithic in northern Italy (Biagi pers. comm.).

Conclusions

Various lessons can be learned from the foregoing review of radiocarbon dating.[20] We have come a long way from the 1960s. We still have much work to do. In closing, I would like to offer several suggestions about ^{14}C dating and its relation to the question of the Mesolithic-Neolithic transition in Italy. First, we still need to convert some traditional archaeologists to using radiocarbon dating—particularly so that radiocarbon samples can be obtained for poorly dated areas. Second, it is clear that all of us must employ a more rigorous and critical approach to radiocarbon samples—from their initial recovery in the field through their publication and on to their eventual acceptance or rejection. The production and analysis of larger series of determinations for individual sites will obviously help, but only if they too are dealt with in a rigorous and critical manner. Third, in dating the Mesolithic-Neolithic transition, we need to be explicit about what cultural features we are discussing: forest clearance, the exchange of rare commodities, the intensification of foraging strategies, food production, technological innovation or the digging of ditches around aggregation sites. Fourth, in order to advance our dating and interpretation of the Mesolithic-Neolithic transition, we need to shift the focus of our research away from the so-called earliest Neolithic to the late Mesolithic. The benefits of this shift in emphasis are apparent in northern Italy, where research on the Mesolithic has been ongoing for over 20 years and where the gap between the latest Mesolithic and the earliest Neolithic dates on a provincial scale is on average down to about 500 calendar years (as opposed to the south and the western islands where late Mesolithic sites are poorly known and where the gap is on average about 1,150 years). Only when this gap has been closed will we be able to begin to compare dates for the Mesolithic and Neolithic in a meaningful way. Fifth, as this chapter shows, patterns of radiocarbon dates and their interpretations will continue to change over time as more determinations become available and as calibration curves become more refined. Because of this, we must make use of the full range of available dates for any particular region or period (rather than limited, personal selections of them), and we must remain tentative in the conclusions that we draw from them. We still have a long way to go before we can talk of an *absolute* chronology for the Mesolithic-Neolithic transition in Italy.

REFERENCES

Alessio, M., F. Bella, F. Bachechi, and C. Cortesi. 1965. "University of Rome Carbon-14 Dates III." *Radiocarbon* 7:213–22.

———. 1967. "University of Rome Carbon-14 Dates V." *Radiocarbon* 9:346–67.

Alessio, M., F. Bella, C. Cortesi, and B. Graziadei. 1968. "University of Rome Carbon-14 Dates VI." *Radiocarbon* 10:350–64.

Alessio, M., F. Bella, S. Improta, G. Belluomini, C. Cortesi, and B. Turi. 1970. "University of Rome Carbon-14 Dates VIII." *Radiocarbon* 12:599–616.

Alessio, M., L. Allegri, F. Bella, S. Improta, G. Belluomini, G. Calderoni, C. Cortesi, L. Manfra, and B. Turi. 1978. "University of Rome Carbon-14 Dates XVI." *Radiocarbon* 20:79–104.

Alessio, M., L. Allegri, F. Bella, A. Broglio, G. Calderoni, C. Cortesi, S. Improta, M. Preite Martinez, V. Petrone, and B. Turi. 1983. "14C datings of Three Mesolithic Series of Trento Basin in the Adige Valley (Vatte di Zamana, Pradestel, Romagnano) and Comparisons with Mesolithic Series of Other Regions." *Preistoria Alpina* 19:245–54.

Alessio, M., L. Allegri, S. Improta, G. Belluomini, C. Cortesi, L. Manfra, and B. Turi. 1991. "University of Rome Radiocarbon Dates XVII." *Radiocarbon* 33:131–40.

Alessio, M., L. Allegri, A. Ferrari, S. Improta, and A. Pessina. 1995. "Nuovi dati di cronologia sulle prime comunità neolitiche dell'Italia nord-orientale." *Gortania: Atti del Museo Friulano di Storia Naturale* 17:37–55.

Allegri, L., C. Cortesi, and A.M. Radmilli. 1987. "La cronologia neolitica in base al radiocarbonio." In *Atti della XXVI Riunione Scientifica: il Neolitico in Italia,* edited by A. Revedin, 67–77. Florence.

Ambers, J., K. Matthews, and S. Bowman. 1989. "British Museum Natural Radiocarbon Measurements XXI." *Radiocarbon* 31:15–32.

Ammerman, A.J. 1990. Reviews of *Apulia* 1, by G.D.B. Jones; *Passo di Corvo e la civiltà neolitica del Tavoliere,* by S. Tinè; and *Coppa nevigata e il suo territorio,* by S.M. Cassano. *American Journal of Archaeology* 94:493–6.

Ammerman, A.J., and S. Bonardi. 1985–1986. "Ceramica stentinelliana di una struttura a Piana di Curingia (Catanzaro)." *Rivista di Scienze Preistoriche* 40:201–24.

Ammerman, A.J., and L.L. Cavalli Sforza. 1971. "Measuring the Rate of Spread of Early Farming in Europe." *Man* 6:674–88.

———. 1984. *The Neolithic Transition and the Genetics of Populations in Europe.* Princeton.

Awsiuk, R., H. Hereman, and M.F. Pazdur. 1991. "Radiocarbon Dating of Human Habitation." *Preistoria Alpina* 27:151–60.

Azzi, C.M., L. Bigliocca, and E. Piovan. 1973. "Florence Radiocarbon Dates I." *Radiocarbon* 15:479–87.

Bagolini, B. 1987. "Vallée de l'Adige: Naissance de premières communautés paysannes dans un territoire alpin." In *Premières Communautés Paysannes en Méditerranée Occidentale: Colloque International du C.N.R.S., Montpellier, 1983,* edited by J. Guilaine, J. Courtin, J.-L. Roudil, and J.-L. Vernet, 455–9. Paris.

Bagolini, B., and P. Biagi. 1980. "The Mesolithic and Neolithic Settlement of Northern Italy." In *Problèmes de la Néolithisation dans Certaines Régions de l'Europe. Actes du Colloque International,* edited by J.K. Kosłowski and J. Machnik, 9–26. Krakow.

———. 1990. "The Radiocarbon Chronology of the Neolithic and Copper Age of Northern Italy." *Oxford Journal of Archaeology* 9:1–23.

Bagolini, B., and G. Cremonesi. 1987. "Il processo di neolitizzazione in Italia." In *Atti della XXVI Riunione Scientifica dell'Istituto Italiano di Preistoria e Protostoria: Il Neolitico in Italia,* edited by A. Revedin, 21–30. Florence.

Bagolini, B., and P. Von Eles. 1978. "L'insediamento neolitico di Imola e la corrente culturale della Ceramica Impressa nel medio e alto Adriatico." *Preistoria Alpina* 14:33–63.

Banchieri, D. 1988–1989. "Bodio Lomnago (VA), Località Pizzo di Bodio: Abitato spondale neolitico." *Notiziario: Soprintendenza Archeologica della Lombardia* 1988–1989:55–6. Milan.

Barfield, L.H. 1971. *Northern Italy before Rome.* London.

Barker, G.W.W. 1975. "Prehistoric Territories and Economies in Central Italy." In *Palaeoeconomy,* edited by E.S. Higgs, 111–75. Cambridge.

———. 1981. *Landscape and Society: Prehistoric Central Italy.* London.

Baroni, C., P. Biagi, R. Nisbet, and R.G. Scaife. 1990. "Laghetti del Crestoso: A High Altitude Castelnovian Camp in Its Environmental Setting (Brescia—Northern Italy.)" In *The Neolithization of the Alpine Region,* edited by P. Biagi, 43–51. Brescia.

Barra, A., R. Grifoni Cremonesi, F. Mallegni, M. Piancastelli, A. Vitello, and B. Wilkens. 1989–1990. "La Grotta Continenza di Trasacco: I livelli a ceramiche." *Rivista di Scienze Preistoriche* 42:31–100.

Batović, S. 1987. "La néolithisation en Adriatique." In *Premières Communautés Paysannes en Méditerranée Occidentale: Colloque International du C.N.R.S., Montpellier, 1983,* edited by J. Guilaine, J. Courtin, J.-L. Roudil, and J.-L. Vernet, 343–9. Paris.

Belluomini, G., and L. Delitalia. 1983. "Datazione di una sequenza stratigrafica del villaggio neolitico di Santa Tecchia con i metodi del 14C e della racemizzazione dell'acido aspartico." In *Studi sul Tavoliere della Puglia: Indagine Territoriale in un'Area-Campione,* edited by S.M. Cassano and A. Manfredini, 265–8. *BAR-IS* 160. Oxford.

Bernabò Brea, M., M. Cattani, R. Conversi, M. Cremaschi, R. Nisbet, and C. Ricci. 1984. "L'insediamento neolitico della Cassa di Risparmio a Travo (Pc)." *Preistoria Alpina* 20:59–80.

Biagi, P. 1979. "Stazione neolitica a Ostiano (CR), località Dugali Alti: scavi 1980." *Preistoria Alpina* 15:25–38.

———. 1980. "Some Aspects of the Prehistory of Northern Italy from the Final Palaeolithic to the Middle Neolithic: A Reconsideration on The Evidence Available to Date." *Proceedings of the Prehistoric Society* 46:9–18.

———. 1990. "The Prehistory of the Early Atlantic Period along the Ligurian and Adriatic Coasts of Northern Italy in a Mediterranean Perspective." *Rivista di Archeologia* 14:46–54.

———. 1996. "North Eastern Italy in the Seventh Millennium B.P.: A Bridge between the Balkans and the West?" In *The Vinča Culture, its Role and Cultural Connections,* edited by F. Draşovean, 9–22. Timişoara.

Biagi, P., E. Starnini, and B.A. Voytek. 1993. "The Late Mesolithic and Early Neolithic Settlement of Northern Italy: Recent Considerations." *Porocilo o Raziskovanju Paleolita, Neolita in Eneolita v Sloveniji* 21:45–67.

Bowman, S.G.E., J.C. Ambers, and M.N. Leese. 1991. "Re-evaluation of British Museum Radiocarbon Dates Issued between 1980 and 1984." *Radiocarbon* 32:59–79.

Brown, D.F. 1964. "The Chronology of the Northwestern Mediterranean." In *Chronologies in Old World Archaeology*, edited by R.W. Ehrich, 321–42. Chicago.

———. 1992. "Radiocarbon Chronology of Prehistoric Italy." In *Chronologies in Old World Archaeology.* 3rd ed. Vol. 1, edited by R.W. Ehrich, 289–94. London.

Burleigh, R. 1984. "Radiocarbon Dates for the Western Mediterranean Region." In *The Deya Conference of Prehistory: Early Settlement in the Western Mediterranean Islands.* Pt. 1, edited by W.H. Waldren, R. Chapman, J. Lewthwaite, and R.-C. Kennard, 277–90. *BAR-IS* 229. Oxford.

Calderoni, G., I. Caneva, A. Cazzella, M. Frangipane, and V. Petrone. 1994. "Department of Earth Sciences at the University of Rome Radiocarbon Dates III." *Radiocarbon* 36:143–52.

Cassano, S. 1985. "Considerazioni sugli inizi dell'economia produttiva sulle sponde dell'Adriatico." In *Studi di Paletnologia in Onore di Salvatore M. Puglisi*, edited by M. Liverani, A. Palmieri, and R. Peroni, 731–43. Rome.

Castelletti, L., A. Maspero, and C. Tozzi. 1994. "Il popolamento della Valle del Serchio (Toscana settentrionale) durante il Tardiglaciale Würmiano e l'Olocene Antico." In *Highland Zone Exploitation in Southern Europe*, edited by P. Biagi and J. Nandris, 189–204. Brescia.

Cipolloni Sampò, M. 1987. "Problèmes des débuts de l'économie de production en Italie sud-orientale." In *Premières Communautés Paysannes en Méditerranée Occidentale*, edited by J. Guilaine, J. Courtin, J.-L. Roudil, and J.-L. Vernet, 181–8. Paris.

Clark, J.G.D. 1965. "Radiocarbon Dating and the Spread of Farming Economy." *Antiquity* 39:45–8.

Coppola, D. 1987. "L'insediamento neolitico di Scamuso." In *Atti della XXV Riunione Scientifica dell'Istituto Italiano di Preistoria e Protostoria della Puglia centrale*, edited by A. Revedin, 223–38. Monopoli.

Costantini, L., and M. Stancanelli. 1994. "La preistoria agricola dell'Italia centro-meridionale: Il contributo delle indagini archeobotaniche." *Origini* 18:149–244.

Donahue, R.E., D.B. Burroni, G.M. Coles, R.H. Colten, and C.O. Hunt. 1992. "Petriolo III South: Implications for the Transition to Agriculture in Tuscany." *Current Anthropology* 33:328–31.

Evans, J. 1987. "The Development of Neolithic Communities in the Central Mediterranean: Western Greece to Malta." In *Premières Communautés Paysannes en Méditerranée Occidentale*, edited by J. Guilaine, J. Courtin, J.-L. Roudil, and J.-L. Vernet, 321–7. Paris.

Fedele, F. 1992. "Steinzeitliche Jäger in den Zentralalpen: Piano dei Cavalli (Splügenpass)." *Helvetia Archaeologica* 89:2–22.

Ferrara, G., G. Fornaca Rinaldi, and E. Tongiorgi. 1961. "Carbon-14 Dating in Pisa II." *Radiocarbon* 3:99–104.

Follieri, M. 1987. "L'agriculture des plus anciennes communautés rurales d'Italie." In *Premières Communautés Paysannes en Méditerranée Occidentale,* edited by J. Guilaine, J. Courtin, J.-L. Roudil, and J.-L. Vernet, 243–7. Paris.

Fugazzola Delpino, M.A., G. D'Eugenio, and A. Pessina. 1993. "La Marmotta (Anguillara, Sabazia, RM). Scavi 1989. Un abitato perilacustre di età neolitica." *Bullettino di*

Paletnologia Italiana 84:181–304.

Gnesutta, P.U., and A. Bertagnini. 1993. "Grotta delle Settecannelle (Ischia di Castro–Viterbo): Analisi ed inquadrimento della ceramica preistorica." *Rassegna di Archeologia* 11:67–112.

Gowlett, J.A.J., R.E.M. Hedges, I.A. Law, and C. Perry. 1987. "Radiocarbon Dates from the Oxford AMS System: *Archaeometry* Datelist 5." *Archaeometry* 29:125–55.

Guerreschi, G., P. Catalani, G. Longo, and A. Iannone. 1981–1992. "Grotta Bella (Terni). Una sequenza stratigrafica dal Neolitico inferiore all'età imperiale: I livelli preistorici." *Bullettino di Paletnologia Italiana* 83:143–228.

Guilaine, J. 1976. *Premiers Bergers et Paysans de l'Occident Méditerranéen*. Paris.

———. 1979. "The Earliest Neolithic in the West Mediterranean: A New Appraisal." *Antiquity* 53:22–30.

———. 1980. "Problèmes actuels de la néolithisation du Néolithique ancien en Méditerranée Occidentale." In *Interaction and Acculturation in the Mediterranean: Proceedings of the Second International Congress of Mediterranean Pre- and Protohistory*, edited by J.G.P. Best and N.M.W. De Vries, 3–22. Amsterdam.

Guilaine, J., and A. Calvet. 1970. "Nouveaux points de chronologie absolue pour le Néolithique Ancien de la Méditerranée Occidentale." *L'Anthropologie* 74 (1–2):85–92.

Guilaine, J., and G. Cremonesi. 1996. "La chronologie du Néolithique ancien a Trasano (Matera, Basilicata) dans le contexte de la Méditerranée centrale." In *Forme e Tempi della Neolitizzazione in Italia Meridionale e in Sicilia*, edited by V. Tiné, 433–50. Genoa.

Guilaine, J., J. Simone, and Y. Thommeret. 1981. "Datations C14 pour le Néolithique du Tavoliere (Italie)." *Bulletin de la Société Préhistorique Française* 78:154–60.

Hedges, R.E.M., R.A. Housley, I.A. Law, C. Perry, and E. Hendy. 1988. "Radiocarbon Dates from the Oxford AMS System: *Archaeometry* Datelist 8." *Archaeometry* 30:291–305.

Hedges, R.E.M., R.A. Housely, I.A. Law, and C.R. Bronk. 1989. "Radiocarbon Dates from the Oxford AMS System: *Archaeometry* Datelist 9." *Archaeometry* 31:207–34.

———. 1990. "Radiocarbon Dates from the Oxford AMS System: *Archaeometry* datelist 10." *Archaeometry* 32:101–8.

Improta, S., and A. Pessina. 1998. "La neolitizzazione dell'Italia settentrionale: Il nuovo quadro cronologico." In *Settemila Anni fa il Primo Pane: Ambienti e Culture delle Società Neolitiche*, edited by A. Pessina and G. Muscio, 107–15. Udine.

Klein Hofmeijer, G., and P.Y. Sondarr. 1993. "Pleistocene Humans in the Island Environment of Sardinia." In *Sardinia in the Mediterranean: A Footprint in the Sea. Studies in Sardinia Archaeology Presented to Miriam S. Balmuth*, edited by R.H. Tykot and T.K. Andrews, 49–56. Sheffield.

Klein Hofmeijer, G., P.Y. Sondaar, C. Alderliesten, K. Van der Borg, and A.F.M. De Jong. 1987. "Indications of Man on Sardinia." *Nuclear Instruments and Methods in Physics Research* B29:166–8.

Lawn, B. 1975. "University of Pennsylvania Radiocarbon Dates XVIII." *Radiocarbon* 17:196–215.

Leighton, R. 1996. "Research Traditions, Chronology and Current Issues: An Introduction." In *Early Societies in Sicily: New Developments in Archaeological Research*, edited by R.

Leighton, 1–19. London.

Lewthwaite, J. 1986. "The Transition to Food Production: A Mediterranean Perspective." In *Hunters in Transition: Mesolithic Societies of Temperate Eurasia and Their Transition to Farming*, edited by M. Zvelebil, 53–66. Cambridge.

———. 1987. "Isolating the Residuals: The Mesolithic Basis on Man-Animal Relationships on the Mediterranean Islands." In *The Mesolithic in Europe*, edited by C. Bonsall, 541–55. Edinburgh.

Linick, T.W. 1980. "La Jolla Natural Radiocarbon Measurements IX." *Radiocarbon* 22:1034–44.

Lo Porto, F.G. 1987. "Problemi cronologici del Neolitico nell'Italia meridionale." In *Atti della XXVI Riunione Scientifica: il Neolitico in Italia*, edited by A. Revedin, 119–31. Florence.

Lubell, D., and M. Mussi. 1995. "Upper Palaeolithic to Neolithic in Abruzzo: Preliminary Data from the 1989–1994 Field Seasons." *Old World Archaeology Newsletter* 18(2):31–7.

Meulengracht, A., P. McGovern, and B. Lawn. 1981. "University of Pennsylvania Radiocarbon Dates XXI." *Radiocarbon* 23:227–40.

Milliken, S., and R. Skeates. 1989. "The Alimini Survey: The Mesolithic-Neolithic Transition in the Salento Peninsula (S.E. Italy)." *Bulletin of the Institute of Archaeology, University College London* 26:77–98.

Pessina, A., and M. Rottoli. 1996. "New Evidence on the Earliest Farming Cultures in Northern Italy: Archaeological and Palaeobotanical Data." *Porocilo o Raziskovanju Paleolitika, Neolitika in Eneolitika v Sloveniji* 23:77–103.

Phillips, P. 1975. *Early Farmers of West Mediterranean Europe*. London.

Pluciennik, M. 1994. "Holocene Hunter-Gatherers in Italy." In *Radiocarbon Dating and Italian Prehistory*, edited by R. Skeates and R. Whitehouse, 45–59. London.

———. 1997. "Radiocarbon Determinations and the Mesolithic-Neolithic Transition in Southern Italy." *Journal of Mediterranean Archaeology* 10:115–50.

Renfrew, C. 1970. "The Tree-Ring Calibration of Radiocarbon: An Archaeological Evaluation." *Proceedings of the Prehistoric Society* 36:280–311.

Sargent, A. 1985. "The Carbon-14 Chronology of the Early and Middle Neolithic of Southern Italy." *Proceedings of the Prehistoric Society* 51:31–40.

Sarti, L., C. Corridi, F. Martini, and P. Pallecchi. 1994. "Mileto: Un insediamento neolitico della ceramica a linee incise." *Rivista di Scienze Preistoriche* 43:73–154.

Shotton, F.W., D.J. Blundell, and R.E.G. Williams. 1970. "Birmingham University Radiocarbon Dates IV." *Radiocarbon* 12:385–99.

Skeates, R. 1992. "The Neolithic and Copper Age of the Abruzzo-Marche Region, Central Italy." D.Phil. thesis, Oxford.

———. 1994a. "Towards an Absolute Chronology for the Neolithic in Central Italy." In *Radiocarbon Dating and Italian Prehistory*, edited by R. Whitehouse and R. Skeates, 61–72. London.

———. 1994b. "A Radiocarbon Date-List for Prehistoric Italy (c. 46,400 B.P.–2,450 B.P./400 cal. B.C.)." In *Radiocarbon Dating and Italian Prehistory*, edited by R. Whitehouse and R. Skeates, 147–288. London.

———. 1999. "Unveiling Inequality: Social Life and Social Change in the Mesolithic and

Early Neolithic of East-Central Italy." In *Social Dynamics of the Prehistoric Central Mediterranean*, edited by R.H. Tykot, J. Morter, and J.E. Robb, 15–45. London.

Skeates, R., and R. Whitehouse. 1994. "New Radiocarbon Dates for Prehistoric Italy 1." *The Accordia Research Papers* 5:137–50.

———. 1995–1996. "New Radiocarbon Dates for Prehistoric Italy 2." *The Accordia Research Papers* 6:179–91.

———. 1997–1998. "New Radiocarbon Dates for Prehistoric Italy 3." *The Accordia Research Papers* 7:149–62.

———, eds. 1994. *Radiocarbon Dating and Italian Prehistory*. London.

Spataro, M. 1997–1998. "La Caverna dell'Edera di Aurisina (TS): Studio archeometrico delle ceramiche." *Atti della Società per la Preistoria e Protostoria della Regione Friuli-Venezia Giulia* 11:63–89.

Stuiver, M., and P.J. Reimer. 1986. "A Computer Program for Radiocarbon Age Calibration." *Radiocarbon* 28:1022–30.

———. 1993. "Extended 14C Data Base and Revised CALIB 3.0 14C Age Calibration Program." *Radiocarbon* 35:215–30.

Tiné, S. 1976. "La neolitizzazione dell'Italia peninsulare." In *IXe Congrès de l'Union Internazionale des Sciences Préhistoriques et Protohistoriques*, edited by G. Bailloud, 74–88. Nice.

Tiné, V. 1996. "Favella." In *Forme e Tempi della Neolitizzazione in Italia Meridionale e in Sicilia*, edited by V. Tiné, 451–2. Genoa.

Tirabassi, J. 1981–1982. "Pozzo neolitico di Via Rivoluzione d'ottobre: I materiali." *Emilia Preromana* 9–10:47–71.

Tongiorgi, E., A.M. Radmilli, G. Rinaldi Fornaca, and G. Ferrara. 1959. *Programma di Datazione delle Culture Italiane della Preistoria Recente in Collaborazione con l'Istituto Italiano di Preistoria e Protostoria: Laboratorio di Geologia Nucleare.* Unpublished Mimeograph. Pisa.

Trump, D.H. 1966. *Central and Southern Italy before Rome*. London.

———. 1983. *La Grotta di Filiestru a Bonu Ighinu, Mara (SS).* Sassari

Tusa, S. 1985. "The Beginnings of Farming Communities in Sicily: The Evidence of Uzzo Cave." In *Papers in Italian Archaeology IV*, edited by C. Malone and S. Stoddart, 61–82. *BAR-IS* 244. Oxford.

———. 1987. "Il Neolitico della Sicilia." In *Atti della XXVI Riunione Scientifica: Il Neolitico in Italia*, edited by A. Revedin, 361–80. Florence.

———. 1996. "From Hunter-Gatherers to Farmers in Western Sicily." In *Early Societies in Sicily: New Developments in Archaeological Research*, edited by R. Leighton, 41–55. London.

Tykot, R.H. 1994. "Radiocarbon Dating and Absolute Chronology in Sardinia and Corsica." In *Radiocarbon Dating and Italian Prehistory*, edited by R. Skeates and R. Whitehouse, 115–45. London.

Ucelli Gnesutta, P., and A. Bertagnini. 1993. "Grotta delle Settecannelle (Ischia di Castro-Viterbo)." *Rassegna di Archeologia* 11:67–112.

Whitehouse, R. 1968. "The Early Neolithic of Southern Italy." *Antiquity* 42:188–93.

———. 1969. "The Neolithic Pottery Sequence in Southern Italy." *Proceedings of the Prehistoric Society* 35:267–310.

———. 1978. "Italian Prehistory, Carbon 14 and the Tree-Ring Calibration." In *Papers in Italian Archaeology*. Vol. 1, *The Lancaster Seminar: Recent Research in Prehistoric, Classical and Medieval Archaeology*, edited by H. Blake, T.W. Potter, and D.B. Whitehouse, 71–91. *BAR* Suppl. 41. Oxford.

———. 1984. "Social Organisation in the Neolithic of Southeast Italy." In *The Deya Conference of Prehistory: Early Settlement in the Western Mediterranean Islands and the Peripheral Areas*, edited by W.H. Waldren, R. Chapman, J. Lewthwaite, and R.-C. Kennard, 1109–33. *BAR-IS* 229. Oxford.

———. 1985. "New Radiocarbon Dates for the Neolithic of Eastern Italy." In *Lancaster in Italy: Archaeological Research Undertaken in Italy by the Department of Classics and Archaeology in 1984*, edited by H. Blake and R. Whitehouse, 35–9. Lancaster.

———. 1987. "New Radiocarbon Dates for Prehistoric Italy." In *Lancaster in Italy: Archaeological Research Undertaken in Italy by the Department of Classics and Archaeology in 1986*, edited by H. Blake and R. Whitehouse, 13–20. Lancaster.

———. 1994. "The British Museum 14C Programme for Italian Prehistory." In *Radiocarbon Dating and Italian Prehistory*, edited by R. Skeates and R. Whitehouse, 85–98. London.

Appendix: "Latest" Mesolithic and "Earliest" Neolithic Radiocarbon Determinations for Italian Provinces

This appendix comprises a list of the single "latest" acceptable Mesolithic and single "earliest" acceptable Neolithic radiocarbon determinations (lying in the seventh and sixth millennia Cal B.C.) that currently exist for Italy. It is therefore not a list of all Mesolithic and early Neolithic dates. It is arranged on a province-by-province basis. The determinations have been calibrated according to the intercepts method at the 2σ confidence level, using the CALIB 4.2 computer program (Stuiver and Reimer 1993).

Twenty radiocarbon determinations have been rejected for various sound reasons (Skeates 1994b, 271–2; Tykot 1994, 121; Whitehouse 1994, 86–7; Pluciennik 1997). The stratigraphic inversion of series of determinations in cave and rock shelter deposits: R-1941α and R-1936 from Mondeval de Sora in the Belluno province (Skeates 1994b); UtC-22 from Grotta Corbeddu in the Nuoro province (Klein Hofmeijer et al. 1987, 167); Rome-446, Rome-447, Rome-449, Rome-450, and Rome-451 from Grotta III di Latronico in the Potenza province (Skeates and Whitehouse 1995–1996); Gd-6153, Gd-5639, Gd-6154 from Grotta d'Ernesto in the Trento province (Awsiuk et al.

1991); R-1175 from Grotta di Monte Venere in the Viterbo province (Alessio et al. 1991). Excessively large standard deviations: GX-19568 from Grotta dell'Edera in the Trieste province; I-11444 from Campo Ceresole in the Cremona province (Biagi 1979); BM-2557 from Coppa Nevigata in the Foggia province (Ambers et al. 1989). The misattribution of a sample to Neolithic instead of Mesolithic deposits: R-285 from Grotta del Santuario della Madonna (Alessio et al. 1967, 354). An unpublished determination of marine shell: Pi-? from Coppa Nevigata in the Foggia province (Tongiorgi et al. 1959). A sample of *mixed* bones: the determination from Masseria Santa Tecchia in the Foggia province (Belluomini and Delitalia 1983). A sample of mixed charcoal and soil from disturbed deposits: P-1999 from Casa San Paolo in the Bari province (Lawn 1975). Another sample of charcoal, OxA-1363, from Petriolo III South, must also be rejected for the moment due to contradictions in the published details of its cultural context, which ascribe it both to the Final Epigravettian (Hedges et al. 1988) and to a "ceramic Mesolithic" (i.e., the early Neolithic; Donahue et al. 1992). A sample of bone, OZB-653, from an early Neolithic burial at Piancada should also be rejected, since it has produced a date in the early eighth millennium Cal B.C., which is far too early, both for the Neolithic in Italy, and in relation to a second determination on bone from this context (Improta and Pessina 1998).

I have chosen to accept all other radiocarbon determinations until proven and published otherwise. To my mind, this is the only way in which to retain a degree of objectivity in the analysis of the radiocarbon determinations available. I have therefore resisted pleas to remove, for example, the "early" dates for north Italian contexts containing Square Mouthed Pottery, which scholars have traditionally assigned to a middle phase of the Neolithic. I have also retained the "early" date for Molino Casarotto, which some scholars consider to be as much as 1,000 years too early. In this respect, then, the list below is provisional and potentially open to alteration, but it aspires to objectivity.

Savona	
Arma dello Stefanin (Final Upper Paleolithic/ Epipaleolithic) R-109 7800 ± 100 b.p. (Alessio et al. 1965) 7039(6641)6441 Cal B.C.	Caverna delle Arene Candide (Impressed Ware) UB-2423 6980 ± 115 b.p. (Bagolini and Biagi 1990) 6062(5840,5816,5815)5639 Cal B.C.

Varese	
	Pizzo di Bodio (Isolino) B-5090 6320±80 b.p. (Banchieri 1988–1989) 5473(5303)5062 Cal B.C.

Cremona	
	Campo Ceresole (Vhò) I-11445 6170 ± 110 b.p. (Biagi 1979) 5363(5204,5179,5137,5131,5073) 4800 Cal B.C.
Sondrio	
Pian dei Cavalli 1 (Castelnovian Mesolithic) NA-192 7540 ± 210 b.p. (Fedele 1992) 7027(6417) 5934 Cal B.C.	
Brescia	
Laghetti del Crestoso (Castelnovian late Mesolithic) HAR-8871 6790 ± 120 b.p. (Baroni et al. 1990) 5964(5707,5685,5667) 5481 Cal B.C.	
Trento	
Riparo di Pradestel (Castelnovian) R-1148 6870 ± 50 b.p. (Alessio et al. 1978) 5840(5730) 5643 Cal B.C.	Riparo di Romagnano III (Late Castelnovian and Early Neolithic) R-1136 6480 ± 50 b.p. (Alessio et al. 1978) 5514(5473) 5323 Cal B.C.
Verona	
	Lugo di Grezzana (Fiorano) R-2745 6524 ± 76 b.p. (Improta and Pessina 1998) 5620(5478) 5321 Cal B.C.
Vicenza	
Grottina dei Covoloni del Broion (Castelnovian) R-892 6930 ± 60 b.p. (Alessio et al. 1978) 5978(5796) 5711 Cal B.C.	Site 3, Molino Casarotto (Neolithic) Birm-175 6450 ± 110 b.p. (Shotton et al. 1970) 5619(5468,5438,5422,5399) 5150 Cal B.C.
Belluno	
Mondeval de Sora (Castelnovian Mesolithic) R-1939 7330 ± 50 b.p. (Skeates and Whitehouse 1994) 6331(6216,6167,6164) 6033 Cal B.C.	
Pordenone	
	Fagnigola (Early Neolithic) R-2550 6570 ± 75 b.p. (Alessio et al. 1995) 5635(5509,5501,5484) 5371 Cal B.C.
Udine	
	Piancada (Early Neolithic) R-2705 6751 ± 108 b.p. (Improta and Pessina 1998) 5839(5659,5651,5640) 5479 Cal B.C.
Trieste	
Grotta Benussi (Castelnovian) R-1043 7050 ± 60 b.p. (Alessio et al. 1978) 6021(5975,5950,5916) 5787 Cal B.C.	Caverna dell'Edera (ceramic Mesolithic) GX-19569 6700 ± 130 b.p. (Spataro 1997–1998) 5838(5623) 5381Cal B.C.
Piacenza	
	Casa di Risparmio (Square-Mouthed Pottery) I-12585 6580 ± 150 b.p. (Bernabò Brea et al. 1984:79) 5733(5512,5498,5485) 5264 Cal B.C.

Reggio Emilia	
Lama Lite (Late Castelnovian) Rome-394 6620 ± 80 b.p. (Skeates and Whitehouse 1995–1996) 5705(5606,5591,5557) 5390 Cal B.C.	Rivaltella-Ca'Romensini (Square-Mouthed Pottery) I-12519 6070 ± 110 b.p. (Tirabassi 1981–1982) 5298(4949) 4715 Cal B.C.
Modena	
	Fiorano Modenese (Fiorano) GrN-19838 6690 ± 180 b.p. (Improta and Pessina 1998) 5976(5621) 5304 Cal B.C.
Ravenna	
	Lugo di Ravenna R-2747 6626 ± 110 b.p. (Improta and Pessina 1998) 5727(5608,5590,5558) 5367 Cal B.C.
Lucca	
Piazzana (Castelnovian) R-400 7330 ± 85 b.p. (Castelletti et al. 1994) 6393(6216,6167,6164) 6012 Cal B.C.	Pian di Cerreto (Early Neolithic) Rome-548 6680 ± 80 b.p. (Skeates and Whitehouse 1995–1996) 5727(5620,5569,5565) 5478 Cal B.C.
Firenze	
	Mileto (Linear Incised Ware) Beta-44114 6180 ± 80 b.p. (Sarti et al. 1994) 5317(5206,5177,5139,5128,5077) 4854 Cal B.C.
Livorno	
Isola Santa (Mesolithic) R-1525 7380 ± 130 b.p. (Alessio et al. 1983) 6460(6228) 5933 Cal B.C.	
Siena	
	Petriolo III South (ceramic Mesolithic) OxA-2630 7480 ± 100 b.p. (Donahue et al. 1992) 6473(6382,6277,6273) 6086 Cal B.C.
Viterbo	
	Grotta delle Settecannelle (Cardial Impressed) GrN-14543 7830 ± 150 b.p. (Ucelli Gnesutta and Bertagnini 1993:69) 7079(6647) 6405 Cal B.C.
Roma	
	La Marmotta (Impressed and Sasso) R-2311 6370 ± 95 b.p. (Fugazzola Delpino et al. 1993) 5506(5336,5334,5322) 5076 Cal B.C.
Perugia	
	San Marco (Impressed and Incised Ware) OxA-1853 6430 ± 80 b.p. (Hedges et al. 1990) 5527(5464,5448,5418,5403,5376) 5261 Cal B.C.
Terni	
	Grotta Bella (Sasso/Early Ripoli) Gif-4375 6450 ± 90 b.p. (Guerreschi et al. 1981–1992) 5610(5468,5438,5422,5399,5387) 5261 Cal B.C.

Ancona		
		Ripabianca di Monterado (Impressed and Incised Ware) R-599 6260 ± 85 b.p. (Alessio et al. 1970) 5465(5278,5273,5261,5216,5215) 4963 Cal B.C.
Macerata		
		Maddalena di Muccia (Impressed and Incised Ware) R-643 6580 ± 75 b.p. (Alessio et al. 1970) 5656(5512,5498,5485) 5374 Cal B.C.
L'Aquila		
	Grotta di Pozzo (Sauveterrian) TO-3420 8110 ±90 b.p. (Lubell and Mussi 1995) 7448(7076) 6714 Cal B.C.	Grotta Continenza (initial Neolithic) Rome-549 7230 ± 90 b.p. (Skeates and Whitehouse 1995–1996) 6238(6068,6035,6035) 5915 Cal B.C.
Pescara		
		Villaggio Leopardi (Impressed and Incised Ware) Pi-101 6578 ± 135 b.p. (Ferrara et al. 1961) 5727(5511,5498,5484) 5300 Cal B.C.
Chieti		
		Marcianese (Impressed and Incised Ware) BM-2250R 6590 ± 130 b.p. (Bowman et al. 1991) 5728(5525,5522,5515) 5304 Cal B.C.
Campobasso		
		Monte Maulo (Impressed and Painted) OxA-651 6540 ± 80 b.p. (Gowlett et al. 1987) 5624(5480) 5323 Cal B.C.
Foggia		
		Masseria Giuffreda (Impressed and Painted Ware) MC-2292 7125 ± 200 b.p. (Guilaine et al. 1981) 6404(5992) 5635 Cal B.C.
Salerno		
	Grotta delle Soppressate (Mesolithic) F-33 7540 ± 135 b.p. (Azzi et al. 1973) 6643(6417) 6084 Cal B.C.	
Potenza		
		Rendina (Impressed and Painted Ware) LJ-4548 7110 ± 140 b.p. (Linick 1980) 6231(5990,5938,5932) 5718 Cal B.C.
Matera		
		Trasano (Impressed Ware) Ly-5297 7030 ± 160 b.p. (Guilaine and Cremonesi 1996) 6222(5961,5954,5892) 5624 Cal B.C.
Bari		
		Scamuso (Impressed Ware) Gif-6339 7290 ± 110 b.p. (Coppola 1987) 6398(6199,6193,6160,6138,6093) 5920 Cal B.C.

Brindisi	
	Sant'Anna III (Middle Neolithic) Rome-322 6780 ± 90 b.p. (Calderoni et al. 1994) 5839(5664) 5528 Cal B.C.
Lecce	
	Torre Sabea (Impressed Ware) ? 6960 ± 130 b.p. (Costantini and Stancanelli 1994) 6065(5838,5822,5809) 5623 Cal B.C.
Cosenza	
	Favella (Impressed Ware) Beta-7163 6860 ± 60 b.p. (Tiné 1996) 5841(5727) 5637 Cal B.C.
Catanzaro	
	Area H, Piana di Curingia (Cardial and Stentinello Impressed Ware) P-2946 6930 ± 60 b.p. (Ammerman and Bonardi 1985–1986) 5978(5796) 5711 Cal B.C.
Agrigento	
	Poggio/Piano Vento (Impressed Ware) A-4474 6130 ± 90 b.p. (Skeates 1994b) 5302(5048) 4799 Cal B.C.
Trapani	
Grotta dell'Uzzo ('transitional' Epipalaeolithic) P-2734 7910 ± 70 b.p. (Meulengracht et al. 1981) 7057(6748,6721,6702) 6592 Cal B.C.	Grotta dell'Uzzo (Impressed Ware) P-2733 6750 ± 70 b.p. (Meulengracht et al. 1981) 5739(5658,5651,5640) 5532 Cal B.C.
Nuoro	
Grotta Corbeddu (pre-Neolithic) UtC-301 7860 ± 130 b.p. (Klein Hofmeijer et al. 1987) and 7078(6679,6671,6657) 6441 Cal B.C.	Grotta Corbeddu (Cardial Impressed Ware) UtC-1251 6690 ± 80 b.p. (Klein Hofmeijer Sondaar 1993) 5729(5621) 5479 Cal B.C.
Sassari	
	Grotta Filiestru (Cardial Impressed Ware) Q-3020 6710 ± 75 b.p. (Trump 1983) 5731(5625) 5482 Cal B.C.

NOTES

[1] Two on carbonized grains of *Hordeum* from Coppa Nevigata (Foggia) and three on grains of *Triticum aestivum*, *Triticum compactum,* and *Hordeum vulgare* from San Marco near Perugia (Hedges et al. 1989, 1990).

[2] Caverna delle Arene Candide (Savona), Villaggio Leopardi and Grotta dei Piccioni (Pescara), Grotta La Porta (Salerno), and Coppa Nevigata (Foggia).

[3] Grotta Arma dello Stefanin (Savona) and Grotta del Santuario della Madonna and Grotta del Romito (Cosenza).

[4] Riparo Arma di Nasino (Savona), Riparo di Vatte di Zambana (Trento), Maddalena di Muccia (Macerata), Ripabianca di Monterado (Ancona), Riparo Blanc (Latina), and Villaggio A, Scaramella (Foggia).

[5] R-350 for Villaggio A, Scaramella; R-643 for Maddalena di Muccia; Pi-101 for Villaggio Leopardi; and Birm-172 and Birm-177 for Site 4, Molino Casarotto.

[6] Mesolithic Grotta delle Soppressate (Salerno); and the Neolithic sites of Caverna delle Arene Candide (Savona), Grotta del'Orso (Siena), Passo di Corvo (Foggia), and Casa San Paolo (Bari).

[7] The new Radiocarbon Dating Laboratory of the University of Florence (F), the Rome Laboratory, the Laboratory of Applied Radioactivity of the Scientific Centre in Monaco (MC), the Radiocarbon Laboratory of the University of Pennsylvania (P), and the Mount Soledad Radiocarbon Laboratory of the University of California at La Jolla (LJ).

[8] Riparo di Pradestel and Riparo di Romagnano III (Trento), Grottina dei Covoloni del Broion (Vicenza), and Grotta Benussi and Grotta dei Ciclami (Trieste).

[9] Laboratories: the Florence Laboratory, Teledyne Isotopes of New Jersey (I), the Central Institute for Ancient History and Archaeology in Berlin (Bln), the La Jolla Laboratory, the Birmingham Laboratory, the Pennsylvania Laboratory, and the Monaco Laboratory. Sites: Caverna delle Arene Candide (Savona), Campo Ceresole (Cremona), Monte Bagioletto Alto and Passo della Comunella (Reggio Emilia), Masseria Giuffreda and Villa Comunale (Foggia), Rendina (Potenza), Grotta di Porto Badisco (Lecce), and Grotta dell'Uzzo (Trapani).

[10] Grotta Arma dello Stefanin (Imperia), Grotta dell'Edera (Savona), Cassa di Risparmio and Casa Gazza (Piacenza), Dugali Alti (Cremona), Fienile Rossino (Brescia), Fornace dei Cappuccini and Lugo (Ravenna), Sammardénchia (Udine), Isola Santa (Lucca), Marcianese (Chieti), Monte Maulo (Campobasso), Coppa Nevigata, Grotta Scaloria and Lagnano da Piede, Masseria Santa Tecchia, and Masseria Fontanarosa Uliveto (Foggia), Torre Canne, Scamuso, and Santa Barbara (Bari), Acconia (Cosenza), Piana di Curingia and Bevilacqua (Catanzaro), Grotta di Molara (Palermo), Poggio Vento (Agrigento), Grotta Filiestru (Sassari), and Grotta Corbeddu (Nuoro).

[11] The Godwin Laboratory of the University of Cambridge (Q), the Rome Laboratory, the La Jolla Laboratory, Teledyne Isotopes, the Monaco Laboratory, the Centre for Isotope Research of the University of Groningen (GrN), the Laboratory of Isotope Geochemistry of the University of Arizona (A), the Pennsylvania Laboratory, the Berlin Laboratory, the Palaeoecology Laboratory of the Queen's University in Belfast (UB), the California Laboratory, and the Research Laboratory of the British Museum (BM).

[12] The Laboratory of the Centre for Radioactive Traces at Gif-sur-Yvette (Gif), the Robert Van der Graff Laboratory of the University of Utrecht (UtC), and the Oxford Radiocarbon Accelerator Unit of Oxford University (OxA).

[13] He also suggested that the earliest appearance of pottery and domesticated caprines preceded the adoption of other farming practices in the west Mediterranean, but soon changed his mind for southern Italy, where he regarded the first domestic sheep, goat, cattle, and pig as having been introduced simultaneously as an integrated "package."

[14] Caverna delle Arene Candide and Arma dell'Aquila (Savona), Pizzo di Bodio (Varese),

Laghetti di Crestoso (Brescia), Pian dei Cavalli I (Sondrio), Riparo di Romagnano III and Grotta d'Ernesto (Trento), Grotta dell'Edera (Trieste), Isola Santa (Lucca), Petriolo III South (Siena), Grotta di Monte Venere (Viterbo), Grotta Bella (Terni), San Marco (Perugia), Fonterossi (Chieti), Scamuso (Bari), Coppa Nevigata and Defensola (Foggia), Trasano (Matera), Perriere Sottano (Catania), Grotta dell'Uzzo (Trapani).

[15] The Physics Institute of the University of Bern (B), the Rome Laboratory, the Belfast Laboratory, the Berlin Laboratory, the Isotope Measurements Laboratory at Harwell (HAR), the Radiocarbon Laboratory of the University of Lyon (Ly), the Radiocarbon Laboratory of the Centre of Applied Research and Documentation at Udine (UD), the Gliwice Laboratory (Gd), and the British Museum.

[16] The Laboratories at Oxford, Gif-sur-Yvette, and Utrecht.

[17] Arene Candide (Savona), Laghetti del Crestoso (Brescia), Lago delle Buse (Trento), Plan de Frea II (Bolzano), Mondeval de Sora (Belluno), Fagnigola, Pordenone, Valer, Aquileia, and Sammardenchia (Udine), Grotta dell'Edera (Trieste), Lama Lite (Reggio Emilia), Piazzana, Il Muraccio, and Pian di Cerreto (Lucca), Mileto (Firenze), Grotta delle Settecannelle (Viterbo), La Marmotta (Roma), Colle Santo Stefano, Fonte Chiarano, Grotta Continenza, Grotta di Pozzo, and Rio Tana (L'Aquila), Catignano (Pescara), Masseria Candelaro and Ripatetta (Foggia), St. Anna III (Brindisi), Grotta III di Latronico (Potenza), Trasano (Matera), Stretto (Palermo), Grotta del Cavallo (Trapani), Grotta Corbeddu (Nuoro).

[18] The Groningen Laboratory, the Rome Laboratory (R), the Laboratory of the Department of Earth Sciences of the University of Rome (Rome), Beta Analytic of Miami (Beta), Geochron Laboratories in Cambridge, Mass. (GX), the Alberta Environmental Center of Vegreville (AECV), the Heidelberg Laboratory (Hd), the Lyon Laboratory, and the Texas A&M University Laboratory (TAM).

[19] The Utrecht Laboratory, the Oxford Laboratory, and the Isotrace Laboratory of the University of Toronto (TO).

[20] I am very grateful to Paolo Biagi for sending me details of recently published radiocarbon determinations and for commenting in detail on an earlier draft of this paper. I would also like to thank Albert Ammerman for his additional comments and for his invitation to participate in the Venice meeting.

— 10 —

Aspects de la Néolithisation en Méditerranée et en France

Jean Guilaine

Depuis quelque 35 ans, les questions touchant à la néolithisation des terres méditerranéennes constituent l'un des principaux axes de ma réflexion scientifique. En effet dès 1963 je commençais mes recherches sur le gisement stratifié cardial/épicardial de la grotte Gazel, en Languedoc. Dès lors, je n'ai cessé, à la fois par ma réflexion théorique et par mes expériences de terrain, de m'intéresser au problème de la transition de l'économie de chasse à la constitution des premières sociétés de production. Soucieux d'avoir sur cette question une vision élargie et non limitée à une aire spécifique, j'ai essayé de multiplier les points d'ancrage de mon jugement en ouvrant des chantiers dans plusieurs zones de la Méditerranée: à l'Ouest d'abord, en France du Sud, avec l'étude d'une série de sites de grotte et d'abri (grotte Gazel, abri Jean-Cros, abri de Dourgne, abri de Font-Juvénal) ou de plein air (Pont de Roque Haute), voire "sous-marin" (Leucate); dans le Nord-Est de l'Espagne ensuite (Cova del Toll, Balma de l'Espluga; Guilaine et al. 1982), en Andorre aussi (Balma Margineda) tandis qu'en 1970, je rédigeais, en collaboration avec O. da Veiga Ferreira, un essai de synthèse sur le Néolithique ancien au Portugal (Guilaine et Veiga Ferreira 1970). J'ai ensuite ouvert des chantiers de fouilles en Italie du Sud-Est, en collaboration avec mon regretté collègue G. Cremonesi, car cette région me paraissait jouer un rôle capital dans la transmission des influences néolithisantes entre Orient et Occident. J'ai ainsi travaillé sur un site à céramique impressa des Pouilles (Torre Sabea) et sur un établissement stratifié de la région de Matera: Trasano (Basilicate;

Cremonesi et Guilaine 1987; Guilaine et Cremonesi 1987). Enfin, depuis quelques années, je dirige la fouille d'un site pré-céramique de Chypre (Parekklisha-Shillourokambos), qui se trouve être le plus ancien site néolithique connu de l'île (Guilaine et al. 1997-1998; Vigne et al. 1998).

J'ajoute que ces diverses expériences, conduites dans une perspective d'approche collégiale générant des monographies pluridisciplinaires, ont été réalisées dans des contextes écologiques volontairement diversifiés: aires côtières (Torre Sabea, Pont de Roque Haute, Leucate) ou sub-littorales (Shillourokambos), étage collinéen (grotte Gazel), basse ou moyenne montagne (El Toll), voire milieu montagnard (Balma Margineda; Guilaine 1979; Guilaine et al. 1984; Guilaine 1993; Guilaine et Martzluff 1995). J'ai, enfin, sous-tendu cette recherche par une réflexion plus générale jalonnée par des ouvrages concernant tout ou partie du domaine méditerranéen ou de la France (Guilaine 1976; 1980; 1994) ou par l'organisation de colloques et publications collectives (Guilaine et al. 1987).

Analyser ainsi les processus de mise en place du Néolithique par des regards multiples sur des aires géo-culturelles diversifiées offre l'incomparable avantage d'approcher la variété des situations et la complexité des processus qu'une vision théorique trop générale a tendance à réduire à des mécanismes simplifiés. Mes terrains m'ont ainsi amené à réfléchir sur trois des grands ensembles "primaires" du Néolithique méditerranéen (le Pré-céramique de Chypre, l'Impressa italo-adriatique, le Cardial occidental) et c'est sur ces trois zones bien distinctes que je voudrais livrer quelques impressions, avant d'aborder des problèmes plus généraux: mise en place du Néolithique en France, rythmes de la transmission en Europe de l'économie de production.

Chypre et le Proche-Orient

Chypre constitue un excellent exemple de néolithisation en milieu insulaire. En effet l'île n'est distante que d'une centaine de kilomètres du littoral levantin et, faute d'espèces végétales et animales domesticables, le Néolithique pré-céramique (qui en constitue la phase la plus ancienne) ne peut être que le produit d'une diffusion maritime. On est donc ici face à un exemple de transmission du complexe néolithique à peu de distance du "noyau moteur " proche-oriental et l'on peut dès lors essayer de saisir les vitesses de propagation et l'éventuelle altération des fondements économiques et culturels au cours de ce trajet.

Jusqu'à ces dernières années, cette vision est restée embrouillée. En effet tous les sites du néolithique pré-céramique chypriotes fouillés jusqu'au début des années quatre-vingt-dix (Khirokitia, Kalavasos-Tenta, Cap Andreas Kastros) révélaient une forte "distance" culturelle par rapport au néolithique proche-oriental contemporain: usage de constructions circulaires massives—s'opposant aux maisons rectangulaires du PPNB continental—manque de spécificité des industries lithiques, alors qu'à la même époque la zone levantine présente de nombreuses variétés d'armatures sophistiquées. Autre divergence: parmi les espèces domestiquées transmises à Chypre (caprins et porcs), le boeuf était absent alors qu'il faisait partie sur le continent de la panoplie des animaux désormais élevés.

Enfin ces différences en regard du "noyau moteur" étaient renforcées par le caractère récent de ce Pré-céramique, centré, d'après le radiocarbone, sur le VIIe millénaire B.C. (cal.): or, à cette époque, la céramique avait déjà fait son apparition sur le littoral levantin, de la Cilicie au Liban. On était donc face à un Pré-céramique tardif, aux caractères culturels accusés, fort différents du panorama présenté par les groupes humains levantins.

Toutes ces divergences entre une souche supposée et une île peuplée de "colons" sont aujourd'hui devenues plus claires. En effet tous les sites de la phase jusqu'ici évoquée appartiennent à une étape évoluée du Pré-céramique chypriote, à une époque où celui-ci fait preuve d'un certain isolement culturel et s'enracine dans une spécificité insulaire caractérisée.

Les fouilles de Shillourokambos ont démontré l'existence d'une phase plus ancienne, apparue dès la seconde moitié du IXe millénaire et s'étant développée tout au long du VIIIe. En l'état actuel des données, ces populations agricoles de Chypre peuvent—tout au moins pour les plus anciennes—être considérées comme les "colons" ayant implanté sur l'île l'économie pastorale en provenance du Proche-Orient. Dès lors les liens avec celui-ci, peu visibles, on l'a vu, lors de la phase de Khirokitia, sont clairement attestés et tout particulièrement dans le domaine de l'industrie lithique. On retrouve en effet à Shillourokambos des techniques de taille très proches de celles du PPNB: recours à des nucleus naviformes pour la pratique du débitage bipolaire (qui n'élimine pas pour autant le mode unipolaire), fabrication d'armatures de projectiles de qualité. Ces contacts sont renforcés par une ample circulation de l'obsidienne d'origine cappadocienne. A côté du daim (probablement chassé), des chèvres, des moutons et des porcs (élevés), le boeuf est présent dès les débuts du néolithique: il disparaîtra ensuite, dans le courant du VIIIe millénaire, pour des raisons non élucidées. Enfin les seuls bâtiments connus sont au départ des sortes d'enclos palissadés de tracé curviligne. Cette architecture à poteaux de bois sera, dès la première moitié du VIIIe millénaire, remplacée par les constructions circulaires en pierre qui marqueront le début d'une spécificité chypriote, ancrée tout au long du Pré-céramique et résurgente aussi, bien plus tard, au IVe millénaire (Chalcolithique).

On voit donc qu'au départ l'installation à Chypre des producteurs néolithiques s'effectue dans le cadre d'une diffusion de populations originaires du continent, dont la souche géographique n'est pas clairement identifiée. Certains indices pourraient orienter vers la zone du moyen ou du haut Euphrate. En tout cas on est bien dans une "ambiance culturelle" de type PPNB ancien-moyen: les savoirs techniques en particulier montrent un lien direct avec le Proche-Orient.

On signalera enfin combien les recherches sur une île—apparemment lieu de diffusion secondaire à partir d'une aire motrice continentale—peuvent contribuer à modifier notre perception chronologique de la domestication sur les lieux mêmes du continent où elle s'est opérée: tandis que les étapes de la domestication étaient jusqu'à ces derniers temps datées du VIIIe millénaire (chèvres et moutons: 8000–7500; boeuf et porc: 7500–7000), nous avons trouvé ces quatre espèces à Shillourokambos dès 8200. Il faut donc en déduire que toute la chronologie de la domestication animale au Proche-Orient doit être repensée et vieillie vers 8500, soit un demi-millénaire à un millénaire en amont des propositions jusqu'ici formulées.

La Méditerranée Centrale: l'Italie du Sud-Est

On est ici dans le domaine de l'horizon culturel à céramique "impressa," réparti dans un premier temps sur les côtes occidentales de la Grèce, l'Albanie littorale, la Dalmatie, l'Italie du Sud et la Sicile.

Des recherches conduites sur les sites de Torre Sabea (Puglia) et de Trasano (Basilicata) émergent quelques idées-force.

En premier lieu l'ancrage au sol marqué par la multiplication des structures touchant à l'habitat et à ses dépendances. A Trasano par exemple, un grand enclos de plus de 20 m de diamètre, sur lequel venaient se greffer d'autres espaces ceinturés, a été bâti en moellons de forte taille: il pouvait s'agir de lieux de parcage du bétail. Des placages de fumiers montrent la part tenue par l'élevage. Des structures de torchis ont notamment permis d'identifier parmi les plus anciens fours reconnus en Italie.

Etudes palynologiques et anthracologiques montrent d'emblée un environnement anthropisé, où, toutefois, la présence de milieux boisés n'est pas exclue (Trasano). L'analyse des faunes souligne le faible rôle de la chasse (cf. aussi Rendina; Cipolloni Sampò 1977–1982) et un élevage orienté vers l'acquisition de viande et reposant à plus de 60% sur les bovins. Les porcs pouvaient servir de ressource d'appoint, les caprinés étaient utilisés pour la viande et, surtout, pour le lait. On retire l'image d'un système très ordonné et de haute technicité (Vigne 1996). Les céréales cultivées sont, à Torre Sabea et à Trasano, le blé et l'orge. A Trasano, l'engrain et l'amidonnier se maintiennent autour de 25–30% tout au long de la séquence du Néolithique ancien; l'orge distique, bien représentée au départ, se raréfie ensuite tandis que le blé tendre-compact est en augmentation (Marinval 1988).

Le Néolithique ancien semble donc se manifester ici, dès les débuts, par l'introduction de systèmes techniques élaborés. Transferts et emprunts de savoirs ou populations nouvelles? Cette question demeure ici grévée par la connaissance encore très sommaire des horizons du Mésolithique terminal. Par certains caractères, l'outillage lithique de Torre Sabea pourrait indiquer une dérivation à partir d'un fonds castelnovien: lames débitées par pression (pour l'obsidienne seulement), usage du microburin, piquants trièdres, trapèzes. Or ce substrat n'a guère été décrit jusqu'à présent dans le Sud italien. D'autres éléments de cet ensemble (cf. grattoirs circulaires) peuvent dériver d'un fonds épi-romanellien. L'idée d'une tradition castelnovienne est aussi évoquée à propos du site de Terragne (Manduria, Tarente; Gorgoglione et al. 1995; Tiné 1996). Toutefois de nouvelles actions sur le milieu expliquent désormais l'enrichissement de ce fonds technique (outils polis, pièces lustrées).

Si ce modèle se confirme, on serait donc face à un système économique fondamentalement introduit (probablement par des groupes externes) mais adopté, au moins en partie, par un fonds de population indigène, celui-ci perçu à travers la permanence des techniques lithiques. On demeure toutefois surpris par le très petit nombre de sites mésolithiques repérés à ce jour et par les difficultés à les identifier. L'idée que le fonds de population mésolithique devait être faible est vraisemblable. Enfin, en Italie du Sud, le compactage autour de 7000 B.P. de la plupart des datations ^{14}C se rapportant au Néolithique le plus ancien, pour des horizons culturels différenciés et sans doute diachroniques, ne facilite pas la réalisation d'un cadre chronologique détaillé. On peut admettre, en l'état des données, que le Néolithique a commencé de se manifester ici vers 6200/6000 avant J.-C.

Le Sud de la France

L'un des acquis récents de la recherche est certainement la complexité des facteurs intervenus dans la néolithisation des terres méditerranéennes de la France, état qu'exprime notamment la grande variété des styles céramiques aujourd'hui reconnus:

1. le Cardial "tyrrhénien," à décor de triangles et de chevrons, observé en Latium, Toscane, Sardaigne, Corse, Ligurie, Côte d'Azur;
2. le Néolithique "ligurien," dont la céramique porte fréquemment des sillons d'impressions. Centré sur la Ligurie, on le connaît aussi en Bas-Languedoc héraultais et audios;
3. le faciès inférieur de l'abri Pendimoun (Alpes-Maritimes), connu sur ce seul site et dont les affinités, probablement italiques, demandent à être précisées;
4. le "Cardial," à décor de coquille dominant, et dont la Provence centrale et occidentale, le Languedoc et le Roussillon constituent l'aire d'expansion. Son homogénéité est, sans doute, relative, et son affinement devrait permettre, à terme, des subdivisions stylistiques plus poussées dans l'espace et le temps;
5. l'"Epicardial" languedocien à poterie à base de faisceaux de sillons et d'impressions, et dont la répartition va de la Basse-Provence à l'Isère et au Roussillon avec de nettes implantations dans la Montagne Noire (Gazel, Camprafaud, Poussarou, Saint-Pierre-la-Fage). Des affinités avec les horizons "épicardiaux" de la Péninsule ibérique sont manifestes;
6. le "Roucadourien" et autres groupes "peri-cardiaux," développés à la marge géographique du Cardial, forment dans les Causses, l'Aquitaine, ou les Pyrénées une sorte de substitut continental à ce dernier: la céramique montre une maîtrise technique et décorative nettement moins élaborée.

Les relations géographiques et chronologiques entre ces divers groupes ne sont pas claires, sauf pour ce qui concerne l'Epicardial qui se situe en position secondaire (5200–4600) par rapport au Cardial ou au Ligurien. Les datations de ces deux derniers complexes se positionnent globalement entre 5800 et 5200. On ignore toutefois comment s'articulent Cardial et Ligurien. Plusieurs datations ^{14}C placent le dernier nommé assez tôt dans la chronologie (vers 5800–5600); de plus ce faciès véhicule des marqueurs italiques comme l'obsidienne de Lipari ou des îles Pontines, et montre d'emblée une forte orientation vers l'agriculture et l'élevage de caprins et bovins. Ce pourrait donc être dans le Midi un "agent néolithisant" actif. Par contre, le Cardial fait l'objet de datations contradictoires, tantôt "hautes" (Gazel, Leucate), tantôt "basses" (Courthézon, la Draga). Sa connaissance repose encore, pour beaucoup, sur la fouille de grottes et d'abris où la chasse est abondante (grotte Lombard, grotte de l'Aigle), en regard de l'élevage et de l'agriculture, ce qui ne permet pas de saisir tous les aspects du système économique. Si l'on part de l'hypothèse que ces sites sont, pour la plupart, représentatifs de la culture et de son économie, on pourrait voir dans le Cardial un système faisant encore appel à la prédation et à un élevage peu spécialisé. En ce sens grande serait la démarcation avec les sites de Grèce et d'Italie méridionale, caractérisés par un sous-système d'acquisition de viande très specialize (Vigne 1996).

Mais la situation se complique du fait que certains sites de grotte à Cardial (Fontbregoua et, en Espagne, les grottes de l'Or et de la Sarsa notamment) montreraient un faible taux de chasse au grand gibier et une contribution essentielle des bovins et caprinés. J.-D. Vigne y voit des sites "colons," établis par déplacements maritimes à longue distance. Si l'idée de sites-colons est vraisemblable, il faut la considérer comme mettant en jeu des "sauts" maritimes d'assez faible ampleur (100–200 km au maximum), mais non des déplacements transméditerranéens. En effet l'examen de la culture matérielle ne permet pas de corroborer l'existence de telles "enjambées," le Cardial ibérique ne trouvant guère d'affinités dans les céramiques de Méditerranée centrale. Peut-être faut-il davantage penser à une plus franche hétérogénéité, voire une dichotomie, à l'intérieur de la culture elle-même avec des communautés à stratégies différentes, un gradient intervenant entre quelques sites à haute spécialisation technique et d'autres où l'exploitation de certaines niches associerait une chasse active à des contributions variables d'espèces domestiques (cf. modèle "Jean Cros"). Ce sont d'ailleurs ces derniers sites de grotte ou d'abri qui accréditent parfois l'idée d'une tradition mésolithique dans les styles de vie. La permanence à la fois dans la fréquentation des cavités et dans la collecte des matériaux lithiques a parfois tendance à conforter cette opinion (abri de Dourgne).

Ce problème de la succession (voire de la cohabitation) entre Mésolithique et Néolithique ancien est actuellement grevé par la rareté—qu'il faudra bien tenter d'expliquer un jour—des sites du Mésolithique terminal (absent de stratigraphies comme Fontbregoua ou l'Abeurador). Là où ce Mésolithique est présent (Châteauneuf, Montclus, Gazel, Dourgne), on constate, dès le Néolithique ancien, l'abandon de certaines techniques castelnoviennes (débitage par pression, usage du microburin) et un redéploiement rapide de la technique des armatures (dans le bassin de l'Aude, les pointes triangulaires disparaissent au profit de la technique "Jean-Cros"; Briois 1997).

Enfin la question du mouton "mésolithique" semble mal posée. Les vestiges retrouvés dans certains niveaux sans céramique sont, plus vraisemblablement, le produit d'éleveurs néolithiques occupant certains abris, au cours de tournées "pionnières" d'exploitation de la basse, moyenne ou haute montagne. Ainsi apparaît mieux le rôle de l'élevage comme facteur d'expansion du système néolithique dans les terres moins aptes à l'agriculture.

On observera également le développement particulier, parmi les céréales cultivées au cours du Cardial, de Triticum aestivo-compactum et de l'orge nue (*Hordeum vulgare var. nudum*). En ce sens le Cardial se distingue des horizons néolithiques primaires de la Méditerranée égéenne ou adriatique; il occupe, en regard de ceux-ci, une position chronologique secondaire (Marinval 1988).

Questions Ouvertes sur la Néolithisation de la France

Pendant longtemps la France a été considérée comme un espace géographique néolithisé par le biais des deux grands axes géoculturels de propagation du néolithique en Europe: la voie méditerranéenne et la voie danubienne. On a même associé à cette double diffusion l'image de deux modèles fort différents. En Méditerranée aurait eu lieu une propagation maritime par acculturation, fondée sur un simple emprunt de techniques par des populations de chasseurs-collecteurs denses et bien implantées. Un certain émiettement culturel du Néolithique ancien, la persistance de la chasse et de la pêche, seraient le reflet de cette permanence autochtone, également sensible dans les industries, voire dans certaines résistances à la pénétration continentale du Néolithique (cf. Portugal).

Par contre la moitié nord de la France serait perçue comme un espace néolithisé sous l'effet d'un processus de diffusion, celui de colons à poterie rubanée, complexe stéréotypé à longues maisons rectangulaires établies en vallée. Autres caractères: une forte unification des styles céramiques, une vocation agricole ou horticole complétée par un élevage préférentiellement axé sur les bovins et se distinguant de ce fait de l'élevage méditerranéen plus orienté vers les caprins.

Les recherches récentes ont beaucoup nuancé l'orthodoxie de ces deux modèles et souligné, en fait, la complexité des situations. On se limitera ici à quelques unes de ces constatations:

1. il n'est pas sûr que le fonds mésolithique terminal du Sud de la France ait été très dense. On peine à mettre en évidence les gisements de cette époque. Cette situation ne conforte guère l'idée d'une population mésolithique importante qui aurait acquis les savoirs néolithiques par simple processus d'emprunt;
2. de clairs mouvements de diffusion sont perceptibles en Méditerranée. Le cas du Ligurien—bien différencié du Cardial—qui s'étale du golfe de Gènes jusqu'aux limites du Roussillon, c'est-à-dire sur presque toute la façade méditerranéenne française, en est un bon exemple. Il véhicule par ailleurs des obsidiennes italiennes de Palmarola et de Lipari, ce qui non seulement confirme la part des navigations côtières mais souligne l'amplitude de certaines relations d'échanges;
3. on a déjà dit la diversité des données méditerranéennes concernant la nourriture carnée (avec des sites à production spécialisée—avec élevage de haute technicité—et d'autres à acquisition de viande diversifiée avec large recours à la chasse). Les premiers indiquent l'implantation d'un système élaboré d'origine externe, les seconds font une plus large place à l'acculturation;
4. en domaine rubané, la chasse joue parfois un rôle important. Ainsi lors des phases anciennes du village de Cuiry-les-Chaudardes (Aisne) où les espèces chassées atteignent jusqu'à 25% en nombre de restes; en poids de viande, les animaux

sauvages (sanglier, cerf) tiennent une place encore plus forte dans l'alimentation (Hachem 1997). Selon L. Hachem, la chasse peut être "une composante importante de l'identité collective" de la culture à céramique linéaire. Cette proposition nuance l'idée d'un complexe à forte composante agricole;

5. une autre interrogation concerne le rôle éventuel des substrats indigènes dans la néolithisation. Cette place a été souvent considérée comme faible en domaine linéaire, voire inexistante, en raison de l'opposition entre les économies de chasse-collecte et l'économie agricole, les stratégies et les espaces mis en jeu. Aussi a-t-on fréquemment évoqué des contacts sporadiques, sans interconnexion évidente entre chasseurs et agriculteurs. Or l'étude des armatures danubiennes asymétriques semble indiquer la forte possibilité d'une acculturation de certains groupes du Mésolithique final d'Alsace ou de Lorraine en raison des parentés entre les projectiles des premiers producteurs et ceux des populations autochtones. Cette impression est renforcée par une donnée de poids: ces armatures si particulières, malencontreusement appelées "danubiennes," sont rares outre-Rhin et inexistantes sur le Danube (où les têtes de flèches accusent des morphologies différentes). Elles répondraient donc à un modèle élaboré en Occident (Thevenin 1995);
6. le poids de la tradition mésolithique occidentale dans le domaine funéraire trouve quelques manifestations dans le Rubané: position allongée de certains sujets inhumés en Basse-Alsace, individus orientés à l'Ouest dans les nécropoles d'Esloo, Niedermez et Souffelweyersheim (Jeunesse 1997).

Une résurgence d'un vieux fonds mésolithique est également parfois avancée dans la culture de Cerny (autour de 4500 avant J.-C.): recours au dépôt du corps allongé, rôle des trophées de chasse—dents de sanglier ou de cerf—dans les mobiliers funéraires.

Au delà de la présence conjointe de processus de diffusion et d'acculturation dans les deux domaines concernés (Cardial/Rubané), d'autres observations viennent nuancer l'idée d'une néolithisation du Nord-Est de la France par simple mouvement migratoire d'origine danubienne. La découverte, dans des contextes rubanés, de céramique dite de la Hoguette, très particulière dans ses formes et son ornementation, invite à s'interroger sur ce phénomène de cohabitation. Extérieure à la tradition rubanée, cette céramique présente quelques affinités avec les productions du domaine méditerranéen (Cardial/Imprimé). Aussi certains auteurs interprètent-ils ces vestiges comme le résultat d'un premier épanouissement céramique en Europe du Nord-Ouest, par contact précoce avec certaines groupes du Néolithique ancien méditerranéen (Manen 1997). Cette pénétration, peut-être complétée par le transfert de techniques lithiques (cf. flèches de Montclus), voire par des premières tentatives de production, aurait entraîné une sub-néolithisation de cette partie du continent avant même que ne parviennent sur ces territoires les premiers colons rubanés.

Pour corroborer cette hypothèse, il faudrait pouvoir en jalonner la progression spatiale et chronologique. Or, sur l'Atlantique, la néolithisation méditerranéenne semble marquer le pas au Nord de la Loire. Sur l'axe du Rhône, une éventuelle pénétration demande à être plus finement balisée. L'existence des récipients de type Hoguette est assurée jusqu'en Languedoc (grotte Gazel) à des dates contemporaines de l'Epicardial (autour de 5000): elles ne peuvent, en l'état des données, être considérées comme des prototypes aux premières céramiques Hoguette, plus anciennes (Guilaine et Manen 1997).

Chronologie de la Diffusion Agricole en Europe: Pour un Modèle Arythmique

Je voudrais enfin faire quelques commentaires sur les rythmes de la diffusion néolithique tels qu'ils ont été proposés par A. Ammerman et L. Cavalli-Sforza (Ammerman et Cavalli-Sforza 1984; Cavalli-Sforza et Cavalli-Sforza 1994; Cavalli-Sforza 1996). Aujourd'hui où l'on dispose de datations plus nombreuses et où les données archéologiques se sont multipliées, on mesure combien les processus de la propagation ont été complexes et n'ont pu prendre la forme d'un courant homogène et régulier, mais ont sans doute donné lieu à des configurations diverses. De sorte que les rythmes avancés (1 km/an) ne peuvent être que globalement indicatifs sans être performants sur les mécanismes de détail. Plusieurs questions sont dès lors posées.

Pour bien cerner la chronologie d'un tel mécanisme de diffusion, il faudrait bien maîtriser la date d'apparition des civilisations "primaires" européennes: "Pré-céramique," monochrome et Proto-Sesklo grecs, Néolithique balkanique à poterie peinte en rouge et blanc, Danubien, Hoguette, pour les complexes plus importants de la "route" continentale; Néolithique ancien crétois, Impressa adriatique, Cardial pour la zone de propagation méditerranéenne. Or les pôles anciens de ces cultures ne sont pas toujours fixés avec toute la précision nécessaire, comme on a pu le voir lors du récent Colloque "C-14 et Archéologie" de Lyon (avril 1998). Un gros effort d'affinement de la chronologie des sites et de la diffusion des productions "domestiques" (céréales, caprins) reste à faire.

Il est probable qu'en fonction des environnements aptes à être colonisés et de la dynamique des populations agricoles, la chronologie de la diffusion n'a pas répondu à un modèle homogène et régulier mais à ce que l'on peut appeler un modèle général arythmique, marqué par des accélérations ou des tassements (fig. 10.1). Ces tassements se sont opérés dans les aires de mutation culturelle. On peut identifier au minimum trois de ces zones: en Anatolie, dans la zone frontière du PPNB; en Grèce occidentale, où se met en place la culture adriatique à impressa; au Nord des Balkans, là où le Néolithique égéo-balkanique amorce une transformation qui donnera naissance au Rubané. Il en résulte plusieurs modèles en fonction des aires géographiques considérées.

Le Modèle Originel de Diffusion Primaire

Il concerne notamment l'aire de maturation et de première propagation de l'économie néolithique (Syrie, Palestine, Anatolie orientale). Le point de départ du pôle émetteur au Proche-Orient pourrait être fixé autour de 8500 c'est-à-dire au cours du PPNB ancien. En effet c'est vers cette date que les céréales sont peu ou prou "domestiquées" (on sait que la "domestication" végétale s'amorce au cours du PPNA). Quant aux animaux, plutôt que d'évoquer de façon catégorique un statut soit sauvage soit domestique que les critères morphologiques ont du mal à exprimer, on peut estimer que leur "appropriation" par l'homme est devenue plus ou moins irréversible vers 8500 comme en témoigne peu après le transfert de ces espèces dans un milieu insulaire comme Chypre. Ce dernier exemple permet d'ailleurs de dater l'un des premiers moments indiscutables de la diffusion de l'économie agro-pastorale. Il a donc valeur d'exemple.

Si l'on prend donc 8500 point moyen des premières diffusions, l'on conviendra que la progression du premier néolithique a d'abord été un

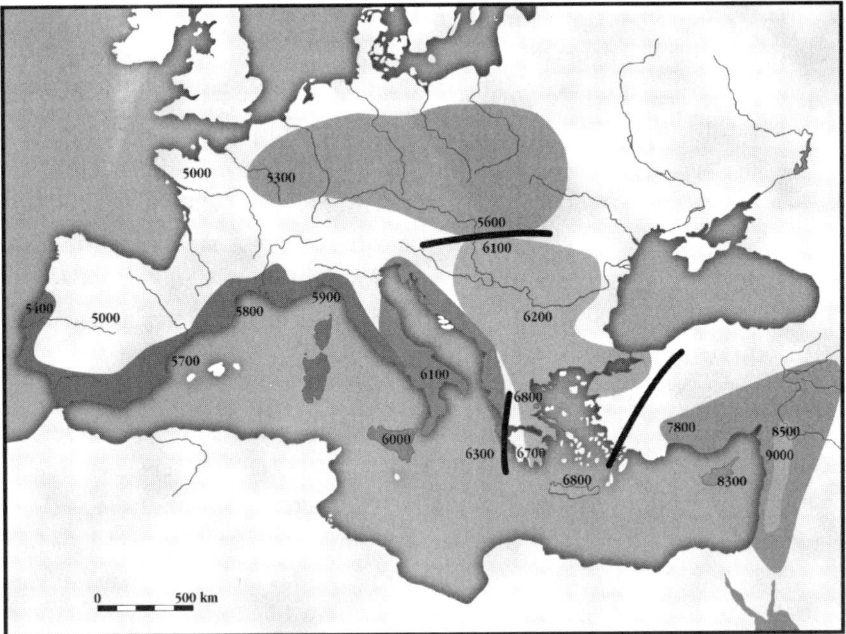

Fig. 10.1. Diffusion chronologique de l'agriculture en Europe selon des processus arythmiques liés à la diversité culturelle des premières civilisations néolithiques: au Proche-Orient: en clair: le PPNA dans la zone du corridor levantin en sombre: l'aire d'extension des PPNB; en Europe du Sud-Est: extension géochronologique des néolithiques égéens et balkaniques; en Méditerranée centrale: extension géochronologique du néolithique adriatique à céramique impressa; en Méditerranée occidentale: extension géochronologique du néolithique cardial franco-ibérique; en Europe tempérée: extension géochronologique du néolithique à céramique linéaire; en traits noirs: principales frontières chronoculturelles.

processus plutôt lent. En effet, si l'on se fie au radiocarbone, les premiers villages du Sud-Est de l'Europe (Thessalie, Crète) sont datés entre 6800 et 6500 B.C. (cal.). Il a donc fallu au minimum 1,800 ans (et peut-être 2,000) pour que l'économie de production diffuse depuis le Haut-Euphrate (sorte de focus qui montre parmi les plus anciennes traces de néolithisation au Proche-Orient) et franchisse les quelque 1,500 km qui séparent cette région de la Grèce du Nord. Et le processus a été encore plus lent si l'on rapproche le "point de départ" en le plaçant en Anatolie centro-orientale. Cette "lenteur" semble pouvoir être expliquée par un essoufflement du PPNB sur les marges occidentales de sa diffusion. Dès lors le redémarrage de la vague d'avancée fut-il tributaire d'une "refonte" culturelle ayant nécessité un certain temps.

Le Modèle Maritime de Diffusion Rapide

A l'opposé, il existe des rythmes de diffusion qui semblent avoir été rapides. Ainsi de la propagation en Méditerranée, à l'Ouest de la Grèce, encore que certaines difficultés de datation touchant au Néolithique ancien du Sud de l'Italie ne soient pas levées. Les plus anciennes datations du Néolithique ancien au Portugal se plaçant aux alentours de 4500 B.P. soit vers 5400–5300 B.C., on peut estimer que la vitesse de propagation maritime de l'Egée au Portugal est d'environ 2 km/an (3,000 km en 1,500 ans). Un gradient chronologique de diffusion est donc globalement observable, même si beaucoup de datations doivent être "manipulées" avec précaution.

On ignore si cette diffusion s'est faite par cabotage ou s'il a existé dès le Néolithique ancien des parcours trans-méditerranéens évitant certaines régions: par exemple de Sicile ou de Sardaigne directement vers la Péninsule ibérique, en laissant de côté le grand arc de la France méridionale. Les différences de style observables entre les céramiques de l'impressa sud-italienne et celles du cardial occidental ne militent pas en faveur de contacts évidents sauf rares incursions exploratoires, toujours possibles. Par ailleurs l'absence d'un néolithique ancien caractérisé aux Baléares ne favorise pas ce type de réflexion.

Si l'on s'en tient à la thèse de la progression littorale par cabotage, on s'aperçoit que sur les 3,000 km de côtes qui séparent les Pouilles du Portugal, la diffusion pourrait s'inscrire entre 6200/6000 et 5400/5300 avant notre ère soit une fourchette de 700/800 ans, donc un rythme supérieur à 3,5 km/an. Mais il est vrai que la propagation maritime a pu être très rapide tandis que l'extension à partir des littoraux vers l'intérieur des terres aurait nécessité des mécanismes plus lents. De plus, comme on l'a déjà évoqué, des ponts maritimes ont pu contribuer à abréger les distances.

Une autre difficulté concerne la datation précise des premiers impacts anthropiques. C'est pourquoi ce point ne sera évoqué que pour mémoire. Ainsi, en France méridionale et en Catalogne, l'approche palynologique révèle, sur des sites d'étangs ou de dépressions fermées, des interventions humaines entraînant précocement l'ouverture du milieu et, parfois, leur mise en culture. Dans l'étang de Capestang (Hérault) on note vers 6300/6200 la présence de pollens de céréales et d'indicateurs de milieux ouverts (*Buxus, Calluna*). Des plantes considérées comme des marqueurs d'une ouverture du milieu (*Plantago, Rumex, Artemisia, Buxus*) sont signalées vers la base de la séquence du Petit Castelou à Narbonne (Aude) vers 6300. A Embouchac, près de Lattes (Hérault), plantain et pollens de céréales sont notés entre 6500 et 6000. Des pollens de céréales sont aussi observés près de l'étang de Berre, vers 6100. En Catalogne, une phase d'incendies est signalée sur le site de Drassanes 1, dans la plaine barcelonaise, entre 6400 et 6000 (éclaircissements "mésolithiques" de la forêt? activités pionnières d'élevage?). L'ensemble de ces manifestations se situe dans un contexte pré-cardial, les premières manifestations du Cardial se positionnant vers 5800. Si ces observations se confirment, il faudra définir l'identité "archéologique" de leurs auteurs.

Une semblable situation a été reconnue dans le Jura français. Pollens de cérales, plantes rudérales et messicoles, chutes d'*Ulmus*, hausses fortes de *Corylus* montrent au Pré Mourey (Villers-le-Lac, Doubs) une première agression humaine entre 5938 et 5677 B.C., bientôt suivie, à Chalain 3, par une seconde poussée vers 5500. Cette première colonisation de la montagne s'inscrit donc dans un créneau antérieur à la diffusion du Rubané centre-européen (s'agit-il d'interventions des populations de la Hoguette?).

Arythmie de la Diffusion Continental: Des Propagations Rapides ou Freinées

Autant la propagation maritime a pu être rapide, autant certaines régions on pu être le siège d'une diffusion lente. Ainsi, dans la péninsule Ibérique, hormis les cas de quelques pénétrations sub-continentales (grotte de Chaves, Cariguela de Piñar), la néolithisation des terres intérieures ne s'effectuera qu'au stade de l'Epicardial ("cultura de las cuevas"), autour de 5000. C'est dire qu'il aura fallu autant de siècles pour parvenir à néolithiser les pays de la Meseta que ceux nécessités par les navigateurs cardiaux pour s'implanter de la Ligurie jusqu'au Portugal. Ce sont précisément les étapes de cette progression continentale, à partir des aires côtières précocement néolithisées, qui devraient être mieux mesurées à l'échelle du temps. On peut parler de lenteur dans la mesure où ce terme traduit une certaine inertie par rapport à l'espace limité à coloniser.

Dans le même esprit, on devrait s'interroger sur les temps de "latence" qui ont pu marquer localement la progression en latitude des fermiers pionniers. Ceci vaut, par exemple, pour la culture cardiale *stricto sensu* dont la propagation semble stoppée en limite de la zone de climat méditerranéen, c'est-à-dire, en gros, au Portugal moyen, vers Figueira da Foz. Il est en effet intéressant d'observer que les datations du "Cardial atlantique" français sont sensiblement plus jeunes que celles du Cardial de l'aire ouest-méditerranéenne. Si la technique du décor à la coquille, connue de la Gironde à la Loire, s'inscrit dans la sphère de la céramique imprimée, on est peut-être ici dans un autre contexte culturel, avec un sensible décrochement chronologique en regard du Cardial franco-ibérique.

Une même réflexion amène à s'interroger sur d'éventuelles dysharmonies dans la transmission du Néolithique à travers l'Europe tempérée. Ainsi le développement, dans les Balkans, des horizons à poterie peinte (Karanovo, Starcevo) semble connaître son plus grand impact vers la fin du VIIe millénaire avant notre ère. La progression à partir du domaine thessalien semble donc avoir été régulière et sans à-coup. Par contre l'élaboration du Rubané (Hongrie, Slovaquie), qui correspond au remodelage du système néolithique méditerranéen pour s'adapter au contexte écologique de l'Europe tempérée, n'a pu subvenir qu'après un délai de "fermentation," autour de 5700/5600. Il conviendrait de vérifier à la fois l'hypothèse de ce temps de latence (phase d'arythmie) et ses éventuelles conséquences sur la progression spatiale de l'économie de production. On sait comment, une fois constitué, le système danubien occidental répercutera jusqu'à l'embouchure du Rhin et au Bassin parisien, et cette fois de façon accélérée, les acquis de la "révolution" néolithique. On est ici, entre Danube moyen et bassin de Paris, face à un processus de diffusion continentale rapide.

Ces réflexions devront certes être soumises au test d'analyses radiocarbone plus affinées. Mais ce modèle de la "diffusion arythmique" a l'avantage, par rapport à celui de la vague d'avancée régulière, de montrer l'existence d'à-coups et d'essoufflements du front pionnier en périphérie des cultures primaires (PPNB en Anatolie, Néolithique égéo-balkanique sur le Danube et en Grèce de l'Ouest, Cardial en limite de son extension méditerranéenne au Portugal). Ces essoufflements signifient que les cultures sont alors parvenues aux limites de leur adaptation écologique et structurelle. Un temps de "fermentation" est donc à nouveau nécessaire sur ces frontières culturelles: Anatolie occidentale où s'effectue la transition du PPNB aux groupes néolithiques égéo-balkaniques, Grèce de l'Ouest où, après que la céramique ait fait son apparition (cf. Sidari), émergeront après une pause les horizons à céramique impressa, cours moyen du Danube où le Néolithique ancien balkanique servira de stimulus à une mutation culturelle profonde donnant naissance à la culture rubanée.

L'on constatera qu'une fois chacun de ces handicaps franchis, les nouvelles cultures émergentes—Impressa/Cardial et Rubané—font preuve d'une rapidité toute particulière dans leur diffusion. C'est sur leur périphérie, en Europe occidentale ou en Europe du Nord, que l'ancrage au sol semble avoir été plus laborieux, plus modéré, dans sa dynamique conquérante.

REFERENCES

Ammerman, A.J., et Cavalli-Sforza, F. 1984. *The Neolithic Transition and the Genetics of Populations in Europe.* Princeton.

Briois, F. 1997. *Les industries lithiques en Languedoc méditerranéen (6000–2000 avant J.-C.).* Thése EHESS, 3 vols.

Cavalli-Sforza, L. 1996. *Gènes, peuples et langues.* Paris.

Cavalli-Sforza, L., et F. Cavalli-Sforza. 1994. *Qui sommes-nous? Une histoire de la diversité humaine.* Paris.

Cipolloni Sampò, M. 1977–1982. "Scavi nel villaggio neolitico di Rendina (1970–1976). Relazione preliminare." *Origini* 11:183–323.

Cremonesi, G., et J. Guilaine. 1987. "L'habitat de Torre Sabea (Gallipoli, Puglia) dans le cadre du Néolithique ancien de l'Italie du Sud-Est." *Premières communautés paysannes en Méditerranée occidentale,* 377–85. Paris.

Gorgoglione, A., S. Di Lernia, et G. Fiorentino. 1995. *L'insediamento preistorico di Terragna (Manduria, Taranto). Nuovi dati sulle processo di neolitizzazione nel Sud-Est italiano.* Manduria.

Guilaine, J. 1976. *Premiers bergers et paysans de l'Occident méditerranéen.* Paris.

———, ed. 1979. *L'abri Jean Cros: Essai d'approche d'un groupe humain du Néolithique ancien dans son environnement.* Toulouse.

———. 1980. *La France d'avant la France.* Paris.

———, ed. 1993. *Dourgne: Derniers chasseurs-collecteurs et premiers éleveurs de la Haute Vallée de l'Aude.* Carcassonne.

———. 1994. *La mer partagée: La Méditerranée avant l'écriture, 7000–2000 avant J.-C.* Paris.

Guilaine, J., F., Brois, J. Coularou, J.D. Vigne, et I. Carrére. 1997–1998. "Les débuts du Néolithique à Chypre." *L'Archéologue* 33:35–40.

Guilaine, J., J.-L. Courtin, J.-L. Roudil, et J.-L. Vernet, eds. 1987. P*remières communautés paysannes en Méditerranée occidentale.* Paris.

Guilaine, J., et G. Cremonesi. 1987. "L'habitat néolithique de Trasano (Matera, Basilicate). Premiers résultats." En *Il Neolitico in Italia,* 707–19. Florence.

Guilaine, J., A. Freises, et R. Montjardin. 1984. *Leucate-Corrège: Habitat noyé du Néolithique cardial.* Toulouse.

Guilaine, J., et O.V. Ferreira. 1970. "Le Néolithique ancien au Portugal." *Bulletin de la Société Préhistorique Française* 67:304–22.

Guilaine, J., D.M. Geddés, M. Barbaza, J.L. Vernet, M. Llongueras, et M. Hopf. 1982. "Prehistoric Human Adaptations in Catalonia." *Journal of Field Archaeology* 9:407–16.

Guilaine, J., et C. Manen. 1997. "Contacts sud-nord au Néolithique ancien: témoignages de la grotte Gazel en Languedoc." In *Le Néolithique danubien et ses marges entre Rhin et Seine,* 301–6. Strasbourg.

Guilaine, J., et M. Martzluff. 1995. *Las excavaciones a la Balma de la Margineda (1979–1991).* Andorra.

Guilaine, J., et O. da Veiga Ferreira. 1970. Le Néolithique ancien au Portugal." *Bulletin de la Société Préhistorique Francaise* 67:304–22.

Hachem, L. 1997. "Structuration spatiale d'un village du Rubané récent, Cuiry-les-Chaudardes (Aisne). Analyse d'une catégorie de rejets domestiques: la faune." En *Espaces physiques, espaces sociaux dans l'analyse interne des sites du Néolithique à l'Age du Fer,* 245–61. Paris.

Jeunesse, C. 1997. *Pratiques funéraires au Néolithique ancien: Sépultures et nécropoles danubiennes (5500–4900 avant J.-C.).* Paris.

Manen, C. 1997. *L'axe rhodano-jurassien dans le problème des relations sud-nord au Néolithique ancien.* British Archaeological Reports, International Series 665. Oxford.

Marinval, P. 1988. *Cueillette, agriculture et alimentation végétale de l'Epipaléolithique jusqu'au deuxième Age du Fer en France méridionale. Apports palethnographiques de la carpologie.* Thèse EHESS, 2 vols.

Thevenin, A. 1995. "Mésolithique récent, Mésolithique final, Néolithique ancien dans le quart nord-est de la France: pour une reinterprétation des données." *Revue Archéologique de Picardie.* Spécial no. 9:3–15.

Tiné, V., ed. 1996. *Forme e tempi della Neolitizzazione in Italia meridionale e in Sicilia.* Genoa.

Vigne, J.D. 1996 (sous presse). "Faciès culturels et sous-système technique de l'acquisition des ressources animales. Applications au Néolithique ancien méditerranéen." En *Deuxième Rencontres Méridionales de Préhistoire récente.* Arles.

Vigne, J.-D., I. Carrére, J.-F. Salliége, A. Person, J. Guilaine, et F. Briois. 1998 (sous presse). "Domestic Cattle, Sheep, Goat and Pig during the 8th Millenium cal. B.C. on Cyprus: Preliminary Results of Shillourokambos (Parekklisha, Limassol)." Paris.

— 11 —

The Neolithic Transition in Portugal and the Role of Demic Diffusion in the Spread of Agriculture across West Mediterranean Europe

João Zilhão

My own involvement in the debate on the origins of agriculture in Europe is to a large extent accidental. A consideration of my empirical, theoretical, and social background may go a long way toward explaining not only why it happened at all, but also why I have come to advance the ideas on the subject that I currently sustain.

The focus of my archaeological research and the subject of my Ph.D. dissertation have been the Upper Paleolithic of Portugal (Zilhão 1995, 1997a). In going through past literature and in looking for and excavating new cave sites with deposits from this time period, several regularities quickly became apparent. For instance, although some examples are now known of sites where the surface of the sediment fill is Upper Paleolithic, most well-preserved contexts tend to be buried under thick Holocene sequences. In almost every single case, however, the early Holocene is represented by a hiatus, and there is direct contact between the early Neolithic and the late Upper Paleolithic. It is also almost always the case that Mousterian deposits underlie the Upper Paleolithic.

In this context, my dealing with the latter inevitably led to collecting empirical data of relevance for the understanding of the two major processes of European prehistory: the biological extinction of Neanderthals and their replacement by anatomically modern humans and the cultural extinction of hunter-gatherer systems and their replacement by agropastoral economies. Since elements of an east-west diffusion are involved in both cases (regardless of whether what diffused were ideas, artifacts, genes, or people), the geographical position of Portugal in the far western end of Europe means that a good understanding of the latest Mousterian and the earliest Neolithic of our country is critical to a good understanding of such processes—namely with regard to the establishment of their chronological boundaries and of the rates of spread of the diffused elements. Given that these processes created the basic frame for the subsequent history of the continent and, as such, have occupied, in one way or another, the front stage of research in European prehistory, I could not avoid being pulled into the discussion of the Mesolithic–Neolithic transition in the west Mediterranean (Zilhão 1990, 1992, 1993, 1997b, 1998, 2000; Zilhão and Carvalho 1996).

The early 1980s, when I was excavating the early Neolithic levels of the cave site of Caldeirão (Zilhão 1992), were characterized by the major influence of the taphonomic perspective in Paleolithic studies (Binford 1983). Beginning in Lower Pleistocene Africa with the man-the-hunter or man-the-hunted controversy, this perspective quickly found its way into the archaeology of later periods, as exemplified by Joachim Hahn's pioneer work at the early Upper Paleolithic cave sites of southwestern Germany (cf. Hahn 1988). However, with few exceptions (Paola Villa's work at the Baume de Fontbrégoua; Villa and Courtin 1982), this taphonomic revolution stopped at the gates of the Neolithic. As a result, my 1993 article on the impact of site formation processes on then current views of the agricultural transition in west Mediterranean Europe was somewhat of a novelty. Why, at that specific moment in time, following the previous work by Fortea and Martí (1984–1985), I happened to be in the right position to show that the Mesolithic sheep of France and the very early dates for an autochthonous Neolithic in southern and eastern Spain were illusory realities created by the operation of different taphonomic agents that can be explained by many different factors.

In Portuguese universities, the late 1970s and early 1980s brought about the emancipation of archaeology as a scientific discipline separate from history. This was accompanied by an explosion in the number of courses, the number of students, and the number of professionals. Since new ground was being broken, allegiance to old schools of thought or old masters was materially impossible, and theoretical novelties spread fast in a milieu of mostly self-trained practitioners molded in a time of great intellectual freedom following the political and social revolution of 1974–1975. In my own case, the effects of exposure to the literature on site formation processes were enhanced by the personal experience of having been an active speleologist throughout my youth. As such, I became familiar with caves as natural sites even before I began to look at them as potential containers for archaeological remains. Having often camped inside underground galleries and having learned about the ways other animals use caves contributed to the skepticism with which I approached the interpretation of the early Neolithic cave record of the western Mediterranean.

The 1980s also witnessed a renewed interest by both national and international teams in the scientific potential of the major Mesolithic shell-midden sites of central-littoral Portugal (Arnaud 1987, 1989, 1990; González Morales and Arnaud 1990; Jackes et al. 1997a, 1997b; Lubell and Jackes 1988; Lubell et al. 1994). Most of the research in question was processual in inspiration, leading to the collection of data on subsistence, diet, and health parameters of the last Iberian hunter-gatherers that were largely unavailable until then. An important mass of radiometric data was also accumulated. This has made it possible to look at both sides of the transition with a wealth of information that, in Europe, is matched or surpassed only in Scandinavia. As a result, several striking spatio-temporal features of the process became apparent.

First, it is now clear that all late Mesolithic and early Neolithic sites known in Iberia are located along the periphery of the Peninsula, for the most part right along the coast or, in the few cases of more inland locations, sufficiently close to major waterways leading to important estuarine areas. Second, north of the Mondego river, halfway between the northwestern and the southwestern ends of Iberia's Atlantic facade, recent research has confirmed that there are no Neolithic sites earlier than ca. 4700 B.C. (cal.), while elsewhere to the south and to the east the establishment of Neolithic economies may be as early as 5600 B.C. (table 11.1). Third, along the southern half of Portugal's west coast, the latest Mesolithic groups survived for as much as 500 years as hunter-gatherer enclaves with territories centered on the estuaries of the Tagus, the Sado, and the Mira Rivers, surrounded by Neolithic groups occupying territories located in the limestone massifs of central Estremadura, to the north, and the Algarve, to the south (fig. 11.1).

These well-established empirical patterns have major implications for the

Table 11.1. Radiocarbon Dates for the Earliest Neolithic in Iberia

Site	Region	Site Type	Provenience	Culture	Sample
Buraco da Pala	Portugal	Rock shelter	Level IV, base	Impressed Ware	Wood charcoal
			Level IV, base	Impressed Ware	Wood charcoal
Gruta do Caldeirão	Portugal	Cave	Layer Eb	Cardial	Wood charcoal
			Horizon NA2	Cardial	Animal bone collagen
			Horizon NA2	Cardial	Animal bone collagen
			Horizon NA2	Cardial	Human bone collagen
			Horizon NA1	Impressed Ware	Animal bone collagen
			Horizon NA1	Impressed Ware	Animal bone collagen
			Horizon NA1	Impressed Ware	Human bone collagen
Abrigo da Pena d'Água	Portugal	Rock shelter	Layer Eb (base)	Cardial	Wood charcoal
Cabranosa	Portugal	Open air	Neolithic hearth	Cardial	Estuarine shells
Padrão	Portugal	Open air	Neolithic hearth	Cardial	Estuarine shells
			Neolithic hearth	Cardial	Estuarine shells
Balma Margineda	Andorra	Cave	Level 3b base	Cardial	Wood charcoal
			Level 3b	Cardial	Wood charcoal
			Level 3a	Cardial	Wood charcoal
La Draga	Catalonya	Open air	Hearth E-6 (1990)	Late Cardial	Wood charcoal
			Hearth E-6 (1990)	Late Cardial	Wood charcoal
			Hearth E-40 (1991)	Late Cardial	Wood charcoal
			Hearth E-56 (1991)	Late Cardial	Charred cereal seeds
			Hearth E-3 (1991)	Late Cardial	Charred cereal seeds
			Post E-106 (1991)	Late Cardial	Oak wood
			Hearth E-50 (1991)	Late Cardial	Wood charcoal
			Garbage area in H-30 (1991)	Late Cardial	Animal bone collagen
Cova del Frare	Catalonya	Cave	T22-23, level 5c	Cardial	Wood charcoal
			Y35, level 5	Epicardial	Wood charcoal

	Lab Number	Date B.P.	Comments	References
	ICEN-935	5860 ± 140	Earliest context with cereal remains in Portugal	Sanches et al. 1993
	GrN-19104	5860 ± 30		
	ICEN-296	6870 ± 210		Zilhão 1992, 1993
	OxA-1035	6330 ± 80	Direct date of Ovis aries bone	
	OxA-1034	6230 ± 80	Direct date of Ovis aries bone	
	OxA-1033	6130 ± 90		
	OxA-1037	5970 ± 120	Direct date of Bos taurus bone	
	OxA-1036	5870 ± 80	Direct date of Bos taurus bone	
	TO-350	5810 ± 70		
	ICEN-1146	6390 ± 150		Zilhão and Carvalho 1996
	Sac-1321	6930 ± 60	Corrected for reservoir effect: 6550 ± 70	Cardoso et al. 1996
	ICEN-873	6920 ± 60	Corrected for reservoir effect: 6540 ± 70	Gomes 1994
	ICEN-645	6800 ± 50	Corrected for reservoir effect: 6420 ± 60	
	Ly-2839	6670 ± 120		Guilaine and Martzluff 1995
	Ly-3289	6850 ± 150		
	Ly-3288	6640 ± 160		
	GAK-1523	5710 ± 170		Tarrus et al. 1994
	UBAR-245	5920 ± 240		
	UBAR-311	5970 ± 110		
	UBAR-313	6010 ± 70		
	Hd-15451	6060 ± 40		
	UBAR-314	6410 ± 70		
	UBAR-312	6570 ± 460		
	UBAR-315	6700 ± 710		
	I-13030	6380 ± 310		Martin 1986–1989, 1990
	MC-2298	5800 ± 130		

Cueva de Chaves	Aragón	Cave	Level I	Cardial	Wood charcoal
			Level I	Cardial	Wood charcoal
			Level I	Cardial	Wood charcoal
			Level I	Cardial	Wood charcoal
			Level II	Epicardial	Wood charcoal
			Level II	Epicardial	Wood charcoal
Cova de les Cendres	Valencia	Cave	H19a	Cardial	Wood charcoal
			VIIa	Cardial	Animal bone collagen
			H19	Cardial	Wood charcoal
			VII	Cardial	Wood charcoal
			VIe	Cardial	Wood charcoal
			H18	Cardial	Wood charcoal
			H17, fireplace	Cardial	Wood charcoal
			H15a	Epicardial	Wood charcoal
			H15	Epicardial	Wood charcoal
Cova de l'Or	Valencia	Cave	Levels 16–17, 153–163 cm	Cardial	Wood charcoal
			Levels 14–15, 140–153 cm	Cardial	Wood charcoal
			Level 6, 95–100 cm	Epicardial	Wood charcoal
El Retamar	Andalucia	Open air		Cardial	Estuarine shells
Kobaederra	Euskadi	Cave	Level II	Neolithic	Wood charcoal
			Level III	Neolithic	Wood charcoal
			Level IV	Neolithic	Wood charcoal

	GRN-12685	6770 ± 70		Baldellou e Castán 1985; Baldellou e Utrilla 1985; Bernabeu 1989
	GRN-12683	6650 ± 80		
	CSIC-378	6460 ± 70		
	GRN-12686	5210 ± 340	Unacceptable, too young	
	CSIC-379	6230 ± 70		
	CSIC-380	6120 ± 70		
	Beta-116624	8310 ± 80	AMS on Quercus; contamination	Bernabeu et al. 1999
	Beta-107405	6280 ± 80	AMS on Ovis aries bone	
	Beta-116625	20,430 ± 170	AMS on Pinus nigra; contamination	
	Beta-75220	6730 ± 80	Selected (P. nigra + Juniperus excluded)	
	Ly-4302	7540 ± 140	Non-selected, contaminants included	
	Beta-75219	6420 ± 80	Selected (P. nigra + Juniperus excluded)	
	Beta-75218	6260 ± 80	Non-selected, only from fireplace	
	Beta-75217	6150 ± 80	Selected (P. nigra + Juniperus excluded)	
	Beta-75216	6010 ± 80	Selected (P. nigra + Juniperus excluded)	
	GANOP-C13	6720 ± 380	Contains abundant remains of cereals and domestic ovicaprids	Martí 1978; Martí et al. 1980; Bernabeu 1989
	GANOP-C12	6630 ± 290		
	GANOP-C11	5980 ± 260		
	Beta-90122	6780 ± 80	Corrected for reservoir effect: 6400 ± 90	Lazarich et al. 1997
	UBAR-472	5200 ± 110	Earliest context with domesticates in Cantabria	Zapata et al. 1997
	UBAR-471	5820 ± 240		
	UBAR-470	5630 ± 100		

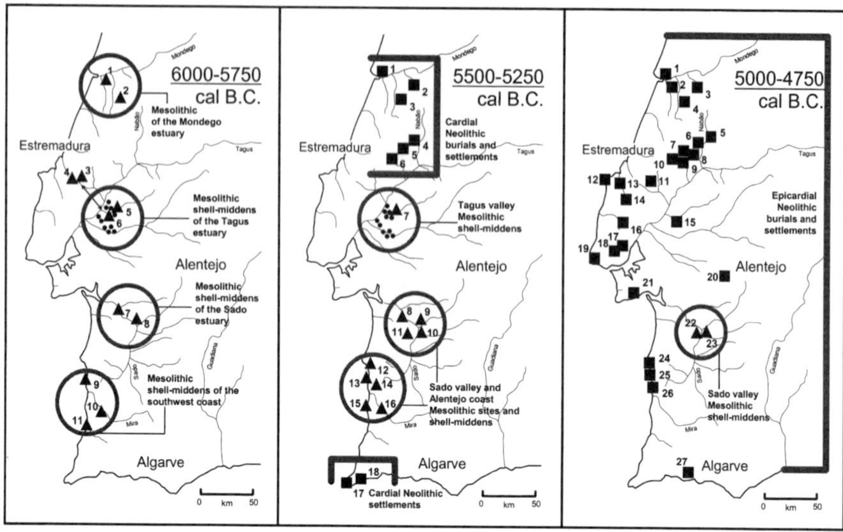

Fig. 11.1. Chronology and distribution of late Mesolithic and early Neolithic sites across the transition to agriculture in Portugal

interpretation of the transition in Iberia (Zilhão 2000). At the onset, it would seem that the interior regions of the peninsula (where no lakes exist and where large sections of even the largest rivers are susceptible to drying up in the summer months) became devoid of archaeologically visible populations after the end of the Pleistocene. By the time of the Atlantic climatic optimum, subsistence systems had become extremely dependent on aquatic resources, explaining current site distribution patterns as well as the consistently nitrogen-enriched isotopic composition of all late Mesolithic skeletons that have been analyzed. This may have been a result of interior forests having become too dense and too poor in resources to sustain year-round human occupation of the interior Meseta. In Portugal, this retreat to the waterside seems to have taken place even within the 50 km wide strip of coastal hills and lowlands between the Tagus and the Mondego Rivers. Inhabited during Boreal times, the interior plateaus of the limestone massif of central Estremadura seem to have become deserted at the onset of the Atlantic—with only a few caves and rock shelters at the mouth of major karstic springs located along the periphery of the massif containing some evidence of a highly logistical late Mesolithic use.

The earliest artifactually defined Neolithic settlement of Portugal is found in exactly those limestone areas that had become abandoned by late Mesolithic hunter-gatherer groups. There are several reasons to believe that this earliest Neolithic settlement is associated with, or triggered by, the arrival of agropastoral seafaring colonist groups originating in Mediterranean Spain. Baroquely decorated Cardial wares similar to those from Valencia (Bernabeu 1989; Bernabeu et al. 1993; Martí et al. 1980, 1987; Martí and Juan-Cabanilles 1987; Martí and Hernandez 1988) are known in several Portuguese sites. The associated lithics feature polished axes and heat pre-treated flint, both technologies being unknown in the late Mesolithic but well documented in the earliest Neolithic of Valencia. The dead were collectively buried in caves not used for habitation, and the artifacts associated with the burials include ceramic vessels and beads made of *Cypraea* and *Glycymeris* shells. These religious practices clearly set the earliest Neolithic groups of Portugal apart from contemporary hunter-gatherers living in the estuaries of the Tagus and the Sado, where individual burial in pits excavated into the shell-midden debris of habitation sites is the norm. Furthermore, neither the ceramic vessels nor the shell-beads characteristic of Cardial burials have ever been found in the graves of the Mesolithic. Domesticates are unknown in the latter, but sheep are present in the artifactually defined early Neolithic burial contexts, which, from the point of view of human skeletal isotopy, are also characterized by markedly terrestrial diets—very different from those of contemporary Mesolithic people.

Coupled with the spatial patterning of the process, these contrasts in material culture and ideology make it extremely difficult to accept the alternative explanation that, in Portugal, the transition is the local adoption of agriculture by hunter-gatherers that had imported the several elements of the Neolithic package through long-distance exchange networks. This may well be the case, however, in northwestern Iberia and in the Cantabrian strip, where the transition takes place several hundred years later—at a time when the expansion of farming groups had already begun the reoccupation of the interior areas of the peninsula, and where some degree of continuity in settlement and burial practices seems to exist (Arias 1991, 1992, 1994; Arias et al. 2000; González Morales 1992, 1996; Zapata et al. 1997). In southern Portugal, this expansion is signaled by the coincidental extinction, ca. 4800 B.C., of the last hunter-gatherer groups. Documented in the Sado valley site of Amoreira, their interaction with encroaching farmers is suggested by the presence of a few Epicardial sherds in the otherwise purely Mesolithic shell midden deposits (Arnaud 1986).

At the continental scale, the Portuguese pattern has several implications. First, given the currently accepted radiometric results for the advent of the Neolithic in Liguria, it permits the calculation of the rate of spread of agro-pastoral systems across the west Mediterranean (fig. 11.2). Taking the earliest radiometric results currently available for Portugal—ca. 5400 B.C. cal.—as trustworthy indicators of the arrival of the first Neolithic settlers to the far western end of their European range, that rate is ca. 5 km/year. If, however, future dating shows that such a first arrival actually happened a couple hundred years earlier, or if the radiocarbon chronology available for the eastern half of the region (almost entirely obtained on bulk charcoal samples) is shown to be skewed by an old wood effect (Whittle 1990; Zilhão 1993), the rate may actually be twice as high: ca. 10 km/year. Given that ceramic vessels do not appear in the Middle East until ca. 6800 B.C. cal., the rate at which pottery spread across the northern shores of the whole of the Mediterranean basin, in turn, can be calculated, in km/year, to lie between 3.8 and 4.5, depending on whether one accepts a date of 5400 or 5600 B.C. cal. for the earliest Portuguese ceramics. Even the lowest value, however, is far above the 1 km/year suggested 30 years ago by Ammerman and Cavalli-Sforza in their wave of advance model for the spread of the Neolithic as a process of demic diffusion (Ammerman and Cavalli-Sforza 1973, 1984).

This recognition has several implications. Figure 11.3 shows the values obtained for the different parameters involved in the calculation of the annual rate of advance of the front *r*, using Ammerman and Cavalli-Sforza's equation $a=6r^2/M$, where *a* is the annual rate of population growth and *M* the generational rate of migratory activity caused by the settlement relocation of individuals over short distances. Ethnographic observations indicate that, in agricultural societies, the maximum value of *M* is ca. 2,000 km^2/generation. Demographic models indicate, in turn, that the logistic growth curves necessary to trigger a process of population expansion cannot occur with values of $a < 0.3\%$. Population growth rates with values above 3% (which would already imply that population doubled every 23 years) are practically impossible; the most likely rate of population growth experienced by farmers at the front of the wave is in the range of 1–2%.

Given that, west of the Balkans, the Neolithic always appears as the full package of domesticates plus pottery, let us assume the most conservative estimate of ca. 4 km/year obtained for the spread of ceramics from the Middle East to Portugal as the best approximation for the rate of spread of

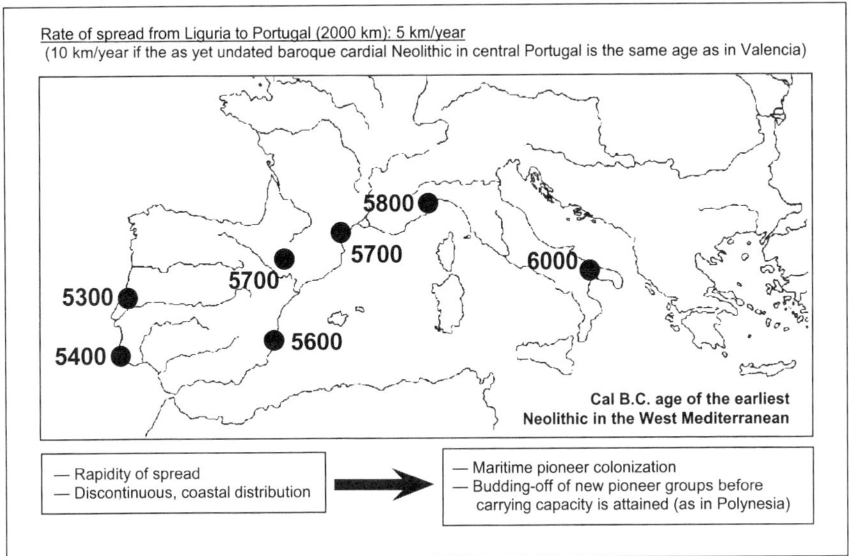

Fig. 11.2. Calibrated radiocarbon age of the earliest Neolithic in west Mediterranean Europe

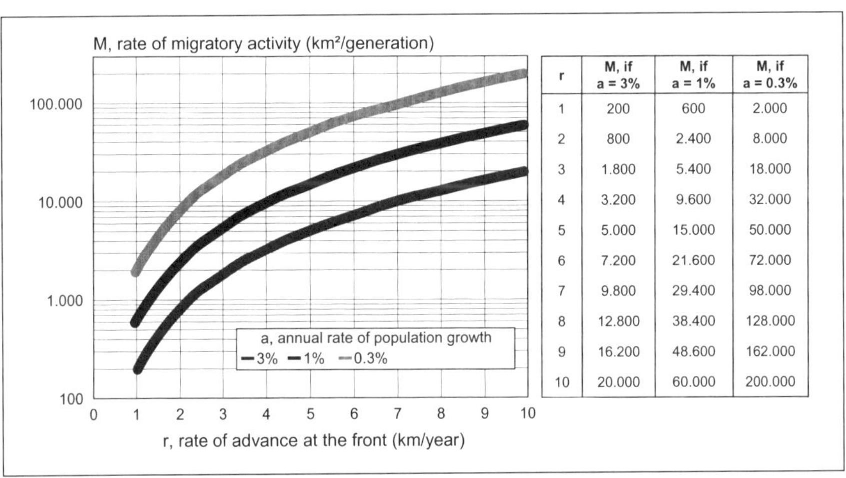

r	M, if a = 3%	M, if a = 1%	M, if a = 0.3%
1	200	600	2.000
2	800	2.400	8.000
3	1.800	5.400	18.000
4	3.200	9.600	32.000
5	5.000	15.000	50.000
6	7.200	21.600	72.000
7	9.800	29.400	98.000
8	12.800	38.400	128.000
9	16.200	48.600	162.000
10	20.000	60.000	200.000

Fig. 11.3. Rates of migratory activity implied by observed rates of advance of the Neolithic package for different values of annual growth of the frontier populations

agropastoral systems across the west Mediterranean. The combination of this rate with the ethnographically observed maximum migratory activity would imply an annual rate of population growth of 4.8%, well above the limit of practical or possible human growth rates (more so, if the rate of advance is set at 10 km/year, in which case the corresponding value would be 30%). If values of 1% are used, then the rate of migratory activity will range between 9,600 and 60,000 km^2/generation depending on whether the rate of advance is estimated at 4 or 10 km/year.

The implication of these values is that either agriculture did not spread through demic diffusion processes or that demic diffusion did not proceed through a mechanism of short-distance relocation of settlements, as previously occupied places became saturated and adjacent land had to be incorporated in the system to accommodate population growth. Since the evidence from Portugal reviewed above seems to refute the first term of this alternative, it remains that different mechanisms of demic diffusion must have been at work.

It is interesting to note that, with a population growth of 1% a year, the system would expand throughout an area of ca. 60,000 km^2 in six generations, if the rate of advance of the front was 4 km/year, or in one generation only, if the front advanced at a rate of 10 km/year (fig. 11.3). Those 60,000 km^2 are precisely the area covered by the 40–50 km wide coastal strip between the Mondego River, in Portugal, and the Cape of Nao, the southern limit of the Gulf of Valencia. The similarities in the style of pottery decoration observed between the two extremes of this range can be taken as evidence of a swift expansion with maintenance of cultural traditions over the relatively short time involved. Given the low resolution of radiocarbon, chances are, anyway, that, when the 95% confidence intervals of the dates are considered, the process will tend to appear to our eyes as an event, regardless of whether its duration was only 25 years or as much as 150 years.

In order for such a considerable area to be covered in such a short time, however, migratory activity must be much more intense than one is allowed to model under the gradualist assumptions of the wave of advance model. Under reasonable estimates of annual population growth, the observed rate of spread of the Neolithic across Mediterranean Europe requires the operation of long-distance settlement relocation episodes. At the same time, it implies such low population densities across the whole of the settled range that considerably large voids must be postulated between each node of the settlement network as, in fact, is made apparent in the mutually exclusive, "enclave" nature of contemporary late Mesolithic and early Neolithic territories observed in central Portugal. The coastal placement of the earliest Neolithic and its discontinuous geographical distribution—most apparent in the Portuguese case but which has also been suggested for eastern Provence, France (Binder 2000)—suggests that the most likely alternative mechanism of demic diffusion is maritime pioneer colonization.

At the same time, the rapidity of spread indicates that individual long-distance colonization events were taking place well before saturation levels were attained at the point of origin. Why this is so remains to be clarified but it is likely that the answer is purely historical: that is, it must be sought in the concrete features of the events that triggered the eastward and westward expansion of agropastoral economies from their Middle Eastern core areas. In Anatolia, the process seems to have coincided with the collapse of the PPNB, for which there are strong indications of a stratified society and a strong influence of cult practices. In contrast, the succeeding Neolithic societies of western Anatolia and southeastern Europe lack any archaeological evidence of specially built temples and of social ranking in settlement or in burial (Özdogan 1997).

Özdogan also suggested that these egalitarian rural societies have their origins in the migration of groups carrying to the west all aspects of their culture except central authority. This hypothesis can be extended to suggest that the driving force behind the expansion of the Neolithic across the Mediterranean may have been a social imperative rooted in a tradition going back to the events leading to the collapse of the PPNB: the imperative to fission before groups get too large and conditions arise for the development of social inequality. In short, agriculture may have been brought to Europe by pioneers escaping from dominance in ranked societies and striving to maintain egalitarianism through the application of strict controls to group size. Along the shores of the north Mediterranean, this tendency to fission and move on would have been further reinforced because opportunities for settlement and expansion around initial enclaves were limited by physical geography and by the presence of local hunter-gatherer groups.

Social and cultural imperatives have also been postulated to explain the similar process of colonization of the Pacific islands by seafaring agriculturalists spreading out to occupy new territory well before carrying capacity was attained at the point of origin—in this case, in the framework of ranked societies (Irwin 1992; Kirch 1984). The greatest merit of these kinds of explanations lies in that explanatory value is given to human volition, and history is represented as the outcome of the interplay of actions carried out by individuals and social groups, not as the simple product of the operation of abstract laws or as the inevitable result of some kind of absolute environmental or demographical pressures.

To conclude in the spirit of the introduction, I must also state that my preference for these kinds of high-level explanations also needs to be understood in the framework of a personal and social background. Mention was made before of the environment of great intellectual freedom that existed in Portugal during my formative years. I was fortunate enough to belong to a generation that, at the time, was sufficiently old to participate actively in the events that radically changed our country in the years 1974–1975, but also sufficiently young to avoid being imprinted by the Stalinist/Maoist caricatures of Marxism that so much influenced previous generations. If that experience taught me an enduring lesson, it was that of the validity of the most basic and most radical contribution of Karl Marx to social sciences: people make their own history, and they must have made their own prehistory too.

REFERENCES

Ammerman, A.J., and L.L. Cavalli-Sforza. 1973. "A Population Model for the Diffusion of Early Farming in Europe." In *The Explanation of Culture Change: Models in Prehistory*, edited by C. Renfrew, 343–57. London.

———. 1984. *The Neolithic Transition and the Genetics of Population in Europe*. Princeton.

Arias, P. 1991. *De cazadores a campesinos: La transición al neolítico en la región cantábrica*. Santander.

———. 1992. "Estrategias económicas de las poblaciones del Epipaleolítico avanzado y el Neolítico en la región cantábrica." In *Elefantes, ciervos y ovicaprinos: Economía y aprovechamiento del medio en la Prehistoria de España y Portugal*, edited by A. Moure Romanillo, 163–84. Santander.

———. 1994. "El Neolitico de la región cantábrica: Nuevas perspectivas." *Trabalhos de Antropologia e Etnologia* 34(1–2):93–118.

Arias, P., J. Altuna, A. Armendariz, J.E. González, J.J. Ibañez, R. Ontañon, and L. Zapata. 2000. "La transición al Neolítico en la región cantábrica. Estado de la cuestión." In *3º Congresso de Arqueologia Peninsular. Actas*. Vol. 3, *Neolitização e Megalitismo da Península Ibérica*, edited by P. Arias, P. Bueno, D. Cruz, J. Enriquez, J. Oliveira, and M.J. Sanches, 115–33. Porto.

Arnaud, J.M. 1986. "Cabeço das Amoreiras—São Romão do Sado." *Informação Arqueológica* 7:80–2.

———. 1987. "Os concheiros mesolíticos dos vales do Tejo e do Sado: Semelhanças e diferenças." *Arqueologia* 15:53–64.

———. 1989. "The Mesolithic Communities of the Sado Valley, Portugal, in Their Ecological

Setting." In *The Mesolithic in Europe*, edited by C. Bonsall, 614–31. Edinburgh.

———. 1990. "Le substrat mésolithique et le processus de néolithisation dans le sud du Portugal." *In Rubané et Cardial*, edited by D. Cahen and M. Otte, 437–46. Liège.

Baldellou, V., and A. Castán. 1983. "Excavaciones en la Cueva de Chaves de Bastaras (Casbas—Huesca)." *Bolskan* 1:9–37.

Baldellou, V., and P. Utrilla. 1985. "Nuevas dataciones de radiocarbono de la prehistoria oscense." *Trabajos de Prehistoria* 42:83–95.

Bernabeu, J. 1989. *La tradición cultural de las cerámicas impresas en la zona oriental de la Península Ibérica.* Valencia.

Bernabeu, J., J.E. Aura, and E. Badal. 1993. *Al Oeste del Eden: Las primeras sociedades agrícolas en la Europa Mediterránea.* Madrid.

Bernabeu, J., M. Pérez,, and R.M. Martínez. 1999. "Huesos, Neolitización y Contextos Arqueológicos Aparentes." In *Actes del II Congrés del Neolític a la Península Ibèrica, Universitat de València*, edited by J. Bernabeu and T. Orozco, 589–96. Valencia.

Binder, D. 2000. "Mesolithic and Neolithic Interaction in Southern France and Northern Italy: Current Research and Hypotheses." In *Europe's First Farmers*, edited by T.D. Price, 117–82. Cambridge.

Binford, L. 1983. *In Pursuit of the Past.* London.

Cardoso, J.L., J.R. Carreira, and O.V. Ferreira. 1996. "Novos elementos para o estudo do Neolítico antigo da região de Lisboa." *Estudos Arqueológicos de Oeiras* 6:9–26.

Fortea, J., and B. Martí. 1984–1985. "Consideraciones sobre los inicios del Neolítico en el Mediterráneo español." *Zephyrus* 37–38:167–99.

Gomes, M.V. 1994. "Menires e cromeleques no complexo cultural megalítico português—trabalhos recentes e estado da questão." In *O Megalitismo no Centro de Portugal: Actas do Seminário*, 317–42. Viseu.

González Morales, M. 1992. "Mesolíticos y Megalíticos: La evidencia arqueologica de los cambios en las formas productivas en el paso al megalistismo en la Costa Cantábrica." In *Elefantes, ciervos y ovicaprinos: Economía y aprovechamiento del medio en la Prehistoria de España y Portugal*, edited by A. Moure Romanillo, 185–202. Santander.

———. 1996. "La transición al Neolítico en la Costa Cantábrica: La evidencia arqueológica." In *Actes. I Congrés del Neolític a la Península Ibérica* 2:879–85. Gavà.

González Morales, M., and J. Arnaud. 1990. "Recent Research on the Mesolithic of the Iberian Peninsula." In *Contributions to the Mesolithic in Europe*, edited by P.M. Vermeersch and P. Van Peer, 451–61. Leuven.

Guilaine, J., and M. Martzluff. 1995. *Les excavacions a la balma de la Margineda (1979–1991).* Andorra.

Hahn, J. 1988. *Das Geissenklösterle I.* Stuttgart.

Irwin, G. 1992. *The Prehistoric Exploration and Colonisation of the Pacific.* Cambridge.

Jackes, M., D. Lubell, and C. Meiklejohn. 1997. "Healthy But Mortal: Human Biology and the First Farmers of Western Europe." *Antiquity* 71:639–58.

———. 1997. "On Physical Anthropological Aspects of the Mesolithic-Neolithic Transition in the Iberian Peninsula." *Current Anthropology* 38:839–46.

Kirch, P. V. 1984. *The Evolution of the Polynesian Chiefdoms.* Cambridge.

Lazarich, M., J. Ramos, V. Castañeda, M. Pérez, N. Herrero, J. M. Lozano, E. García, S. Aguilar, M. Montañés, and C. Blanes. 1997. "El Retamar (Puerto Real, Cádiz): Un asentamiento neolítico especializado en la pesca y el marisqueo." In *II Congreso de Arqueología Peninsular: Tomo I–Paleolítico y Epipaleolítico*, edited by R. Balbín and P. Bueno, 49–58. Zamora.

Lubell, D., and M. Jackes. 1988. "Portuguese Mesolithic–Neolithic Subsistence and Settlement." *Rivista di Antropologia* Suppl. 66:231–48.

Lubell, D., M. Jackes, H. Schwarcz, M. Knyf, and C. Meiklejohn. 1994. "The Mesolithic–Neolithic Transition in Portugal: Isotopic and Dental Evidence of Diet." *Journal of Archaeological Science* 21:201–6.

Martí, B., V. Pascual, and M.D. Gallart. 1980. *Cova de l'Or (Beniarres, Alicante)*. Valencia.

Martí, B., J. Fortea, J. Bernabeu, M. Perez, J.D. Acuna, F. Robles, and M.-D. Gallart. 1987. "El Neolitico antiguo en la zona oriental de la Peninsula Iberica." In *Premières communautés paysannes en Méditerranée Occidentale*, edited by J. Guilaine, J. Courtin, J.-L. Roudil, and J.-L. Vernet, 607–19. Paris.

Martí, B., and M.S. Hernandez. 1988. *El Neolític Valencià: Art rupestre i cultura material.* Valencia.

Martí, B., and J. Juan-Cabanilles. 1987. *El Neolític Valenciá: Els primers agricultors i ramaders.* Valencia.

Martin Colliga, A. 1986–1989. "Reflexión sobre el estado de la investigación del Neolítico en Cataluña y su reflejo en la cronología radiométrica." *Empúries* 48–50:84–102.

———. 1990. "El Neolitico antiguo en Cataluña: Trayectoria de su investigacion." In *Autour de Jean Arnal*, edited by J. Guilaine and X. Gutherz, 37–54. Montpellier.

Özdogan, M. 1997. "The Beginning of Neolithic Economies in Southeastern Europe: An Anatolian Perspective." *Journal of Mediterranean Archaeology* 5:1–33.

Sanches, M.J., A.M. Soares, and F.M. Alonso. 1993. "Buraco da Pala (Mirandela): Datas de carbono 14 calibradas e seu poder de resolução: Algumas reflexões." *Trabalhos de Antropologia e Etnologia* 33(1–2):223–43.

Tarrus, J., J. Chinchilla, and A. Bosch. 1994. "La Draga (Banyoles): Un site lacustre du Néolithique ancien cardial en Catalogne." *Bulletin de la Société Préhistorique Française* 91:449–56.

Villa, P., and J. Courtin. 1982. "Une expérience de piétinement." *Bulletin de la Société Préhistorique Française* 79:117–23.

Whittle, A. 1990. "Radiocarbon Dating of the Linear Pottery Culture: The Contribution of Cereal and Bone Samples." *Antiquity* 64:297–302.

Zapata, L., J.J. Ibañez, and J. González. 1997. "El yacimiento de la cueva de Kobaederra (Oma, Kortzubi, Bizkaia): Resultados preliminares de las campañas de excavación de 1995–1997." *Munibe* 49:51–63.

Zilhão, J. 1990. "Le processus de néolithisation dans le centre du Portugal." In *Rubané et Cardial*, edited by D. Cahen and M. Otte, 447–59. Liège.

———. 1992. *Gruta do Caldeirão: O Neolítico Antigo*. Trabalhos de Arqueologia 6. Lisbon.

———. 1993. "The Spread of Agro-Pastoral Economies across Mediterranean Europe: A View from the Far West." *Journal of Mediterranean Archaeology* 6:5–63.

———. 1995. "O Paleolítico Superior da Estremadura portuguesa." Ph.D. thesis, Universidade de Lisboa.

———. 1997a. *O Paleolítico Superior da Estremadura portuguesa*, 2 vols. Lisbon.

———. 1997b. "Maritime Pioneer Colonization in the Early Neolithic of the West Mediterranean: Testing the Model Against the Evidence." *Porocilo o raziskovanju paleolitika, neolitika in eneolitika v Sloveniji* 24:19–42.

———. 1998. "On Logical and Empirical Aspects of the Mesolithic–Neolithic Transition in the Iberian Peninsula." *Current Anthropology* 39:690–8.

———. 2000. "From the Mesolithic to the Neolithic in the Iberian Peninsula." In *Europe's First Farmers*, edited by T.D. Price, 144–82. Cambridge.

Zilhão, J., and A.M.F. Carvalho. 1996. "O Neolítico do Maciço Calcário Estremenho. Crono-estratigrafia e povoamento." In *Actes. I Congrés del Neolític a la Península Ibérica*, 659–71. Gavà.

PART IV

The Transition in Central and Northern Europe

— 12 —

Origins of the Linear Pottery Complex and the Neolithic Transition in Central Europe

Malgorzata Kaczanowska and Janusz K. Kozłowski

The transition from the Anatolian-Balkan model of a food producing society with painted pottery to the Danubian model represented by the Linear Pottery Complex was of essential importance for the adaptation of the new economy to the environmental conditions in central Europe. For the explanation of this process, one has to look into the relations between the early Neolithic Balkan units with painted ware and the cultural units with Linear Pottery as well as the relationship between these two complexes and the local Mesolithic. The vital regions for the investigation into these relationships are the northern Balkans, the Tisza Basin, and Transdanubia.

In studying a process of economic and cultural transformation, all of the material correlates of human behavior must be taken into account. Of special significance are archaeobotanical and archaeozoological materials, which constitute direct evidence for a specific method of obtaining food. Equally important is the study of lithic assemblages, which are the main link between the Mesolithic and the Neolithic in the sphere of a material culture and a direct expression of the interregional relationships and contacts between groups. Answers to such questions have to be based on paleobiological studies that make it possible to define the relationships between the Mesolithic and the Neolithic populations, as well as the internal relationships within the Neolithic groups, in terms of physical anthropology and paleogenetics.

Leaving the study of archaeobotany, archaeozoology, or paleobiology to those with more experience in these fields, the authors have concentrated on the study of lithic industries. The objective of our work is the description of technological and stylistic traditions and the functional differentiation of the lithic artifacts. In the case of the Mesolithic and Neolithic of the northern Balkans, such studies were almost totally neglected in the 1970s. All the hypotheses about cultural change were based on the examination of pottery in the case of the Neolithic and on a very superficial knowledge of the Mesolithic (defined as the Tardenoisian whenever microliths were present).

The Balkan Confrontation

In the 1970s, the sensational discovery of the Lepenski Vir culture in the region of the Iron Gates compelled a number of researchers to put forward a hypothesis about the existence of a local center of Neolithization in the Balkans. This discovery seemed to have confirmed the earlier speculations about the presence of the preceramic Neolithic in southeastern Europe (Milojčič et al. 1962; Teocharis 1973). It seemed at that time that the stone architecture and the monumental sculpture found at Lepenski Vir could not be reconciled with the Mesolithic economy. All these issues encouraged S.K. Kozłowski and J. K. Kozłowski to study the lithic industries of the classical Iron Gate sites of Vlasac and Lepenski Vir. The late D. Srejovič provided access to the materials from these sites. At the start of our investigations, we were guided by the conviction that the relation between the local sequence of the pre-Neolithic assemblages and the Neolithic could only be explained on the basis of lithic inventories. Such inventories reflect both the technological traditions, the innovations associated with changes in human activities, and the interregional links based on systems of raw material procurement. Our monographs on the lithic industries from Vlasac (Kozłowski and Kozłowski 1982) and Lepenski Vir (Kozłowski and Kozłowski 1983) offered the following answers to the fundamental questions arising from the discoveries in the Iron Gates region. We have: (a) demonstrated the continuation of the Epigravettian technological tradition from the Late Pleistocene until the Atlantic period; (b) showed the isolation of Mesolithic groups in the Early Holocene as seen in the exploitation of exclusively local raw materials; (c) confirmed the persistence of the Mesolithic technological tradition down to the end of the sixth millennium B.C. suggested by radiocarbon dates (initially these dates had been rejected not only by D. Srejovič but by others as well). This implied that the late phase of the Lepenski Vir Mesolithic was contemporaneous with the evolution of the Starčevo culture in eastern Serbia and the Criş culture in Oltenia and Banat; (d) confirmed the possibility of contacts between the Mesolithic groups of the late phase of the Lepenski Vir culture and the population of the Starčevo-Körös-Criş complex on the basis of flint imports from the pre-Balkan platform discovered in the upper portion of layer I at Lepenski Vir; (e) pointed simultaneously to the deep discontinuity or even cultural hiatus separating the Mesolithic industries and the early Neolithic industries at Lepenski Vir and Padina.

When these data were published in the 1980s, the greatest controversy was over the question of whether or not the Starčevo and the Mesolithic settlement were contemporaneous. D. Srejovič, starting from the assumption that the monochrome ceramics of Lepenski Vir phase III corresponded to the earliest phase of the Starčevo culture, rejected all the relatively late radiocarbon determinations (between 6600 and 6500 B.P. for Lepenski Vir phase I_2–II; Srejovič 1969). Srejovič suggested that the real age of the Starčevo culture from the Iron Gates region was earlier than 7000 B.P. dating, at the same time, the end of the Mesolithic in that region. This was the most controversial issue in the discussion between D. Srejovič and the authors—the issue which, at the price of a compromise, was left out of our joint lecture at the conference at Mogilany (Srejovič et al. 1980).

Although the studies on the lithic assemblages from Vlasac and Lepenski Vir were overlooked, a number of researchers attempted to resolve the same questions by resorting to much the same arguments (Voytek and Tringham 1989). Others interpreted the relation between the form and the function of stone artifacts in a naive way (Prinz 1987), while still others offered new arguments based on reviewing the stratigraphical sequences and contextual data (Radovanovič 1994, 1996). The new approaches have not, however, essentially changed the conclusions that we formulated about the lithic industries from Vlasac and Lepenski Vir. Nor have new answers been given to two questions that are vital for understanding the mechanisms of the interrelations between the Mesolithic and the Neolithic in the northern Balkans. These questions are:

1. Could the communities of the late phase of the Early Holocene Epigravettian—having reached a high degree of

sedentism, having intensified the exploitation of food resources, having created excellent and original art (rooted, nonetheless, in the Epigravettian symbolic tradition)—abruptly changed not only their economy and behavior but also their technological traditions to become the champions of the Starčevo-Körös-Criş complex (Kozłowski 1992)? How can the hiatus between Lepenski Vir I/II and III be explained? In fact, the only link between these two worlds are the remains of the stone architecture of Lepenski Vir style known from Padina in the Starčevo context (Radovanovič 1994, 1996).

2. If we assume that population replacement is the explanation of the Mesolithic–Neolithic hiatus, then what happened to the Mesolithic population? Should we fall back, as an explanation, on the possibility of violent confrontation between the indigenous population and the Neolithic invaders, as suggested by F. Magoşeanu (1978) and V. Boroneanţ (1980) on the basis of burial remains from the Romanian side of the Iron Gates?

It is much easier to show the long-term coexistence of the two cultural units—the Epigravettian and the Starčevo-Körös-Criş—than to account for the cultural hiatus between them, the disruption of technological, economic, and symbolic traditions.

It should be added that the scenario for the Iron Gates is but one of the possible forms of the relationship between the Mesolithic and the Neolithic in the Balkans. Examples can be given of the transfer of economy without changes in the material culture (at Franchthi, lithic phase X; Perlès 1990), ceramic transfer without changes in the economy or modifications in the sphere of lithic technology (at Odmut; Kozłowski et al. 1994), and the transfer of the economy and the lithic technology without the ceramics (at Dendra).

When we analyze the problem of the relationship between the Starçevo-Körös cultures and the Mesolithic units, we should stress that the range of the Starçevo-Körös complex covers, first of all, the territory occupied by various Epigravettian (or Epitardigravettian) groups. Only its northeast regions reach the territories occupied by the Pontic cultures (Grebenikian). The differences between the early Neolithic Painted Pottery complex industries and the Mesolithic assemblages in terms of raw material procurement, technology, morphology, and tool functions have been emphasized in the literature (Kozłowski and Kozłowski 1983; Gatsov 1993). But there are also differences within the lithic industries of the early Neolithic Painted Pottery complex itself—namely the occurrence of trapezes, which are the main, if not the only, link between the Mesolithic and the Neolithic. Trapezes are not known in the Kremikovci-Chavdar or the Karanovo cultures, but they occur in varying percentages in the Starçevo and the Körös-Criş culture assemblages. The stratigraphical position of the trapezes at Cuina Turcului (associated with phases Iib/IIIa and IIIb of the Starçevo culture; Paunescu 1985) suggests that they appear mainly toward the end of the Starçevo-Körös culture. As we have pointed out elsewhere (Kozłowski 1982), the occurrence of trapezes in the Starçevo-Körös assemblages (produced from the typical Starçevo macro-blades without employing the microburin technique) could be the consequence of both increasingly intensive contacts between the Neolithic and the Mesolithic groups and adaptation of the Neolithic groups to local environmental conditions (e.g., conditions that called for a greater reliance on hunting weapons).

The Eastern Linear Model: Continuity of the Basic Balkan Traditions

The relatively few radiocarbon dates available for the territory of eastern Hungary and eastern Slovakia allow us to draw the conclusion that, during the interval from 6700 to 6400 years B.P. (5500–5300 B.C.), complex processes of cultural development were taking place that were of key importance for the further evolution of the Neolithic. Simultaneously with the development of the late phase of the Starçevo-Körös culture (described as the spiraloidial A phase; Dimitrijeviç 1969) local cultural variants with painted ware, representing the Northern Körös tradition, made their appearance: Méthelek in the Hungarian part of the upper Tisza Basin and Zastavne-Mala Hora in Transcarpathian Ukraina (Kalicz and Makkay 1976; Potushniak 1994). These variants passed smoothly into the earliest Linear phase represented by the Proto-Linear Szatmár group (Kalicz and Makkay 1977) and the Proto-Kopčany group in eastern Slovakia (Kozłowski 1998). At the same time, the "Vinča impuls" appears in southern Hungary; this is confirmed by the dates for the "Proto-Vinča" phase from Deszk-Oljakút and the "Vinča A" phase from Oszentiván VII (Horvath and Hertelendi 1994).

Despite the complexity of the cultural events between 6700 and 6400 years B.P., a single, clear evolutionary trend led from the Körös-Starčevo to the Eastern Linear Culture. It can be seen in the continuation of the Szatmár group and the early AVK in Hungary as well as the Kopčany and Barca III group in eastern Slovakia and also in the Rinve-Kish Meze II group in Transcarpathian Ukraina. Today this trend can best be traced through the evolution of lithic technology, raw material procurement, morphology, and the function of stone tools. Transformations in these spheres have a gradual character and the ties of cultural tradition in between the particular evolutionary phases do not become severed despite the fact that the process of transformation is connected with considerable expansion of the early Neolithic *oikumene* (only one quarter of the territory occupied by the Szatmár group was inhabited by Körös settlement; cf. Kalicz 1995, map 22a).

We attempted to elucidate the nature of such cultural transformations in our publications in the 1980s and 1990s. A monograph on the Eastern Linear Pottery by S. Šiška (1989) and an article dealing with the lithic industries from Čečejovce and Kopčany (Kozłowski 1989) drew attention to the persistence of technological and morphological traditions (e.g., the domination of marginal retouches) along the continuum from the classical Körös culture, through the Méhtelek group, to the Eastern Linear groups of Szatmár and Kopčany This continuity was maintained although, gradually, the Starčevo system of procurement of final products or blanks of northern Balkan flint was abandoned and the transition to "on-site," full cycle processing from core reduction to tool shaping took place (Kozłowski forthcoming). The best example of the first stage of such modification, with concurrent stability of the Starčevo-Körös culture, is the study of the material from the site of Méthelek, which was used by Kozłowski (1989). These materials were published later by E. Starnini (1993), whose conclusions are closer to our own position (about the internal evolution from Körös to Eastern Linear) than to the model of change (linked with the local Mesolithic substratum) suggested by J. Chapman (1988).

Subsequent publications of the lithic industries of the Eastern Linear Pottery culture in Slovakia have confirmed the maintenance of local traditions (Kozłowski 1998). At the same time, further work on adaptative systems has demonstrated the limited ability of the Eastern Pottery culture to adjust to ecological conditions totally different from those in the basin of the Tisza and its northern tributaries. The constraints were imposed by a more conservative model of stockbreeding in the Eastern Linear culture, where the proportion of sheep/goat and cattle were similar (e.g., Zemlinské Kopčany, Tiszavasvar). Even when cattle dominated quantitatively, the proportion of sheep/goat was still fairly high. Discoveries at some of the younger Eastern Linear settlements reveal the important role of pig breeding (Folyas-Szilmeg, Šarišske Michalany), which may reflect an attempt by Linear groups to adapt to local conditions.

Already in the earliest phases of the Eastern Linear Pottery culture, two types of dwellings have been recorded. The first type—as N. Kalicz and J. Koós (1997) noticed—originated from the Körös house. This is a structure built on the plan of a slightly elongated rectangle. Timber beams strengthened the clay walls, but no traces of rows of inner posts supporting the roof have been recorded. Similar houses were built in the Šariš Basin as late as the Bükk culture. Remains of such dwellings indicate that they were covered with light, low roofs. Other structures were also built in the same territory, however; these were long post houses similar to the dwellings known in the Western Linear Pottery culture (Domboróczky 1997).

The relations between the Eastern and the Western Linear Pottery culture, whose territories were separated by a gap in settlement between the Tisza and the Danube, are generally only weakly documented in the archaeological record, especially for the early phase of the two complexes. The presence of small quantities of Szentgál radiolarite at Méhtelek and obsidian at Budapest-Aranyhegy shows, nonetheless, that such contacts did exist (Biró 1991). Obviously, in the Notenkopf phase of the Western Linear Pottery culture and the Tiszadob phase of the Eastern Linear Pottery culture, when western settlement from the Vistula Basin entered the Poprad and Špis Basin (nearing the northern boundary of the Eastern complex at Šariš), the bilateral contacts between the two complexes gained in intensity (Kaczanowska and Kozłowski 1994).

The Western Linear Model: Breaking Away from the Early Balkan Tradition

Did the change here arise from adaptation, acculturation, or migration? Unlike in the Eastern Linear complex, the disruption of links with the traditions of the Early Balkan Neolithic is much more clearly evident in the Western Linear Complex (LPC).

In the territory of southern Transdanubia, the range of the earliest LPC phase overlaps with the distribution of the Starčevo culture. But this is less than 3% of the spatial range of the early LPC phase. The explanation of such an extensive and relatively fast spread (between 5450 to 5200 B.C.) of the Western Linear Complex in terms of demic diffusion from the territories of southern Transdanubia does not seem to be very realistic (see, however, Bogucki in this volume). Therefore, attempts are made to explain the Western Linear Complex on the basis of the acculturation of the local Mesolithic communities in the Danube Basin and the territories beyond (Tillmann 1993). But the strongest argument against this interpretation is the absence of Mesolithic heritage in the material culture of the early Neolithic communities. The only exceptions here would be the territories farthest to the west in the upper Danube and the Rhein Basin (Gronenborn 1994; Jeunesse 1995). It is worth adding here that the Mesolithic substratum was highly varied throughout the territory covered by the Linear Pottery expansion (in the west the Lower Rhein and the Beuron-Coincy cultures; in the east the Janisławice, the Chojnice-Pieńki and Oldesloe cultures; not to mention the Epigravettian and the Castelnovian units in the middle Danube Basin).

This variability of the Mesolithic substratum has manifested itself only in a very limited degree in the lithic industries, especially in the early phase of the LPC. In the early sites in Lower Austria, a few backed implements occur that show some similarities with Epigravettian elements (Neckenmarket, Brunn II; excavations by P. Stadler and E. Lenneis), whereas in the earliest westernmost sites (Groenborn 1990, 1994) some technological features in common with the Beuronian tradition are present as are some flat retouched microliths (at Bruchenbrucken). In the remaining early LPC territories, no technological features or typological elements common to the Mesolithic have been recorded.

In recent years, attempts have been made in the literature to resolve or at least to take a new outlook on the problem of discontinuity between the Mesolithic and the Neolithic in terms of lithic production. To this end, studies of the production technologies of blanks have been carried out (Gronenborn 1990, 1994). For example, the analysis of the LBP material from the site of Bruchenbrucken indicated that the technologies used for the production of blade blanks and flake blanks differ from one another, and that there is considerable similarity in the methods employed to obtain blade blanks between the Mesolithic assemblages from Jägerhaushöhle and the early LBP lithics from Bruchenbrucken. The similarities are seen above all in platform edge regularization.

The first point was already made in the literature based on the study of Neolithic technology (Kaczanowska and Kozłowski 1986) as well as experimental investigations (Weiner 1985). On the other hand, new arguments on the question have to be based on more detailed statistical comparison of series of late Mesolithic and early Neolithic finds. In a recent study that we have carried out, several selected blade series have been compared using the following parameters: (1) specimen dimensions (length, width, thickness), (2) their proportions expressed in indices (width/length × 100, platform width/blade width × 100), (3) dorsal pattern of blades, (4) platform type, (5) platform shape, (6) platform width, (7) the shape of the proximal part of blades, (8) regularization of the platform edge, (9) the presence of the percussion point, (10) bulb type, (11) shape of the lateral edges, (12) blade cross section, (13) blade profile, (14) type of blade tip.

The analysis was performed on 90 randomly chosen blades from two late Mesolithic sites in Poland, representing the Janisławice culture (Tomaszów and Wieliszew), a similar blade series from the site of Brunn II (kindly placed at our disposal by P. Stadler of the Natural History Museum in Vienna) representing the oldest LBP phase, and 50 specimens from the similarly dated site of Mogiła 62 (all the blades from that site). Inventories from other central European sites of the oldest LBP phase (such as Budapest-Aranyhegy) could not be taken into account since the series are very small.

A detailed analysis revealed the following:

1. There is considerable variability in the examined features between the two Mesolithic sites of the Janisławice culture (e.g., the platform angles as shown in fig. 12.1, the thickness of the proximal part in relation to the distal thickness of the blade as shown in fig. 12.2, and the frequency of the various platform types as shown in fig. 12.3).
2. There is fairly advanced standardization in the production of Neolithic blades despite the fact that various raw materials and differently shaped concretions were used (the material inventory at the Mogiła site is based on Jurassic flint in the form of spherical or ovaloid concretions; at Brunn II local and non-local, tabular-shaped radiolarites were used). Standardization is

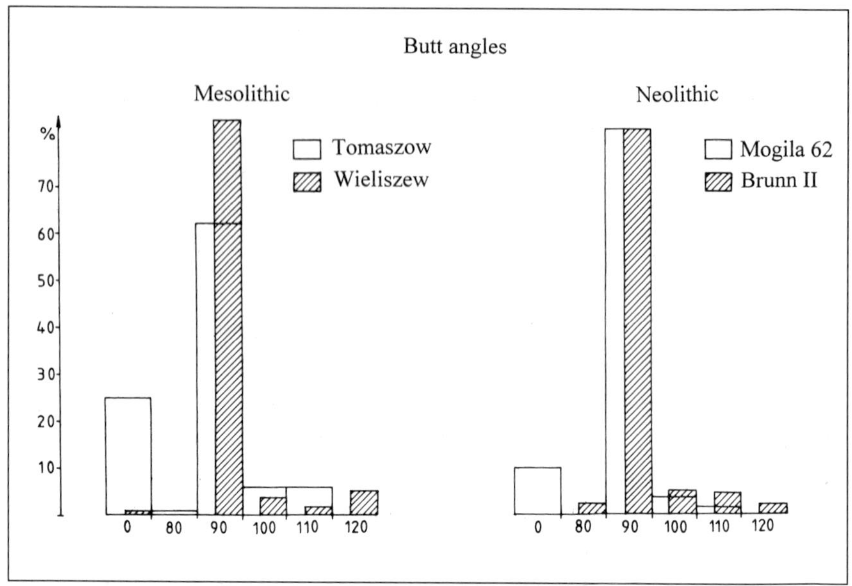

Fig. 12.1. Technological features of the Mesolithic and Neolithic blades: platform angle

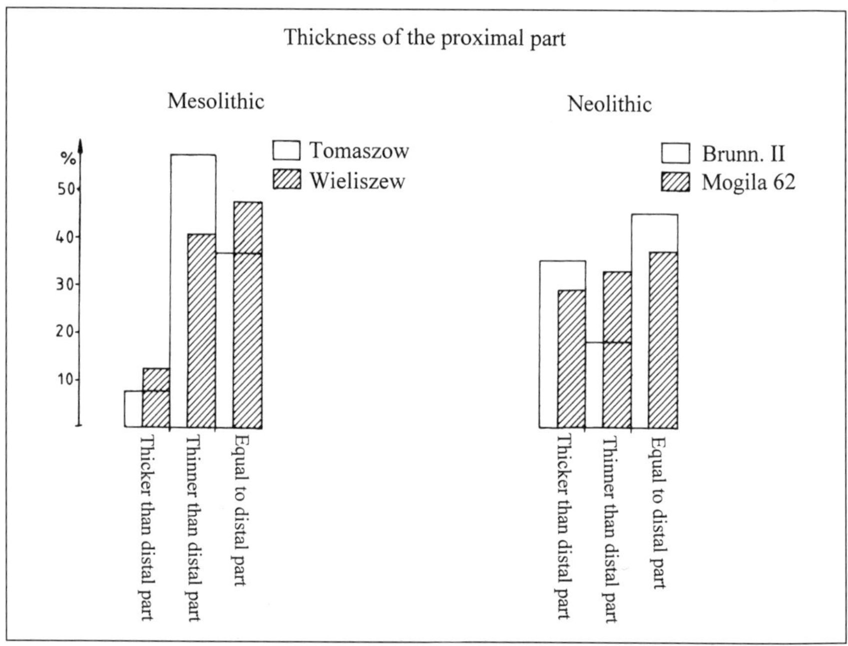

Fig. 12.2. Technological features of the Mesolithic and Neolithic blades: thickness of the proximal part in relation to the distal part

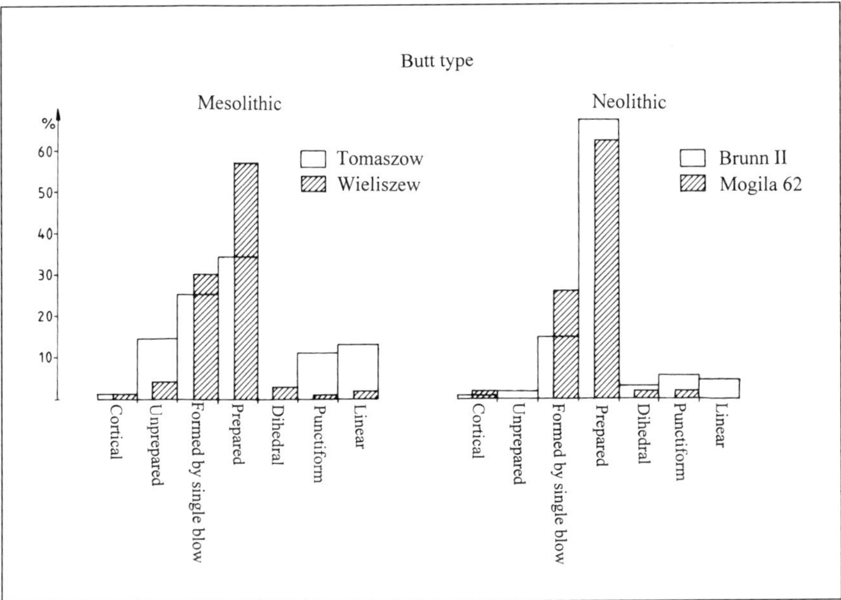

Fig. 12.3. Technological features of the Mesolithic and Neolithic blades: platform types

also seen in the methods of platform preparation; the Neolithic inventories invariably show practically the same proportion of faceted platforms (62–68%; see fig. 12.3).

3. The Mesolithic materials show that core platforms were prepared by means of some operations connected with the pressure technique, such as platform abrasion. That a procedure like this occurs in the context of the pressure technique has already been mentioned in the literature on the subject (Tixier et al. 1980).
4. The most striking feature that distinguishes the Neolithic assemblages is the rare use of platform edge regularization (18%), whereas in the Mesolithic inventories this operation has been recorded for 50% of all blades (fig. 12.4). The fact that, in the Mesolithic blades, the mesial part is thicker than the proximal part (while in the Neolithic the blades are generally thicker in the proximal part), can also be linked with platform regularization. Moreover, the pressure technique was virtually unknown in the oldest phase of the LBP culture.
5. In the Mesolithic inventories, the use of a direct blow with a

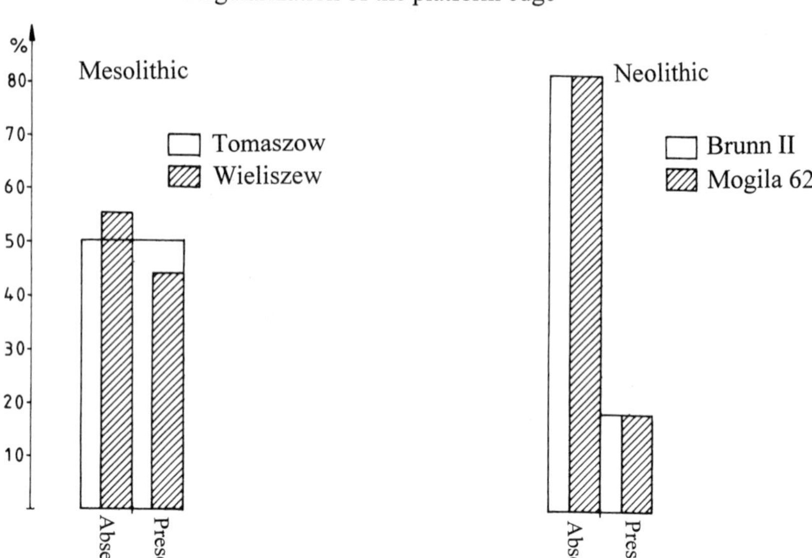

Fig. 12.3. Technological features of the Mesolithic and Neolithic blades: platform types

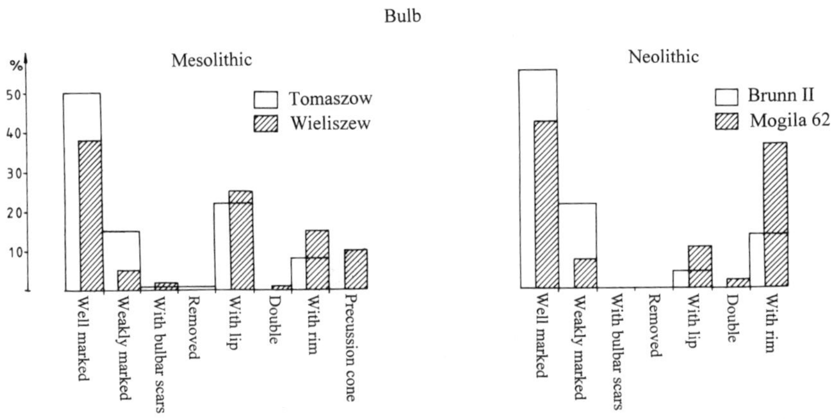

Fig. 12.5. Technological features of the Mesolithic and Neolithic blades: bulb types

soft hammer, shown by the presence of a "lip," is more frequent than in the LBP blades (fig. 12.5).

6. The Mesolithic blades are, as a rule, thinner than the Neolithic specimens. On average they are narrower in the proximal part, whereas the more robust LBP blades have similar width throughout (fig. 12.6).

In conclusion, we can say that the LBP inventories show a single method of blade production, whereas the Mesolithic inventories reveal a variety of techniques that are connected with local traditions of core reduction. So far, blade features do not indicate that the LBP people were acquainted with blade production based on the pressure technique. Various Mesolithic communities, on the other hand, certainly knew and employed this technique. In the Mesolithic and LBP inventories we examined, neither mutual exclusion nor correlation between any of the examined features has been established. For this reason, the presence or absence of a specific feature does not constitute an argument in itself for continuity or discontinuity of technological tradition between the Mesolithic and the LBP cultures.

The comparative analyses of lithic industries that we have performed for two regions of the Western Linear Complex near its eastern and western periphery (Belgium and Poland) have revealed a surprising uniformity for the whole territory with respect to technology, morphology, and function (Caspar et al. 1989). Differences in the raw materials are more marked, especially in the oldest phase. With only a few exceptions (Brunn II in Lower Austria), the oldest Western Linear sites contain either few traces of lithic processing or else comparatively few stone tools (the Silesian and western Slovakian sites). This suggests that, at first, the Linear people had poor knowledge of local raw material resources.

From the point of view of variability in raw material procurement, a division into two zones can clearly be seen (fig. 12.6). One is the Danubian zone covering the territories from Transdanubia to Lower Austria as far as southern Germany, where the raw materials from Transdanubia such as the Szentgál type radiolarite or hornstones from the Danube Basin predominated (Gronenborn 1995). The other zone embraces Moravia, lesser Poland, and the Basin of the lower Vistula with predominance of Jurassic flints from the region of Kraków and middle Polish Świeciechów and "chocolate" flints. Bohemia and upper and central Silesia, which also used the raw material

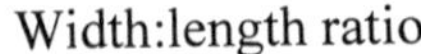

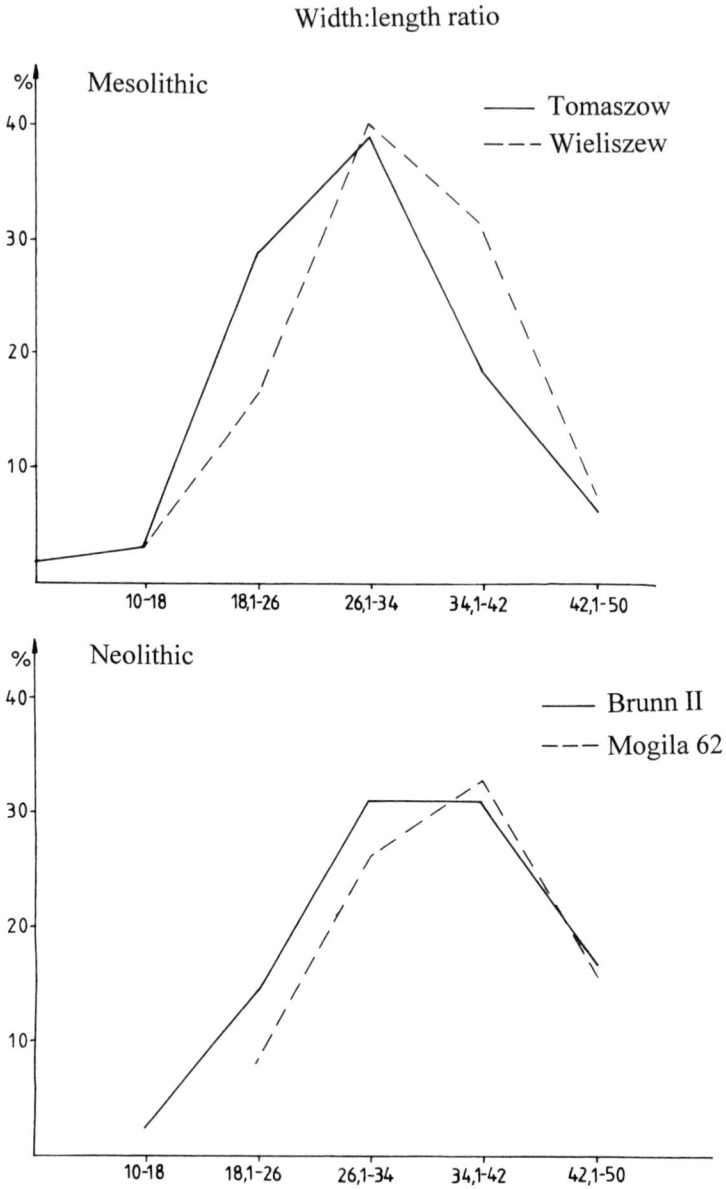

Fig. 12.6. Morphometric features of the Mesolithic and Neolithic blades: width to length ratio

deposits in the Upper Vistula Basin can be assigned to this zone as well (Lech 1988; 1989). The formation of these two broad zones of raw material procurement contrasts with the marked raw material variability among the Mesolithic groups, which exploited mainly local raw materials (often of low quality).

In addition to the complex problem of the relation of the LPC to the Mesolithic units, numerous controversies have arisen around the relationships between the LPC, the Starčevo and Vinča cultures—notably their chronology and ceramic styles. As far as lithic production is concerned, the LPC/Starčevo hiatus is again evident (Kaczanowska and Kozłowski 1987) as are the close ties between the LPC and Vinča. These trends can be seen in Transdanubia and in Lower Austria as early as the oldest phase of the LBP. In the rich lithic inventories from Brunn II, the macro-blade component with the dominance of lateral retouch, which is so characteristic of the Starčevo culture, is absent. But the high frequency of short, almost microlithic end-scrapers has parallels in the Vinča culture assemblages from the territory of Voivodina and Banat (Kaczanowska and Kozłowski 1983). The numerous perforators (including those of the Fiera type) recorded in the Banat and Voivodina assemblages are also an important component of the Brunn II inventory (Paunescu 1970). At this site, trapezes are found in a large number as well. Their source is more difficult to trace since they are known both in some Starčevo contexts as well as in the Vinča culture, and obviously in the Balkan and middle Danubian Late Epigravettian inventories. The presence of fairly numerous impact fractures on the trapezes from Brunn II indicates that these were arrowheads, whereas trapezes of the Starčevo culture do not always indicate this function.

The evolution of the views of other researchers is also worth noting here. J. Pavuk (1997), who linked the subphases of the Early LPC with the evolution of the major part of the Starčevo culture, has recently emphasized that it is only in the oldest LPC subphase (Hurbanovo-Nitra) that elements in common with the Starčevo culture can be seen, whereas the Vinča influences can be traced first of all to the oldest sites north of the Carpathians. N. Kalicz has also documented numerous new elements in the LPC that are close to the Vinča culture in pottery technology, ceramic form, and ornamentation, although he stresses the stylistic similarity of the oldest LPC and the Starčevo-Körös complex. These elements are often assigned to the proto-Vinča phase, providing evidence for the parallel evolution of the final phase of the Starčevo-Körös complex and the early phase of Vinča stylistics (Kalicz 1995). On the other hand, a comparison of the Proto-Vinča and Vinča ceramics done by W. Schrier has shown that the typical Proto-Vinča elements do not continue in the Vinča culture itself but seem to characterize the late Körös culture (Schrier 1996). The "Vinčaoidal" components, therefore, have an intercultural nature and can be regarded as the effect of another southern (Anatolian?) diffusion that initiated a new stage in the European Neolithic. The fundamental question is: was this only a stylistic diffusion or a new migration?

Conclusion: The Wave of Advance Versus Other Models

The diffusion of the Neolithic in the seventh to sixth millennium B.C. is not a continuous process of expansion of the oikumene of the agricultural-stock breeding communities but is, in fact, made up of two stages whose evolution displays certain regularities. The first stage is the expansion of the First Neolithic with monochrome and white-painted ceramics from the Aegean basin to Transylvania. The second stage is the spread of the early phase of the Western Linear Complex from Transdanubia as far as the Vistula Basin in the east and the Rhein in the west (fig. 12.7).

Each of the stages included several episodes (fig. 12.8):

a. the fast expansion over the territories of nearly the maximum range of a given unit;
b. the formation of local settlement systems and the adaptations which enabled the *oikumene* to widen both internally as well as externally until the full range of a given unit was reached;
c. contacts with the local Mesolithic communities. The nature of these contacts was diverse: from hostility (wars) and the exchange of goods (fostering change in the cultural and economic systems) to acculturation.

It is highly likely that following the first episode, connected with demic diffusion, the role of local populations in each subsequent episode grew in importance, while that of gene flow decreased.

The diagram in figure 12.9 represents, in detail, the diachronic approach to the Neolithization process in central Europe. The first stage—the lowest

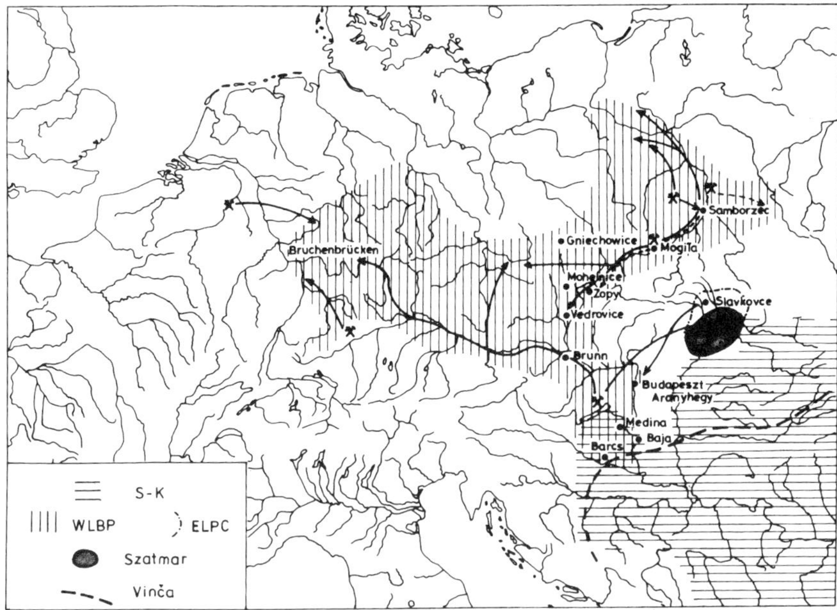

Fig. 12.7. Map of the Western and Eastern Linear Complexes: (S-K) Starčevo-Körös, (WLBP) Western Linear Band Pottery culture, (ELPC) Eastern Linear Pottery Culture, Szatmar and Vinča territories. Arrows indicate the most important WLBP raw material imports with their places of extraction.

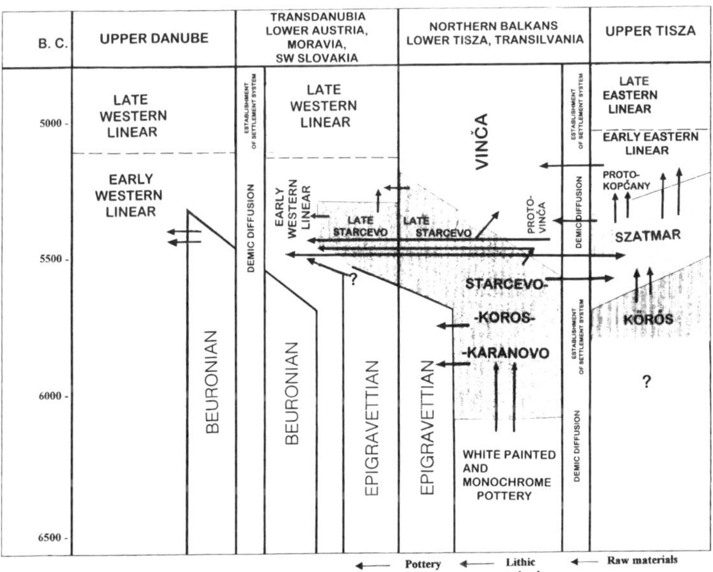

Fig. 12.8. Evolution of the early Neolithic complexes in central Europe

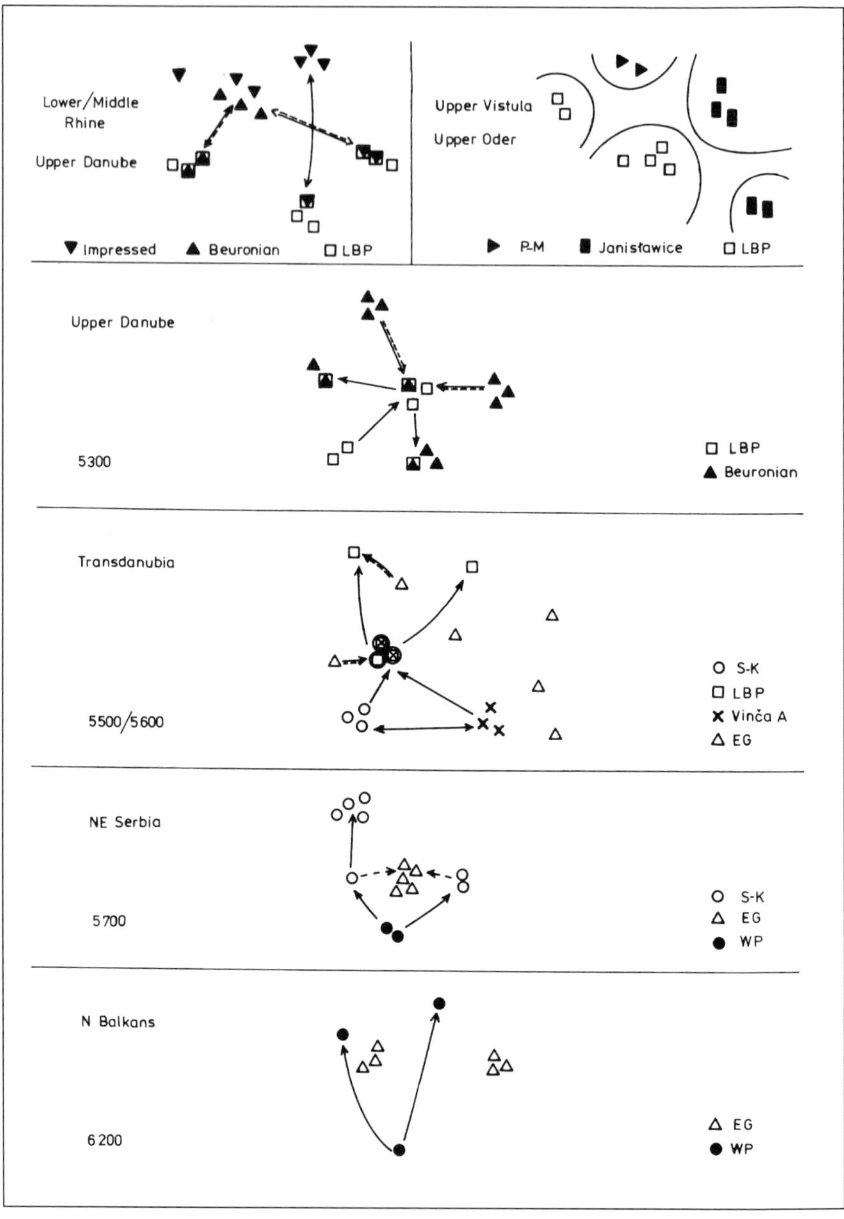

Fig. 12.9. Diagram showing models of Neolithic diffusion and Mesolithic-Neolithic interaction during the Neolithic transition in central Europe: (P-M) Post-Maglemose, (LBP) Linear Band Pottery culture, (S-K) Starčevo-Körös, (EG) Epigravettian, (WP) White-Painted Pottery Complex.

part of the diagram—is the appearance of sites with the monochrome and white-painted ceramics among the rare Epigravettian concentrations scattered like islands in the northern part of the middle Balkans (about 6200 B.C. cal.). The next stage at about 5700 B.C. is the emergence of the painted and barbotine ware complex (the Starčevo-Körös cultures) presented against the background of the special situation in northeastern Serbia, where the Starčevo settlements surrounded the Mesolithic enclave of the Lepenski Vir culture in the Iron Gates. There was no interaction between the Starčevo and the Lepenski Vir communities in the cultural sphere until the end of the Lepenski Vir II phase. It is possible, nevertheless, that the imports of "Balkan" flint in Lepenski Vir phase I and II were brought by the Starčevo population, who exploited deposits of this raw material in western Bulgaria.

The third stage occurs at about 5500–5600 B.C. in the territory of Transdanubia south of Lake Balaton. Here the adaptation of the Starčevo and Vinča groups to local conditions constitutes the foundation of the formation of the first groups of the Western Linear Complex (LBP). At the same time, single Epigravettian sites exist whose population had undergone acculturation, contributing occasional "Mesolithic" elements to the lithic typological and technological repertoire at some of the oldest LBP sites. From centers in Transdanubia and lower Austria, the LBP began to expand into the loess territories of the central European uplands.

The next stage of the Neolithic transition takes place on the Upper Danube at about 5500–5300 B.C. when the LBP penetrates into the range of fairly dense Beuronian settlement. In consequence, all types of interaction are created between the Mesolithic population and the foreign LBP people. For example, Mesolithic sickle inserts appear in LBP assemblages and Neolithic axes now occur on Beuronian sites.

In the period from about 5300 to 5200 B.C., the territories of the Western LBP (upper Danube, Rhine Basin, Netherlands, Belgium) and those of the Eastern LBP (Vistula and Oder Basins) see a diametrically different development. In the west the interaction between the Mesolithic and the Neolithic groups continues to occur not only through exchange but it is also expressed, possibly, in the building of fortifications at the LBP settlements. These contacts are further enriched by the influence of western complexes with impressed ware pottery (La Hougette and Limburg type ceramics). In the east, on the other hand, the local Mesolithic communities and the LBP population live in separate ecological niches and no interaction between them is observed. Interactions occur only later, in the second and third stages of Neolithization of the European lowland—that is, during the expansion of the Lengyel-Polgar cultures into these territories and during the formation of the Funnel Beaker culture.

REFERENCES

Biró, K. 1991. "The Problem of Continuity in the Prehistoric Civilization of Raw Material." *Antaeus* 19–20:41–50.

Boronean_, V. 1980. "Probleme ale culturii Schela Cladovei – Lepenski Vir in lumina noilor cercetarii." *Drobeța* 4:5–25.

Caspar, J.P., M. Kaczanowska, and J.K. Kozłowski. 1989. "Chipped Stone Industries of the Linear Band Pottery Culture (LBP): Techniques, Morphology and Function of the Implements in Belgian and Polish Assemblages." *Helinium* 29:175–205.

Chapman, J. 1988. "Technological and Stylistic Analysis of the Early Neolithic Chipped Stone Assemblage from Méthelek, Hungary." In *International Conference on Prehistoric Flint Mining and Lithic Raw Material Identification in the Carpathian Basin,* 31–52. Budapest.

Dimitrievič, S. 1969. "Starčevačka kultura u Slavonsko-srijemskom prostoru i problem prielaza ranog u srednji neolit u srpskom i hrvatskom Podunavlju." In *Simpozij neolit i eneolit u Slavonji,* 59–121. Vukovar.

Domboróczky, L. 1997. "Füzesabony-Gubakút. Ujkökori falu a Kr. e. VI èvezredböl." In *Utak a Múltha*. Budapest.

Gatsov, I. 1993. *Neolithic Chipped Stone Industries in Western Bulgaria.* Cracow.

Gronenborn, D. 1990. "Mesolithic-Neolithic Interactions: The Lithic Industry of the Earliest Bandkeramik Culture Site at Friedberg-Bruchenbrücken, Wetterau-Kreis (West Germany)." In *Contributions to the Mesolithic in Europe*, edited by P.M. Vermeersch and P. Van Peer, 173–82. Leuven.

———. 1994. "Kommentare zum Beitrag von Andreas Tillmann—Kontinuität oder Diskontinuität? Zur Frage einer Bandkeramischen Landnahe im südlichen Mitteleuropa." *Archäologische Informationen* 17:50–2.

———. 1995. "Überlegung zur Ausbreitung der bäuerlichen Wirtschaft in Mitteleuropa—Versuch einer kulturhistorischen Interpretation ältestbandkeramischer Silexinventare." *Praehistorische Zeitschrift* 69:131–51.

Horváth. F., and E. Hertelendi. 1992. "Contribution to the 14C Based Absolute Chronology of the Early and Middle Neolithic Tisza Region." *Józa András Múzeum Erkönyre* 36:111–33.

Jeunesse, Ch. 1995. "Les groupes régionaux occidentaux du Rubené á travers les pratiques funéraires." *Gallia Préhistoire* 37:115–54.

Kaczanowska, M., and J.K. Kozłowski. 1983. "Einige Bemerkungen zu Fragen der Steinindustrien der Vinča-Kultur." *Istraživanja* 10:253–72.

———. 1986. "Gomolava-Chipped Stone Industries of Vinča Culture." *Prace Archeologiczne* 39:1–138.

———. 1987. "Barbotino" (Starčevo-Körös) and Linear Complex: Evolution or Independent Development of Lithic Industries?" *Arheološki radovi i rasprave* 10:25–52.

———. 1990. "Chipped Stone Industry of the Vinča Culture." In *Vinča and Its World*, edited by D. Srejovič and N. Tasič, 35–47. Belgrade.

———. 1994. "Environment and Highland Zone Exploitation in the Western Carpathians (VII–VI Millennium BP)." In *Highland Zone Exploitation in Southern Europe*, edited by P. Biagi and J. Nandris, 49–71. Monografia di Natura Bresciana 20. Brescia.

Kalicz, N. 1995. "Die älteste transdanubische (mitteleuropäische) Linienbandkeramik. Aspekte Zu Ursprung, Chronologie, und Beziehungen." *Acta Archaeologica Academiae Scientiarum Hungaricae* 47:23–59.

Kalicz, N., and J. Koós. 1997. "Eine Siedlung mit ältestneolitischen Hausresten und Gräber in Nordostungarn." In *Uzdarje Dragoslavu Srejoviču*, 125–35. Belgrade.

Kalicz, N., and J. Makkay. 1976. "Frühneolitische Siedlung in Méhtelek–Nádas." *Mitteilungen des Archäologischen Institutes der Ungarischen Akademie der Wissenschaften* 6:13–24.

———. 1977. *Die Linearbandkeramik in der Großen Ungarischen Tiefebene*. Budapest.

Kozłowski, J.K. 1982. "La néolithisation de la zone balkano-danubienne du point de rue des industries lithiques." In *Origin of the Chipped Stone Industries of the Early Farming Cultures in Balkans*, 131–70. Cracow.

———. 1989. "The Lithic Industry of the Eastern Linear Pottery Culture in Slovakia." *Slovenská Archeológia* 37:377–410.

———. 1992. *L'art de la Préhistoire en Europe orientale*. Paris.

———, ed. 1998. *The Early Linear Pottery Culture in Eastern Slovakia*. Prace Komisji Prehistorii Karpat 1. Cracow.

———. 2001. "Evolution of the Lithic Industries of the Eastern Linear Pottery Culture." In *Szolnok Conference Papers*, edited by R. Kertéz. Budapest.

Kozłowski, J.K., and S.K. Kozłowski. 1982. "Lithic Industries from the Multi-layer Mesolithic Site Vlasac in Yougoslavia." In *Origin of Chipped Stone Industries of the Early Farming Cultures in the Balkans*, edited by J.K. Kozłowski, 11–109. Cracow.

———. 1983. "Chipped Stone Industries from Lepenski Vir, Yugoslavia." *Prehistoria Alpina* 19:37–76.

Kozłowski, J.K., S.K. Kozłowski, and I. Radovanovič. 1994. *Meso- and Neolithic Sequence from the Otmut Cave (Montenegro)*. Warsaw.

Lech, J. 1988. "A Danubian Raw Material Exchange Network: A Case Study from Bylany." In *Bylany Seminar 1987*, 111–20. Prague.

———. 1989. "The Organization of Siliceous Rock Supplies to the Danubian Early Farming Communities (LBK): Central European Examples." In *Rubané et Cardial*, edited by D. Cahen and M. Otte, 253–68. Liège.

Magoşeanu, F. 1978. "Mezoliticul de la Ostrovul Corbului." *Studii si Cercetarii de Istorie Veche si Archeologie* 29:335–51.

Makkay, J. 1982. *A magyarországi neolitikum kutatásának új eredményei*. Budapest.

Milojčič, V., J. Boessneck, and M. Hopf. 1962. "Die deutsche Ausgrabungen auf der Argissa-Magula, Thessalien: Die praekeramische Neolithikum sowie die Tier und Pflanzenreste." In *Beitrage zur Ur und Frühgeschichtlichen Archäologie des Mittelmeer Kulturraumes* 2. Bonn.

Paunescu, A. 1970. *Evolutia uneltelor şi armelor de piatră cioplită pe teritoriul Romnniei.* Biblioteca de Archeologie 15. Bucharest.

———. 1985. "Les industries lithiques du Néolithique ancien de la Roumanie et quelques considérations sur l'inventaire lithique des cultures du Néolithique moyen de cette cointrée." In *Chipped Stone Industries of the Early Farming Cultures in Europe*, edited by J.K. Kozłowski and S.K. Kozłowski, 75–94. Cracow.

Pavuk, J. 1997. "The Vinča Culture and Beginning of the Linear Pottery." In *Uzdarje Dragoslavu Srejoviču*, 167–78. Belgrade.

Perlès, C. 1990. *Les industries lithiques tailleés de Franchthi (Argolide, Grèce).* Vol. 2, *Les industries du Mésolithique et du Néolithique initial.* Bloomington.

Potushniak, M. 1994. "Dejaki naslidki doslidzhenia serednoneolitichnogo sharu na bagatosharovomu poselenni Zastavne-Kovo Domb u Zakarpatti." In *Problemi arkheologij srhidnikh Karpat*, 15–17.

Prinz, B. 1987. *Mesolithic Adaptations on the Lower Danube. BAR-IS* 330. Oxford.

Radovanovič, I. 1994. "The Lepenski Vir Culture: A Contribution to the Interpretation of Its Ideological Aspects." In *Uzdarje Dragoslavu Srejoviču*, 87–92. Belgrade.

———. 1996. "Kulturni identitet djerdapskog mezolita." *Zbornik Radova Narodnog Muzeja* 16:39–47.

Schrier, W. 1996. "Proto-Vinča: Zum Übergang von der Starčevo—zur Vinča-Kultur im Südosten des Karpatenbeckens." In *Uzdarje Dragoslavu Srejoviču*, 155–66. Belgrade.

Srejovič, D. 1969. *Lepenski Vir, Nova praistorijska kultura u Podunavlju*. Belgrade.

Srejovič D., J.K. Kozłowski, and S.K. Kozłowski. 1980. "Les industries lithiques de Vlasac et de Lepenski Vir." In *Problèmes de néolithisation de certaines régions de l'Europe*, edited by J.K. Kozłowski and J. Machnik. Cracow.

Starnini, E. 1993. "Typological and Technological Analyses of the Körös Culture Chipped, Polished and Ground Stone Assemblages of Méthelek-Nádas (North-Eastern Hungary)." *Atti della società per la Preistoria e Protostoria della Regione Friuli-Venezia Giulia* 8:29–96.

Šiška, S. 1989. *Kultura s vychodnou linearnou keramikou na Slovensku.* Bratislava.

Teocharis, D. 1973. *Neolithic Greece.* Athens.

Tillmann, A. 1993. "Kontinuität oder Diskontinuität? Zur Frage einer bandkeramisachen Landnahme im südlichen Mitteleuropa." *Archäologische Informationen* 16:157–87.

Tixier, J., M.L. Inizan, and H. Roche. 1980. *Préhistoire de la pierre taille.* Paris.

Voytek, B., and R.E. Tringham. 1989. "Rethinking the Mesolithic: The Case of South-East Europe." In *The Mesolithic in Europe,* edited by C. Bonsall, 492–500. Edinburgh.

Wiener, J. 1984. "Praktische Versuche zur neolitischen Klingenproduktion: Ein Beiträg zur Frage der Sog. Punchtechnik." *Archäologische Informationen* 8:22–33.

— 13 —

Neolithic Dispersals in Riverine Interior Central Europe

Peter Bogucki

Over the last two decades, I have studied the introduction of agriculture to the region that could be called "riverine interior central Europe." I have chosen to concentrate on this theme and area because I find it to be a fascinating example of the interplay of various cultural processes on several different scales. My own archaeological fieldwork has focused on a 100 km^2 corner of this area along the lower Vistula in the Kuyavia region of Poland. It is also a field of research into which one can introduce many different concepts from social science and allied fields of study that expand the discussion beyond traditional limits of anthropological archaeology. Such "experimentation" is possible because of the superb foundation of data that has been recovered over the last 125 years from Ukraine to France. This body of data continues to expand annually, although the rate at which it is growing has slowed after the virtual explosion that took place between 1950 and 1980. Few other parts of the world can claim such a robust corpus of fundamental data on artifacts and settlements, to which now can be added a growing amount of botanical evidence, although over much of this region animal bones are not well preserved.

Riverine Interior Central Europe

The area of riverine interior central Europe is defined in figure 13.1. A distinctive feature of central Europe south of the coasts of the North and Baltic

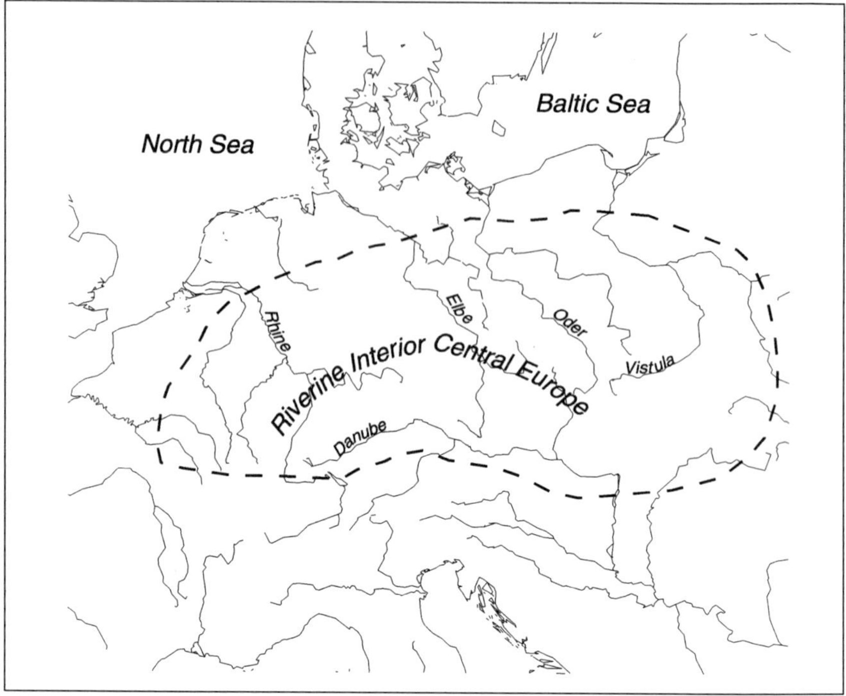

Fig. 13.1. Riverine interior central Europe includes the upper Danube drainage and the middle and upper watersheds of the major rivers that flow into the North Sea and the Baltic Sea

Seas and north and east of the Alps is its dominance by a variety of river basins, catchments, and drainages. The southern part of this zone is dominated by the Danube north of the Hungarian 90° bend and the rivers flowing into it. The northern part is distinguished by the rivers flowing north into the Baltic and North Seas: the Vistula, Oder, Elbe, Weser, Rhine, and Maas, along with their tributaries such as the Main, the Neckar, the Saale, Váh, the Nidzica, the Warta, and also the Aisne. The valleys of these rivers were the foci of early farming settlement for over a millennium after about 5600 B.C. (recalibrated radiocarbon dating). Valley slopes define the distribution of soil types, and drainage patterns define routes of communication and migration. The hill and mountain zones of interior central Europe simply did not play a role in the initial establishment of agrarian communities in this area prior to about 4000 B.C. Riverine interior central Europe embraces about 750,000 km^2, although only a relatively small amount of that area is directly implicated in early farming settlement.

Two decades of study of the earliest farmers of riverine interior central Europe have persuaded me that the primary mechanism of the establishment of farming communities was the dispersal of their initial inhabitants rather than the adoption of agriculture by indigenous populations of foraging peoples. The evidence in favor of this "migrationist" position seems to me to be compelling, as it has been for several generations of archaeologists who have dealt with the early farmers of this region. In contrast to these views are the positions taken by those whom Albert Ammerman (1989) has called the "indigenists." For example, Robin Dennell (1992, 91) maintains that "early farming in Europe always occurred in areas where there were already hunter-forager communities. These cannot be regarded as irrelevant to the pattern of agricultural expansion." Although I disagree with the inclusion of the word "always," it is true that foragers were indeed present throughout pre-agricultural Europe. But it does not follow that the development of agricultural communities was always an in situ process. Alasdair Whittle (1996) has also recently disputed the notion that colonization was a major factor in agricultural dispersal in Europe.

For reasons I will make clear below, I cling to the orthodox position, although there is no reason why indigenous foragers, few as they were in riverine interior central Europe, could not have participated in this process as well. Modern foragers have complicated relationships with their sedentary neighbors, and the nature of these relationships can shift significantly over time. One only has to look at the current debate over the status of the San peoples of the Kalahari to realize that all is not so simple as might have been assumed by archaeologists when they discovered "Man the Hunter" in the late 1960s. The apparent polarity of the migrationist-indigenist debate among archaeologists masks what must have been complicated individual family histories and shifting affiliations among locals and immigrants in Neolithic central Europe.

A Personal Account: 1976–1995

The basis for my interest in the transition to agriculture in riverine central Europe is my fieldwork in north-central Poland at three major Neolithic sites: Brzesc Kujawski, Nowy Mlyn, and Oslonki. Of these three, Brzesc Kujawski was the catalyst. Discovered and initially investigated by Konrad Jazdzewski between 1933 and 1939 (Jazdzewski 1938), my Polish colleague

Ryszard Grygiel and I reopened excavations there in 1976. We were both very young and learned the techniques of large open-site excavation "on the job," since by then Jazdzewski was at an advanced age and did not micro-manage our work. It was clear from the 1930s data, analyzed by Lidia Gabalowna in the 1960s, that there were two major Neolithic components (as they might be called by North American archaeologists) at Brzesc Kujawski (Gabalowna 1966). One belonged to the Linear Pottery culture, which had been dated elsewhere by numerous radiocarbon dates to ca. 5400–5000 B.C. (recalibrated), while the other was later and belonged to what Jazdzewski called the Brzesc Kujawski Group, a late variant of the Lengyel culture (fig. 13.2). We provided the first series of radiocarbon dates for this group, which placed it nearly a millennium after Linear Pottery, between ca. 4500 and 4000 B.C. The Brzesc Kujawski Group was responsible for the striking ground plans of trapezoidal longhouses and the numerous burials with copper and shell ornaments at Brzesc Kujawski, while the Linear Pottery inhabitants left only dense rubbish deposits and some enigmatic structural evidence. In between these two components were vague traces of post-Linear Pottery groups, the so-called Stroke-Ornamented Pottery culture and early Lengyel culture, but apparently not much happened at Brzesc Kujawski between the Linear Pottery settlement and the late phase of the Lengyel culture.

Brzesc Kujawski is situated in the northeastern frontier zone of the Linear Pottery culture in north-central Poland (fig. 13.3). We are now starting to realize the extent to which early Neolithic settlement is found in this region. Not only in Kuyavia where Brzesc Kujawski is located, but also in the area north of Torun along the lower Vistula and in the areas on both side of the lower Oder many new Linear Pottery and Lengyel sites have come to light in the last two decades. An important recent discovery is a trapezoidal longhouse like those at Brzesc Kujawski found at Bukowiec north of Torun (Kukawka and Malecka-Kukawka 1999). A millennium after Linear Pottery, this area still lay along the northern farming frontier in Europe, the most northeastern enclave of settlement in the world of the earliest central European farmers.

The settlements of the Brzesc Kujawski Group reflect an intensity of occupation and land use that was simply not present in this area a millennium earlier. Apart from a newly-reported building at Bozejewice (Czerniak 1998), an under-reported structure at Lojewo (Czerniak 1994), and some

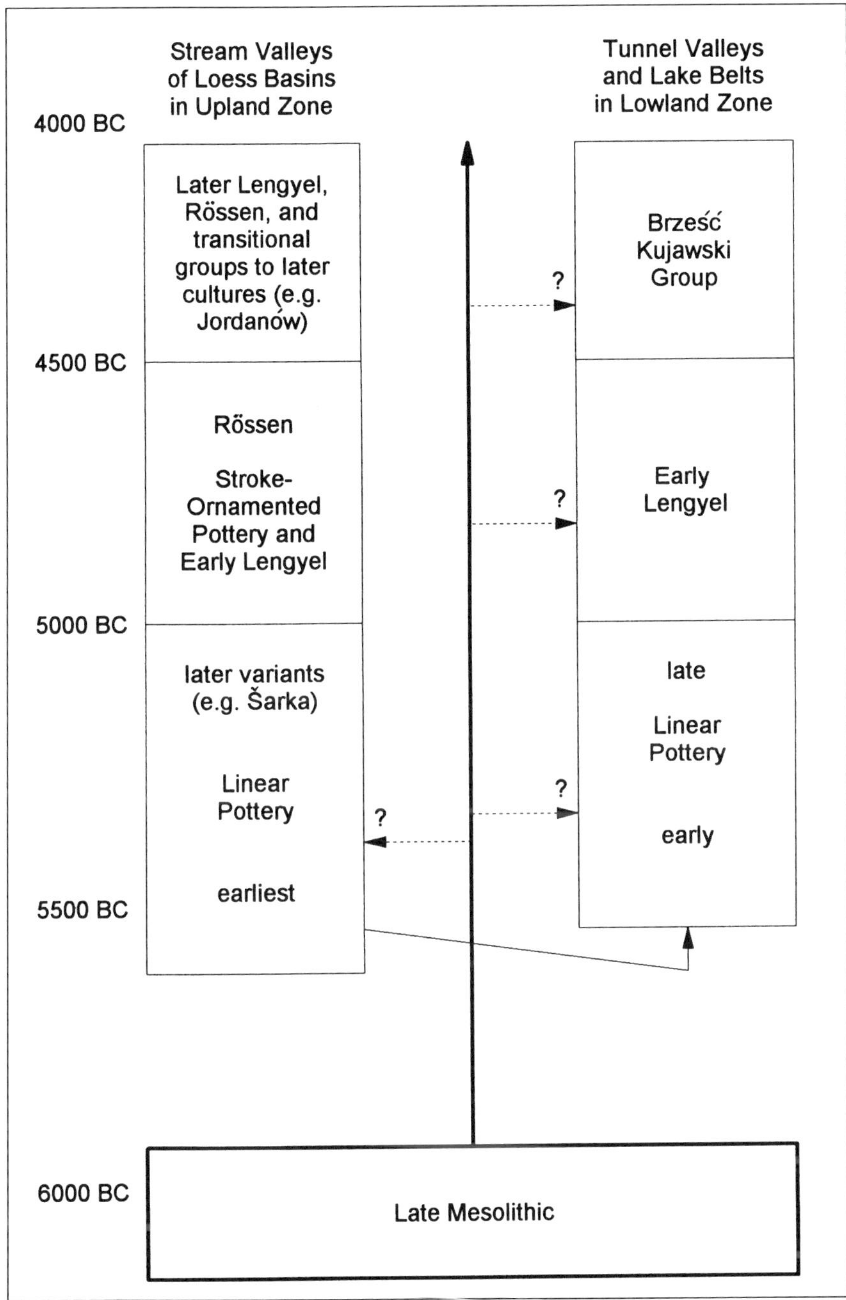

Fig. 13.2. Schematic chronological chart of the principal cultural units discussed in this chapter

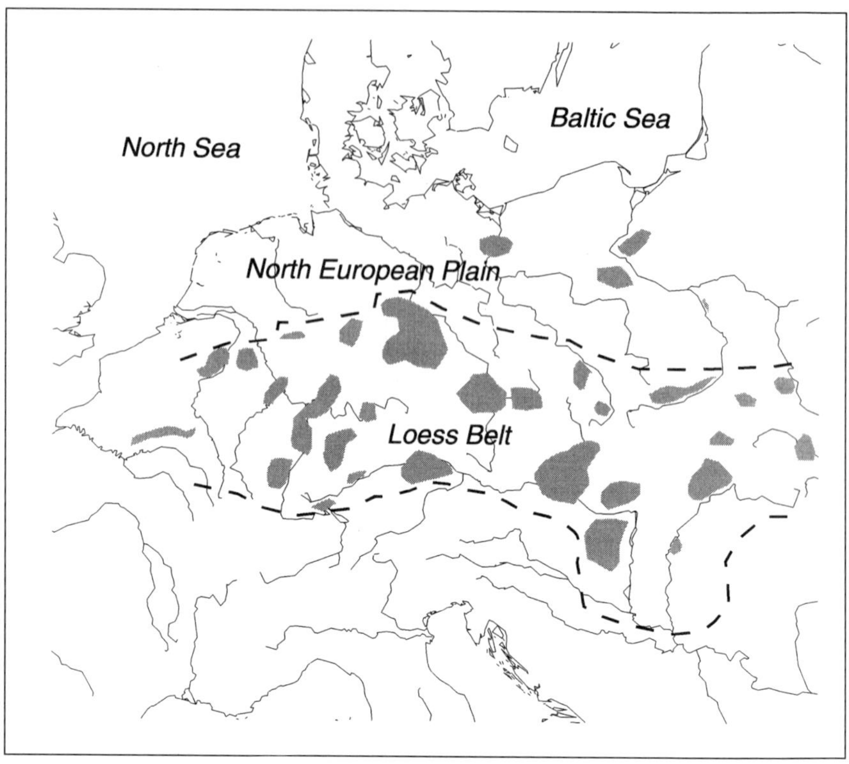

Fig. 13.3. Shaded areas indicate principal cells of Linear Pottery settlement both within the loess belt and on the north European plain. While most maps of Linear Pottery distribution depict it as continuous, this map emphasizes its discontinuous nature, which would be consistent with occasional long-distance colonization events.

enigmatic structural traces at Brzesc Kujawski, we still have almost no post-built Linear Pottery longhouses from the lowlands of northern Poland and Germany. Such structures are a signature of the Linear Pottery culture on the upland loess regions that lay to the south and west. Instead, the Linear Pottery settlements of the lowland zone are small and lack much internal organization other than large oval pits. In this sense, they seem similar to the Linear Pottery settlements reported by Nowak (1997) from the eastern Slovakian lowlands. The Brzesc Kujawski Group was much different. With their sturdy longhouses, Brzesc Kujawski Group settlements were more similar to the Linear Pottery settlements found in the loess uplands a millennium earlier and the longhouses of the Stroke-Ornamented Pottery culture and the Rössen culture, which followed Linear Pottery in many regions.

I focused on this difference between the Linear Pottery culture and Brzesc Kujawski Group settlements early in my study of these communities. One day in the Harvard Coop, I found a copy of an early number of the *Midcontinental Journal of Archaeology* containing an article by R. Berle Clay (1976). This article was a theoretical treatment of Mississippian settlement in the Southeastern United States in which a "tactical" and a "strategic" aspect were described. This was intriguing, and Clay's article made extensive reference to a 1965 article by F.E. Emery and E.L. Trist, entitled "The Causal Texture of Organizational Environments." The article was an abstract sociological analysis of organizational environments, which intrigued me still further. It was much more exotic stuff than we were reading in courses at Harvard, although when he had visited Harvard the previous spring, Robert Adams had encouraged us to read outside our immediate area of interest.

One of Emery and Trist's core ideas was a distinction between environments where the relevant good and bad features had been identified in space and time (what they called "placid-clustered") and those in which they had not (called "placid-randomized"). Organizations respond *strategically* to the former and operate with an eye toward making commitments and setting long-term goals, while they respond *tactically* to the latter by not making decisions, which commit them to any long-term strategy. It is a fairly simple idea, although I should note that Emery and Trist's article was somewhat more profound in sketching the implications of such organizational responses.

From my perspective at Brzesc Kujawski, this concept corresponded to what I was seeing in the Neolithic settlements of the Polish lowlands. The Linear Pottery sites were generally small, and they generally appeared to lack internal organization such as houses and a typology of domestic features such as storage and rubbish pits, burials, and clay extraction pits. They did not seem to me to represent long-term commitments to particular locations, even if they contained ceramics, bones, seeds, and other evidence of sedentary life. The Lengyel sites, in contrast, appeared to represent long-term commitments to particular settlement locations, which had resulted in a full range of domestic features. Primary among these were the trapezoidal longhouses that characterize the Brzesc Kujawski Group, but they also included numerous burials and pits of various sizes, shapes, and presumed functions. In my view, the Linear Pottery sites represented tactical adaptations in the lowland environment by pioneer farmers and stockherders, while the

Lengyel sites were strategic establishments, made with a definite intention to spend a long time on that spot. I wrote up a short paper on this idea in March 1978 during the spring break of my third year in graduate school, and it was published a year later in the *Journal of Anthropological Research* (Bogucki 1979). In this article, I also suggested that we might look at the overall early Neolithic colonization of interior central Europe from this perspective, with the earliest Linear Pottery sites fitting the tactical category (for no Linear Pottery longhouses from the earliest phases had yet been found) and the later ones with many longhouses being more strategic in character.

My 1981 dissertation dealt with the analysis of animal bones from Brzesc Kujawski. Another difference between the Linear Pottery and Lengyel settlements was that the Linear Pottery faunal sample was composed largely of domestic cattle bones, while the Lengyel sample had a greater variety of wild and domestic species. Given the ephemeral nature of Linear Pottery sites in the lowlands (and the earliest Linear Pottery sites in the uplands as well, it seemed then), I suggested that these may have been the product of some form of agropastoral system that led certain members of Linear Pottery communities to range further from their home settlements and thus pioneer the Neolithic settlement of central Europe.

The reaction to this suggestion after it was disseminated more widely in a *BAR* monograph (Bogucki 1982) was skeptical. In retrospect, I have to agree that it was a bit far-fetched to imagine herds of Linear Pottery cattle being driven through the forests of central Europe to meadows and other areas of forage far from the settlements at which their keepers lived, and much less for this to have been an instrument of colonization. Evidence for local manufacture of Linear Pottery ceramics (Cowie 1983) and impressions of carbonized grain in the fired clay points toward greater duration of habitation even in areas like northern Poland which lack great longhouses. The discovery of longhouses at very early Linear Pottery sites such as Brunn in Austria and Schwanfeld and Bruchenbrücken in central Germany has effectively vitiated my "pastoral Linear Pottery" hypothesis and consigned it to the dustheap of prehistory.

While working at Brzesc Kujawski, especially at the area of the site known as Brzesc Kujawski 3, another influence from American archaeology came into play. Kent Flannery's 1976 volume *The Early Mesoamerican Village* made a great impression on many young American archaeologists around this time, particularly the opening chapters about households and household clusters. Brzesc Kujawski 3 was set apart from the main area of the site. A single Brzesc Kujawski Group longhouse, which had been rebuilt once, had been excavated there by Jazdzewski in 1937. We excavated the surrounding area, which contained the clay extraction pits that then became large rubbish dumps, bell-shaped features that appeared to have been storage pits and graves. There was evidence of specific activities by inhabitants of the house, such as a feature that contained the residue of a single session of skinning five beavers 6000 years ago. Taken together with the house, all this seemed very much like a household cluster, so I wrote it up as an article that appeared in *World Archaeology* in 1981. During the summer of 1981 and 1982, when I was not in Poland, Grygiel excavated another isolated house and all its features at Brzesc Kujawski 4. I had sent him a copy of *The Early Mesoamerican Village*; in early 1984, he sent me a copy of his habilitation thesis (published as Grygiel 1986), whose English subtitle could have been *The Early Mesoamerican Village Visits Poland.*

As a result, Grygiel and I became firm adherents of "household archaeology," perhaps among the first in Europe, along with Ruth Tringham, who was thinking along similar lines at Selevac in Yugoslavia (Tringham and Krztic 1990), and others. Our view was that households, as represented by their material remains of houses, features, and graves were fundamental organizational units of Neolithic society. They were the loci of decision making about the acquisition, allocation, and consumption of resources. Although our interest in Neolithic households grew from the study of the Brzesc Kujawski Group component at Brzesc Kujawski, it was not difficult to project this onto the earlier Linear Pottery longhouse communities in other parts of central Europe. Around this same time, I should note that Jens Lüning and his colleagues working on the Aldenhoven Plateau in the Rhineland began to identify clusters of Linear Pottery domestic features that were very much like household clusters (Lüning 1982, 1988). The German archaeologists called these clusters *Hofplätze* (fig. 13.4), although they avoided much deeper paleosociological interpretation of these clusters. Such findings were consistent with our observations in Poland. We concluded that the basic building blocks of Neolithic society in riverine interior central Europe, from Linear Pottery onward, were households that occupied a longhouse.

I developed this idea somewhat further in my 1988 book *Forest Farmers and Stockherders*, specifically the idea that the elemental units of Linear Pottery society were the households that lived in the individual longhouses,

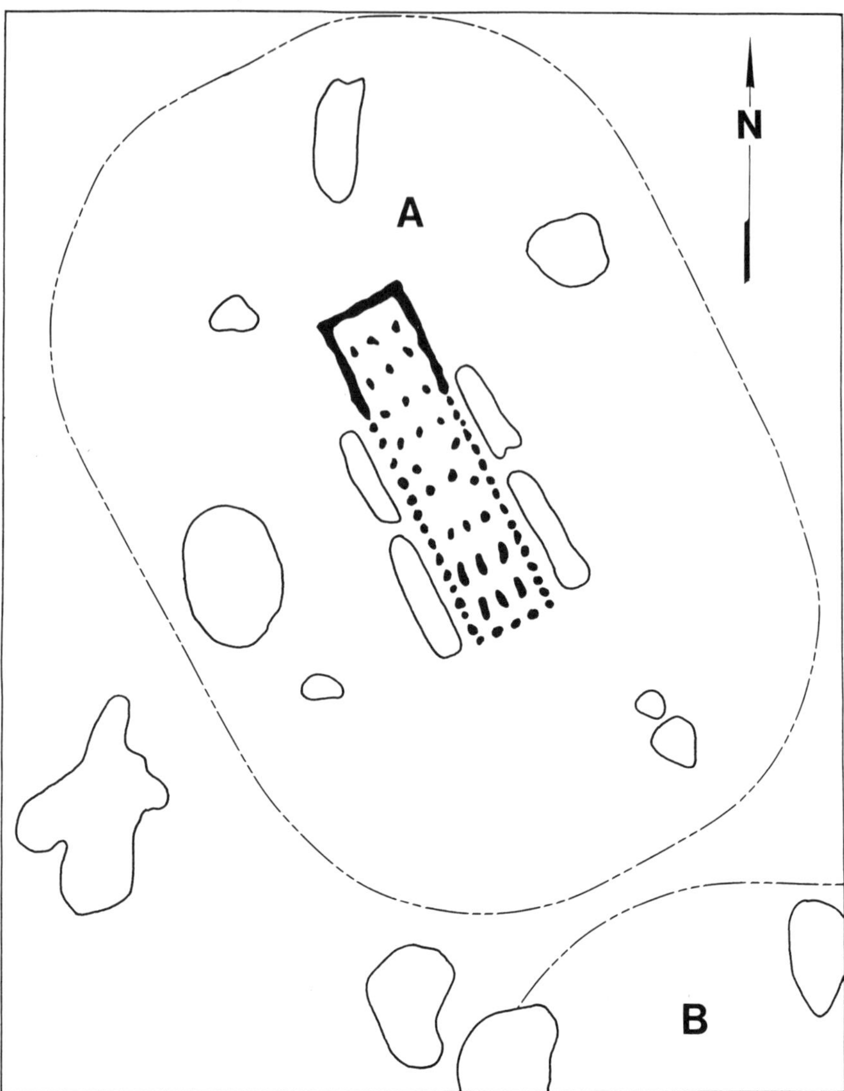

Fig. 13.4. Generalized Linear Pottery Hofplatz consisting of a longhouse and associated pit features and "activity zone" based on examples at Langweiler 8 in the Rhineland (after Lüning 1982, fig. 19). The radius of the "activity zone" is approximately 25 m.

while Grygiel worked along parallel lines and published a large study in 1994. It is especially gratifying that detailed analysis of Linear Pottery households has finally been taken up by others, such as Jonathan Last (1998) in his study of household clusters at Bylany and Miskovice. Seeing households as the basic units of decision-making is crucial for understanding the Neolithic diaspora in central Europe, in my view. If the decision to relocate was made by a household rather than a larger community, then the reasons for the Linear Pottery dispersal probably lay in the motivations and aspirations of these small units rather than in larger phenomena such as soil depletion and population pressure that had previously been invoked as explanations.

Current Thoughts on the Linear Pottery Dispersal

Between 1988 and 1994, I was thoroughly distracted by excavations at Oslonki, a large settlement of the Brzesc Kujawski Group 10 km west of Brzesc Kujawski (Grygiel and Bogucki 1997) and did not worry much further about the introduction of farming to riverine interior central Europe. Not until 1995 did I really revisit the topic. The catalyst for this was an invitation to participate in the symposium organized by Doug Price at the Society for American Archaeology meetings in Minneapolis, which drew together archaeologists working in all parts of Europe to consider the transition to agriculture in their particular regions (Price 2000). The charge for this symposium was to evaluate critically the evidence for the mechanism of agricultural dispersals in each part of Europe, whether it was through the colonization of new habitats by populations of farmers or the adoption of agriculture by indigenous foragers.

In this symposium, I evaluated the Linear Pottery dispersal against the criteria laid down by Irving Rouse (1958) for deciding whether population movement is a better explanation for change in the archaeological record than in situ development:

1. identify the migrating people as an intrusive unit in the region they have penetrated;
2. trace this unit back to its homeland;
3. determine that all occurrences of this unit are contemporaneous;

4. establish the existence of favorable conditions for migration;
5. demonstrate that some other hypothesis, such as independent invention or diffusion of traits, does not better fit the facts.

David Sanger (1975) later added a sixth rule:

6. establish that all cultural subsystems are involved and not just an isolated one (such as burial practice).

Rouse's criteria supported the case that the establishment of early farming communities in this area was the result of colonization rather than a local development, for the following reasons:

1. there is no in situ developmental sequence of pottery in north-central Europe; pottery manufacture appears fully-developed within this area with clear antecedents outside the region; the anomalous La Hoguette and Limburg wares cannot be shown to predate the earliest appearance of Linear Pottery in the region;
2. the house forms, settlement patterns, and burials of the first farmers of this region are completely different from any preceding ones in the region (as poorly known as the latter are);
3. the chipped stone tool types of the first farmers are also different from preceding Mesolithic forms in most Linear Pottery areas, although there are some similarities as well at the earliest sites in central and southern Germany;
4. the key domestic plant species and two of the major livestock species—sheep and goats—have no native, wild, conspecific forms in central Europe and were introduced from southeast Europe and ultimately from southwest Asia;
5. the pottery, house forms, and settlement locations are extraordinarily uniform in their general appearance, although with minor local variations over time, from Slovakia to eastern France;
6. the warm mid-Holocene climate was especially agreeable to cultigens of Near Eastern origin, and the river systems of central Europe presented excellent corridors for the movement of people and livestock;

7. the most remarkable aspect of the Neolithic diaspora in central Europe is its speed; in the course of 500 years, and probably less, farming communities were established between Ukraine and eastern France.

The central European evidence, then, conforms to Rouse's criteria for demonstrating population movement: the earliest agricultural communities are intrusive, have a clear source outside the region, appear across central Europe in a very short span of time, and took advantage of favorable conditions for migration. The real clincher, however, lies in its conformity with Sanger's additional qualifier. All cultural subsystems—technology, subsistence, settlement, mortuary practice—are involved.

The evidence that has been cited in contradiction to this view include certain aspects of lithic technology, faunal remains, and the enigmatic La Hoguette and Limburg pottery of the western part of the Linear Pottery territory (Gronenborn 1994; 1997; 1999; Tillmann 1993; Whittle 1996, 152; Kind 1998). Upon closer scrutiny, it is evident that these are regional, low-frequency phenomena that affect only a few cultural subsystems. Limburg and La Hoguette wares are widely distributed in west-central Europe, but their occurrence on Linear Pottery sites and on nearby sites without Linear Pottery is sporadic compared with the overall quantity of standard Linear Pottery materials. Similarly, the "Mesolithic" elements in stone tool assemblages are confined to a relatively small number of projectile points and blade traits. Moreover, the generic nature of the common blade characteristics suggests that they may be dependent more on raw material than on cultural lithic tradition. They are also confined to the oldest Linear Pottery sites, where one might expect idiosyncratic pioneer adaptations in advance of the main colonization. While the anomalous pottery and stone tools may argue in favor of some interaction between Linear Pottery farmers and local foragers, as Susan Gregg argued in her 1988 book, they do not suddenly transform the locals into the sole agents of the establishment of agricultural communities in riverine interior central Europe.

The argument that the presence of significant numbers of wild animals, especially red deer in the earliest Linear Pottery faunal assemblages in south-central Europe, reflects a Mesolithic pattern (Kind 1998, 13–17) is especially open to criticism. Just because the inhabitants of these sites were farmers and also kept substantial numbers of livestock does not mean that they were incapable of hunting. In fact, the opening of the forest for agriculture would have drawn wild herbivores to cultivated fields, and hunting may have been undertaken as a way of protecting crops and as a ready source of meat. Numerous parallels of the convergence of immigrant and indigenous subsistence strategies are known from other colonizations around the world. For example, a comparison of Native American and Spanish faunal assemblages in colonial Florida has demonstrated the extent to which the colonizers' subsistence patterns took on the characteristics of the native pattern (Reitz 1985).

The fundamental point that the recent indigenist positions ignore is that the Linear Pottery settlements of central Europe were clearly inhabited by people who had made a fundamental and irreversible commitment to agriculture.[1] Hunter-gatherer bands do not make such a commitment overnight, especially not one which would entail such a radical change in settlement, subsistence, technology, and mortuary behavior over such a wide area. Foragers who adopt farming generally do so piecemeal, adopting those elements that make sense to integrate with their existing subsistence and settlement system, ignoring those which do not. The relatively trivial similarities between the sparse late Mesolithic assemblages and the earliest Linear Pottery can be reasonably, and more economically, explained as either the occasional adoption of local technology by the immigrant farmers or the assimilation of indigenous foragers into the farming population.

I remain secure in holding to the traditional colonization model, a view that was reinforced at the Venice conference by data presented by Janusz Kozłowski and Malgorzata Kaczanowska (see ch. 12 in this volume). Moreover, a consistent theme in the discussions in Venice was "where *is* the late Mesolithic?" When viewed continentally, the greatest evidence for late Atlantic (6000–4000 B.C.) forager settlement is found in the rich coastal and estuarine environmental of northern, western, and southern Europe; specific riverine locations in Eastern Europe (e.g. Lepenski Vir and Soroki), sandy outwash areas of the north European plain, and montagne and *karst* regions of south-central Europe. In riverine interior Central Europe, however, late Atlantic forager settlement, either immediately preceding or contemporaneous with Linear Pottery, is virtually invisible, except for sporadic "Mesolithic" elements in Linear Pottery chipped-stone assemblages. Argument on the basis of negative data is viewed as dangerous archaeological practice (although routine in other disciplines), but the more we strain to identify late Atlantic foragers in interior central Europe, the more elusive they become.

Colonizing Logic

Once it can be established that the introduction of agriculture to central Europe was fundamentally the result of colonization through population dispersal, certain aspects of the archaeological record assume a new significance. I discussed these in a symposium on prehistoric migration organized by Dean Snow at the 1997 SAA meetings in Nashville, characterizing them as the "colonizing logic" (a term introduced by Beaton in 1991) of the early agriculturalists. Two characteristics of the Neolithic archaeological record in riverine interior central Europe deserve special attention: the preference of the earliest farmers in this region for very specific habitats to the exclusion of virtually all others and a degree of technological conservatism manifested in pottery, stone tools, and especially houses.

In this connection, I was influenced by a comparison of the Pleistocene colonizations of the Americas and Australia by J.M. Beaton (1991). Beaton discussed the notion of what he refers to as "megapatches" and their implications for the study of colonization and migration. By using the term "megapatch," Beaton means that colonists may not have a sufficiently fine-grained view of the variability contained by virgin territory and see it at a very coarse level of resolution. As a result, a characteristic of a colonizing population might be that it will very clearly demonstrate a preference for gross habitat types, and within these gross habitat types they might locate their settlements in very similar places. Only over time would the colonizers distribute their sites in a variety of habitats as the resolution at which they perceive their environment becomes finer and finer.

This idea resonated with my earlier view of tactical and strategic Neolithic settlements in riverine interior central Europe, or at least it might parallel it to a certain degree. A signature of colonization, it might be argued, would be the focus on one gross habitat type to the virtual exclusion of all others, no matter how similarly attractive or productive the others might eventually prove to be. Within that gross habitat type, moreover, the location of settlements in a specific landscape zone to the exclusion of others would also be a signature characteristic of a colonizing population. By contrast, a settlement pattern that exhibits the use of many different types of gross habitat types and where settlements are located in many different parts of the landscape would be more characteristic of a continued pattern of foraging settlement, even if there were slight changes. The exploitation of many different habitats would suggest a fine-grained familiarity with the environment that might be expected of an indigenous population rather than the coarse-grained perception of a colonizing population.

In central Europe, the concentration by early farmers on a limited number of gross habitat types is obvious. In the upland loess basins, early farmers chose to settle almost exclusively along the valleys of the smaller tributaries of the major river systems, which would have been the most fertile and moist habitats in the loess belt (Kruk et al. 1996; Kruk and Milisauskas 1999). These also may have been the areas of natural breaks in the vegetation, what Verboom (1977) has called "zones of weakness," in the primeval forest where the Linear Pottery farmers might have found it easier to open clearings by felling, or more likely by ring-barking, trees and clearing ground vegetation with fire. The watersheds separating the stream basins, where early agriculture certainly could have been practiced, were generally ignored by the early farmers and were not colonized until later in the Neolithic.[2]

When the early central European farmers entered lowlands of the north European plain, there was a similar concentration on gross habitat types. The lowland analogues of the upland loess basins in northern Poland are patches of ground moraine that are engraved with subglacial meltwater valleys. The earliest farming settlement in northern Poland occurs almost exclusively along these so-called tunnel valleys. These zones would also have been the natural breaks in the forested lowland environment, since they would have been part of the hydrological network during the moist Atlantic period. Intervening areas, despite their suitability for agriculture, are not occupied until later in the Neolithic.

This concentration by the early farmers on a limited number of habitats contrasts markedly with the pattern observed in other parts of northwestern Europe. In the Rhine-Maas delta, for example, early farming sites are known from a variety of habitats and come in varying sizes and functions (Louwe Kooijmans 1993). Such variety appears to reflect a much finer-grained knowledge of the environment than the concentration on a very specific habitat type in the loess zone and in northern Poland. Similarly in Ireland the early farmers used the landscape very widely (e.g., Green and Zvelebil 1993), which appears to reflect the residual knowledge of the terrain from indigenous foraging peoples.

The other dimension of the colonizing logic of the earliest farmers of central Europe that deserves attention lies in their conservative and strikingly remarkably standardized technology. If there had been a mass conversion to agriculture of indigenous foragers over the 750,000 km^2 of central Europe, then one would expect almost immediately to see considerable variation in house forms, pottery, and other aspects of culture. Instead, the initial pattern is strikingly homogeneous on a continental scale, and initial differences from one region to another are trivial. Houses and pottery from Slovakia look much like those from the Paris basin. The expected regional variation did not set in until several centuries later, when local populations developed distinct material identities.

In particular, the use of a standardized, modular form of construction—the longhouse of the early farmers—is consistent with a colonizing population using proven technology to facilitate its expansion. These longhouses, up to 45 m in length, were the largest buildings in the world at this time. They were multipurpose structures, providing shelter for humans and livestock as well as storage and working space. Despite some regional variation in construction details, the dimensions and general organization of the longhouses is remarkably similar throughout central Europe.

New Directions

Early in the 1990s, I became intrigued by the notion that the establishment of agricultural communities belongs to a category of phenomena called "complex adaptive systems." Although I do not recall exactly when I became interested in this, it may have been after seeing an article about the Santa Fe Institute and the work being done there. I also know that a 1991 article in *Scientific American* by Per Bak and Kan Chen on "self-organized criticality" caught my eye. Part of the problem with the establishment of early farming communities in central Europe was the fact that the single-factor explanations such as soil depletion and population pressure that had been advocated by archaeologists beginning with V. Gordon Childe did not account adequately for Neolithic dispersals. Yet I was convinced that a dispersal had occurred throughout riverine interior Central Europe, and somewhere along the way it occurred to me that it might be some sort of large-scale cumulative effect of many small-scale local actions by early farming households.

A complex adaptive system has the following properties (Forrest and Jones 1994):

1. a collection of primitive components, called "agents";
2. interactions among agents and between agents and their environment;
3. unanticipated global properties often result from the interactions;
4. agents adapt their behavior to other agents and environmental constraints;
5. as a consequence, system behavior evolves over time.

Thus, a complex adaptive system is a network of many agents working in parallel and making separate decisions, and these agents make mistakes and learn. The results of individual decisions are not averaged away and forgotten but may be magnified in impact as a result of other decisions and may decide the direction the system takes. At times the system rushes forward, sometimes it stands still, and sometimes it retreats. Many complex adaptive systems are characterized by what economists would call "increasing returns." In increasing-returns systems, small chance events may cause potentially inefficient or suboptimal technology to be chosen and to persist.

I believe that the households that formed the basic decision-making units of early farming society in riverine interior central Europe can be viewed as the primitive agents of a complex adaptive system. Yet they were not completely autonomous, for as Neolithic society developed they would have been connected by a network of kinship affiliations that reached well beyond the immediate environs of the residential location. Moreover, the developmental cycle of household establishment, life-span, and demise would have provided a continual dynamism to this system, which not only generated new agents but also provided them with motivations and aspirations which occasionally caused them to relocate and disperse. Individual household decisions about relocation would have echoed through this web of self-interested agents to result in the agricultural settlement of almost all of north-central Europe within a matter of centuries.[3]

It is also worth noting how increasing-returns phenomena might have played a role in the Neolithic dispersal in riverine interior central Europe. The economist W. Brian Arthur (1990) has pointed out that economic sectors that are subject to increasing returns are largely those that are knowledge-based, for the initial investment in knowledge acquisition is rewarded with a rapid accumulation of experience, which makes the process function more efficiently. The Neolithic settlement of riverine central Europe was effectively a knowledge-based process, for the major challenge would have been the identification of suitable habitats with the best soils, moisture, landforms, and microclimates. The frequency of mistakes, or the need to locate settlements tactically, would have dropped sharply as knowledge increased, and the use of a conservative technology and adaptive strategy also would have simplified the process.

My thoughts on the application of complex adaptive systems theory to the establishment of agricultural communities in riverine interior central Europe are still in the infant stage (see Bogucki 2000). Perhaps the next stage would be to interest colleagues who are more mathematically- and computationally-sophisticated to try to model this process using tools such as *Echo* or *Sugarscape* (Forrest and Jones 1994; Epstein and Axtell 1996). Such agent-based modeling tools are a considerable advance over modeling techniques that were applied to the Linear Pottery dispersal in the 1970s (e.g., Hamond 1981).

An aspect of the dispersal of early farming populations in riverine interior central Europe that has always confounded archaeologists is the apparent speed of this process. When viewed in the perspective of archaeological time, farming communities appear almost instantaneously from the middle Danube to Franconia, and after a short delay, quickly onward to the Rhine and the north European plain. It was not a steady and even diffusion, but rather it seems to have rushed forward in several spurts.

Recent advances in dispersal theory in plant ecology may provide an explanation for the appearance of rapidity (Kot et al. 1996; Pitelka et al. 1997; Clark et al. 1998). Many rapid biological invasions are characterized by great leaps forward and rare (and difficult-to-observe) long strides. Long-distance colonization events are undertaken by a small minority of the population, while the majority take small steps that determine the average dispersal distance. So while average dispersal distances may be small, the range of variation in relocation distances may include a few very long leaps (fig. 13.5). The participants in these long-distance relocations then begin to disperse locally in their new habitat and fill in some of the area that they jumped over. Under such conditions, the dispersal of an organism will not be a smooth and steady diffusion wave that is the cumulative effect of only short-distance moves but rather a rapid—sometimes even accelerating—spread whose rate may vary considerably from one area to another. This

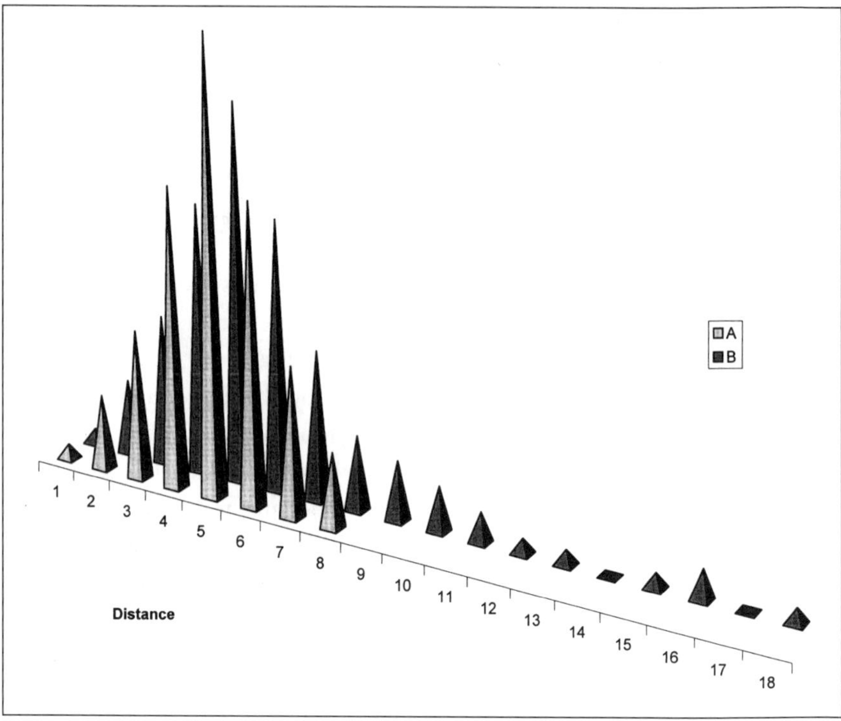

Fig. 13.5. Steady diffusion versus rapid dispersal. Steady and gradual diffusion is based on the assumption that the variation in dispersal distances will follow a normal distribution (A); if, however, the distribution is not normal and there are some examples of long-distance colonization events (B), the dispersal of a population appears, when viewed historically, to have occurred very rapidly.

body of diffusion theory helps to defuse the indigenist argument that the rapid establishment of Linear Pottery communities across central Europe can only be explained through the common adoption of agriculture by in situ peoples. It is also a promising dimension to any future attempts to model the establishment of farming communities in riverine interior central Europe.

Conclusion

The evidence still strongly indicates that the establishment of agricultural communities in central Europe was largely the result of the colonization of this region by migrating farming peoples. Recent indigenist arguments notwithstanding, the case for population movement seems even more compelling in light of the application of Rouse's Rules and the identification of the colonizing logic reflected in the choice of habitats and in technological conservatism. New approaches such as the complex adaptive systems paradigm and dispersal theory may prove useful in understanding this process.

REFERENCES

Ammerman, A.J. 1989. "On the Neolithic Transition in Europe: A Comment on Zvelebil and Zvelebil." *Antiquity* 63:162–5.

Arthur, W.B. 1990. "Positive Feedbacks in the Economy." *Scientific American* 262:92–9.

Bak, P., and K. Chen. 1991. "Self-Organized Criticality." *Scientific American* 264:46–53.

Beaton, J.M. 1991. "Colonizing Continents: Some Problems from Australia and the Americas." In *The First Americans: Search and Research*, edited by T.D. Dillehay and D.J. Meltzer, 209–30. Boca Raton, Fla.

Bogucki, P. 1979. "Tactical and Strategic Settlements in the Early Neolithic of Lowland Poland." *Journal of Anthropological Research* 35:238–46.

———. 1981. "Early Neolithic Economy and Settlement in the Polish Lowlands." Ph.D. diss., Harvard University.

———. 1982. *Early Neolithic Subsistence and Settlement in the Polish Lowlands. BAR-IS* 150. Oxford.

———. 1988. *Forest Farmers and Stockherders: Early Agriculture and Its Consequences in North-Central Europe.* Cambridge.

———. 1995. "The Linear Pottery Culture of Central Europe: Conservative Colonists?" In *The Emergence of Pottery: Technology and Innovation in Ancient Societies*, edited by W.K. Barnett and J.W. Hoopes, 89–97. Washington, D.C.

———. 1996. "The Spread of Early Farming in Europe." *American Scientist* 84:242–53.

———. 2000. "The Neolithic Settlement of Riverine Interior Europe as a Complex Adaptive System." Paper presented at the meeting of the Theoretical Archaeology Group, Oxford, England, December 2000.

Bogucki, P., and R. Grygiel. 1981. "The Household Cluster at Brzesc Kujawski 3: Small-Site Methodology in the Polish Lowlands." *World Archaeology* 13:59–72.

Clark, J.S., C. Fastie, G. Hurtt, S.T. Jackson, C. Johnson, G. King, M. Lewis, J. Lynch, S. Pacala, C. Prentice, E.W. Schupp, T. Webb III, and P. Wyckoff. 1998. "Dispersal Theory Offers Solutions to Reid's Paradox of Rapid Plant Migrations." *BioScience* 48:13–24.

Clay, R.B. 1976. "Tactics, Strategy, and Operations: The Mississippian System Responds to Its Environment." *Midcontinental Journal of Archaeology* 1:137–62.

Cowie, R. 1983. "The Production and Distribution of Linear Pottery Ceramics and Other Neolithic Pottery from Eastern Europe." M.A. thesis, University of Sheffield.

Czerniak, L. 1994. *Wczesny i Środkowej Okres Neolitu na Kujawach, 5400-3650 p.n.e.* Poznan.

———. 1998. “The First Farmers.” In *Pipeline of Archaeological Treasures*, edited by M. Chłodnicki and L. Krzyzaniak, 23–36. Poznan.

Dennell, R. 1992. “The Origins of Crop Agriculture in Europe.” In *The Origins of Agriculture*, edited by C.W. Cowan and P.J. Watson, 71–10. Washington, D.C.

Emery, F.M., and E.L. Trist. 1965. “The Causal Texture of Organizational Environments.” *Human Relations* 18:21–32.

Epstein, J.M., and R. Axtell. 1996. *Growing Artificial Societies: Social Science from the Bottom Up*. Washington, D.C.

Flannery, K.V., ed. 1976. *The Early Mesoamerican Village*. New York.

Forrest, S., and T. Jones. 1994. “Modeling Complex Adaptive Systems with Echo.” In *Complex Systems: Mechanism of Adaptation*, edited by R.J. Stonier and X.H. Yu, 3–20. Amsterdam.

Gabalowna, L. 1966. *Ze Studiów nad Grupa Brzesko-Kujawska Kultury Lendzielskiej*. Acta Archaeologica Lodziensia 14. Lodz.

Green, S.W., and M. Zvelebil. 1993. “Interpreting Ireland’s Prehistoric Landscape: The Bally Lough Archaeological Project.” In *Case Studies in European Prehistory*, edited by P. Bogucki, 1–29. Boca Raton, Fla.

Gronenborn, D. 1994. “Überlegungen zur Ausbreitung der bäuerlichen Wirtschaft in Mitteleuropa—Versuch einer kulturhistorischen Interpretation ältestbandkeramischer Silexinventare.” *Praehistorische Zeitschrift* 69:135–51.

———. 1997. *Silexartifakte der ältestbandkeramischen Kultur*. Bonn.

———. 1999. “A Variation on a Basic Theme: The Transition to Farming in Southern Central Europe.” *Journal of World Prehistory* 13:123–210.

Grygiel, R. 1986. “The Household Cluster as a Fundamental Social Unit of the Brzesc Kujawski Group of the Lengyel Culture.” *Prace i Materialy Museum Archeologicznego i Etnograficznego w Lodzi* 31:43–334.

———. 1994. “Untersuchungen zur Gesellschaftsorganisation des Früh—und Mittelneolithikums in Mitteleuropa.” In *Internationales Symposium Über die Lengyel-Kultur 1888–1988*, edited by P. Kosturik, 43–77. Brno-Lodz.

Grygiel, R., and P. Bogucki. 1997. “Early Farmers in North-Central Europe: 1989–1994 Excavations at Osaonki, Poland.” *Journal of Field Archaeology* 24:161–78.

Hamond, F. 1981. “The Colonisation of Europe: The Analysis of Settlement Process.” In *Pattern of the Past: Studies in Honour of David Clarke*, edited by I. Hodder, G. Isaac, and N. Hammond, 211–78. Cambridge.

Jazdzewski, K. 1938. “Cmentarzyska kultury ceramiki wstegowej i zwiazane z nimi slady osadnictwa w Brzesciu Kujawskim.” *Wiadomości Archeologiczne* 15:1–105.

Kind, C.-J. 1998. “Kompleze Wildbeuter und frühe Ackerbauern.” *Germania* 76:1–23.

Kot, M., M.A. Lewis, and P. van den Driessche. 1996. “Dispersal Data and the Spread of Invading Organisms.” *Ecology* 77:2027–42.

Kreuz, A. 1990. *Die ersten Bauern Mitteleuropas—eine archäobotanische Untersuchung zum Umwelt und Landwirtschaft der ältesten Bandkeramik*. Analecta Praehistorica Leidensia 23. Leiden.

Kruk, J., S.W. Alexandrowicz, S. Milisauskas, and Z. Snieszko. 1996. *Osadnictwo i Zmiany*

Srodowiska Naturalnego Wyzyn Lessowych: Studium Archeologiczne i Palaeogeograficzne nad Neolitem w Dorzeczu Nidzicy. Cracow.

Kruk, J., and S. Milisauskas. 1999. *Rozkwit i Upadek Spoleczenstw Rolniczych Neolitu.* Cracow.

Kukawka, S., and J. Malecka-Kukawka 1999. "The 'Long House' of the Late Band Pottery Culture from Bukowiec, Chelmno Land," *Sbornik Praci Filozofické Fakulty Mazarykovy Univerzity* M4, http://www.phil.muni.cz/archeo/sbornikm4/kukawka.html, viewed 24 April 2002.

Last, J. 1998. "The Residue of Yesterday's Existence: Settlement Space and Discard at Miskovice and Bylany." In *Bylany Varia I,* edited by I. Pavla, 17–46. Prague.

Louwe Kooijmans, L.P. 1993. "The Mesolithic/Neolithic Transition in the Lower Rhine Basin." In *Case Studies in European Prehistory*, edited by P. Bogucki, 95–145. Boca Raton, Fla.

Lüning, J. 1982. "Research into the Bandkeramik Settlement of the Aldenhovener Platte in the Rhineland." *Analecta Praehistorica Leidensia* 15:1–29.

———. 1988. "Frühe Bauren in Mitteleuropa im 6. und 5. Jahrtausend v. Chr." *Jahrbuch des Römisch-Germanischen Zentralmuseums* 35:27–93.

Nowak, M. 1997. "Regional Settlement Patterns of the Early Phases of the Eastern Linear Pottery Culture in the Eastern Slovakian Lowland." In *The Early Linear Pottery Culture in Eastern Slovakia,* edited by J.K. Kozłowski, 15–41. Cracow.

Pitelka, L., and the Plant Migration Workshop Group. 1997. "Plant Migration and Climate Change." *American Scientist* 85:464–73.

Price, T.D., ed. 2000. *Europe's First Farmers.* Cambridge.

Reitz, E.J. 1985. "Comparison of Spanish and Aboriginal Subsistence on the Atlantic Coastal Plain." *Southeastern Archaeology* 4:41–50.

Rouse, I. 1958. "The Inference of Migrations from Anthropological Evidence." In *Migration in New World Culture History*, edited by R.H. Thompson, 63–68. Tucson.

Sanger, D. 1975. "Culture Change as an Adaptive Process in the Maine-Maritimes Region." *Arctic Anthropology* 12:60–75.

Tillmann, A. 1993. "Kontinuität oder Diskontinuität? Zur Frage einer bandkeramischen Landnahme in südlichen Mitteleuropa." *Archäologische Informationen* 16:157–87.

Tringham, R., and D. Krstić. 1990. "Conclusion." In *Selevac: A Neolithic Village in Yugoslavia,* edited by R. Tringham and D. Krstić, 567–616. Los Angeles.

Verboom, W. 1977. "Lines of Weakness in the Forests, and Early Human Settlements." *ITC Journal* 1977 (3):531–44.

Whittle, A. 1996. *Europe in the Neolithic: The Creation of New Worlds.* Cambridge.

NOTES

[1] E.g., Whittle (1996, 150–2) suggests that the substantial Linear Pottery longhouses convey an unwarranted impression of sedentism, although the non-trivial investment of labor in such construction seems inconsistent with a mobile settlement system.

[2] An alternative explanation of this concentration on specific habitat types is presented by Angela Kreuz (1990, 251), who states that "the criteria that seem to have been employed in selecting the settlement sites imply a knowledge of environmental conditions that can be expected only of experienced farmers. Such farmers must have come from areas where crop husbandry and stock breeding were already practiced." This explanation would also support a colonization model rather than the indigenist position.

[3] See Bogucki (1988) for a discussion of the history of explanations for the dispersal of farming in Europe.

— 14 —

The Arrival of Agriculture in Europe as Seen from the North

T. Douglas Price

The call for papers for the conference that produced this volume emphasized an examination of the development of research—a look back over the course of past studies on the beginnings of agriculture in Europe and a look ahead to identify those areas where investigations might focus in the future. Part of such introspection is of course autobiographical. In this article I combine some thoughts on the development of my own ideas with new information pertinent to questions about the transition. I emphasize those aspects of the transition that appear to me most salient in the context of the conference. Finally, I hope to contribute to the more general debate of the introduction of agriculture in Europe by reviewing some of the evidence from southern Scandinavia.[1]

Retrospect

The beginnings of my journey lie in graduate school at the University of Michigan at the end of the 1960s. I was schooled there as part of the second generation of the New Archaeology. A group of Binford's students, freshly molded at the University of Chicago, had come as new faculty to the Museum of Anthropology as I was starting graduate school. There was fervor and conviction that the past could be understood through scientific methods and an emphasis on human ecology. That excitement was absorbed by my cohort of graduate students; we wandered the world in search of problems to solve.

I began my dissertation research in the Netherlands in 1970 with Robert Whallon. I had developed an interest in the Mesolithic in graduate school and the Netherlands provided a good opportunity to get deeper into the subject. I spent the next seven years involved with these investigations. Mesolithic sites in Holland and across the north European plain are normally very small scatters of flint, charcoal, and a hearth or two; few other remains have survived. This experience significantly influenced me. Discussions with Tjalling Waterbolk at Groningen expanded my view of the European Neolithic. My research here solidified my interest in the Mesolithic but, at the same time, made me aware that better preservation was essential for understanding prehistoric hunter-gatherers.

Attending the first International Mesolithic Congress in Warsaw in 1973, I met a number of Mesolithic archaeologists from northern and western Europe who were enthusiastically looking beyond classification and culture history. This group of "young turks" was different from the older typologists and "cartographers" at the congress. They wanted to know about settlement, subsistence, and behavior. I liked that. Among that group was Erik Brinch Petersen of the University of Copenhagen. We stayed in touch and I visited Denmark in 1976 to see some of the Mesolithic evidence there. I was astounded by both the quality and quantity of material. Preservation in the bogs and wetlands of Denmark was remarkable and offered a much bigger window on past Stone Age hunter-gatherers. Brinch and I began a collaborative excavation in the Vedbæk area north of Copenhagen in the early 1980s.

That experience made a strong impression. It was clear that coastal Scandinavia foragers created larger settlements and left behind a much greater variety and amount of material than groups in the inland areas of northwestern Europe. This initial Danish involvement stimulated my interest in more complex hunter-gatherers and was an inspiration for the edited volume, *Prehistoric Hunter-Gatherers: The Emergence of Cultural Complexity* in 1985. At that time, I argued that most of the accepted characteristics of farming populations (including sedentism, subsistence specialization, storage, and hierarchical organization, among others) were also present among more complex hunter-gatherer groups.

I came to the question of agricultural transitions rather late. It was, however, a natural progression. I was interested in the development of foraging society. The final act of that development was, in most parts of the world, the transition to agriculture. This focus on the transition was no doubt influenced by my fascination with Denmark and Danish archaeology. Some of the most visible reminders of prehistory in this area are the marvelous megalithic tombs scattered across the landscape. These tombs are banners of the transition and force one to consider the process.

Discussions with Lisbeth Pedersen of the Kalundborg Museum in western Sjælland, Denmark, and Anders Fischer in the mid 1980s led to plans for a new project in a local region known as the Saltbæk Vig. This project of archaeological survey, testing, and excavation was focused on the transition from the Mesolithic to the Neolithic. Around the same time, I met a Danish archaeologist specializing in the Neolithic, Anne Birgitte Gebauer, and convinced her to work with me.

Gitte and I began the Saltbæk Vig project in 1989. Our work in this area over the last 10 years has brought new insights regarding the transition and several publications on the subject (e.g., Gebauer and Price 1992). This project—the fieldwork and the interaction with other archaeologists in Denmark—has been a constant source of inspiration, of new questions and new ideas. The research has forced me to rethink previous ideas about hunter-gatherers and to be more specific about the nature of change in human society. The following pages then are a condensation of some of what I have learned and how my path and experience has led to my current understanding of the transition to agriculture in southern Scandinavia. I am fortunate to be working in the area where a long history of archaeological research, combined with a richness of both archaeological and natural historical information, makes it possible to examine questions concerning the transition in some detail.

Scandinavia in the Context of Europe

Although southern Scandinavia lies at the margins of cultivation—perhaps the last place agriculture reached in Europe—information from this region reveals much about the nature of the transition across the continent. This region is likely one of the very few places in Europe that supported a substantial indigenous population prior to the transition to agriculture (Price 2000a). This point is important, and I want to digress to emphasize it. With only a few exceptions, most of the continent contains little or no indication of occupation during the period just prior the transition to agriculture. Archaeological evidence for human population is admittedly difficult to obtain, but in recent years a number of systematic surveys have been undertaken in various parts of Europe that provide some sense of the intensity, or lack thereof, of occupation during the Mesolithic.

The Mediterranean islands, with only one or two exceptions, were not colonized until the Neolithic (Cherry 1990). Extensive surveys in Greece and parts of southeastern Europe have failed to record more than a handful of settlements from the Mesolithic (e.g., Bintliff and Snodgrass 1985; Runnels 1996). These surveys have covered hundreds of square kilometers in southern, central, and northern Greece with the specific goal of locating Mesolithic occupation. The possibility exists that Mesolithic settlements now lie deep beneath alluvial plains (Runnels 1997) or submerged along the former shore of the Mediterranean (Shackleton et al. 1984), but these are as yet undiscovered and unknown.

Elsewhere along the Mediterranean, late Mesolithic sites are not at all common. Surveys in southern and central Italy have revealed very little settlement (Martini and Tozzi 1996). Pluciennik (1997) has pointed out that there are very few Mesolithic sites in the south of Italy. A distinct 1,000-year gap in radiocarbon dates exists between the oldest Mesolithic sites and the earliest Neolithic ones. The Acconia Survey in Calabria, Italy was one of the more systematic and intensive survey projects undertaken in Europe (Ammerman 1985; Ammerman and Shaffer 1981). In this region, there have been repeated surveys of the same areas, undertaken in conjunction with land use studies (Ammerman 1995). A large number of Early Neolithic Impressed Ware sites have been recorded, but no Mesolithic materials encountered.

Biagi and others (Biagi 1991, forthcoming; Biagi et al. 1993) have compiled site distribution information for northern Italy. One hundred and ten Mesolithic sites are known from this area, with the majority in the region of the Trieste Karst and the pre-Alpine and Alpine zone. Many of these are small, high altitude camps. The Po valley is practically devoid of Mesolithic remains, and large parts of Liguria to the west contain no evidence of early Holocene occupation (Biagi and Maggi 1994). A similar situation is seen in neighboring areas of Provence in France (Binder 1987; 2000).

This pattern is also observed along the coast of Spain and Portugal where the distribution of late Mesolithic and early Neolithic sites does not overlap (Zilhão 1993, 2000). There are no known late Mesolithic sites in the Estremadura of Portugal, for example, precisely the area where the earliest

Neolithic remains are found. Late Mesolithic sites, best known from large shell middens, are concentrated to the south in the major river valleys of Portugal.

Moving north from the Mediterranean region to the distinctly inland areas of central Europe, surveys confirm a relative absence of later Mesolithic population. In southwestern Germany, for example, there appears to be a substantial decline in human population during the late Mesolithic (Jochim 1998), a few tens of sites compared to several hundred in the early Mesolithic. Surveys in other major river valleys in central and western Europe report Early Neolithic materials, but Mesolithic artifacts are rarely encountered (e.g., Bogucki 1988). Occupation from the later Mesolithic is also not well documented in the United Kingdom and Ireland. Woodman (2000) has noted a substantial decline in the number of radiocarbon dates in the Late Mesolithic of the British Isles, particularly in Ireland.

The Transition to Agriculture in Southern Scandinavia

Elsewhere in Europe it is difficult to view this process of transition. Northern Europe is, in fact, one of the few areas where evidence is found for substantial indigenous population at the time of the introduction of agriculture. The important questions concerning the transition in southern Scandinavia are basic ones: What happened, when did it happen, how did it happen, who was involved, and why did it take place? Because of the rich archaeological record in southern Scandinavia, the what and when questions regarding the transition to agriculture are largely resolved. The answers to who and how are still debated, but the answers seem reasonably clear. The why question, as usual, remains more intractable.

When and What Arrived

The when and what questions are relatively straightforward. Mesolithic foragers were present in northern Europe from approximately 8500 B.C.—shortly after the close of the Pleistocene—about the same time plants and animals were being domesticated in Southwest Asia. By 5000 B.C., and probably earlier, these groups in southern Scandinavia were intensively occupying the coastal areas, living in permanent villages characterized by deep middens of shell and refuse and substantial cemeteries (Larsson 1990; Price 1991). Specialized hunting camps were scattered along the coasts and inland around these residential sites. These groups exploited the resources of both the land and the sea, successfully capturing large quantities of fish, shellfish, marine mammals, large and small game, and birds. In addition, hazel nuts and acorns along with a variety of other plant foods contributed to the diet. The Mesolithic is characterized by an elaborate technology of equipment and facilities, boats and paddles, bows and arrows, art and decoration, and domesticated dogs. Pottery appeared in the late Mesolithic—borrowed from farmers to the south.

Agriculture, in the form of domesticated plants and animals, arrived in southern Scandinavia shortly after 4000 B.C. These crops and herd animals are associated with the arrival of the early Neolithic Funnel Neck Beaker (TRB) culture, which likely originated in Poland ca. 4400 B.C. and spread to northern Germany by 4200 B.C. The TRB is characterized by distinctive pottery styles, the production of polished flint axes, and new forms of burial and sacrifice (Koch 1998). Settlements during the early Neolithic typically are small and ephemeral, and pollen evidence indicates that cereals were not common and land clearance was minor. Substantial evidence for domesticates and land clearance does not appear until later in the Neolithic (Andersen 1992).

How Agriculture Arrived

The question of how agriculture arrived in southern Scandinavia involves both what was introduced and the rate of spread. The elements of the Neolithic—domesticated plants and animals, pottery, polished flint axes, and new burial patterns—appear as a package in Scandinavia.

Evidence from the site of Bjørnsholm in northern Jutland (fig. 14.1) is typical for very early Neolithic sites in southern Scandinavia (Andersen 1991; Andersen and Johansen 1992). The site is of particular importance because of the presence of a shell midden and settlement remains from both the late Mesolithic and early Neolithic periods, along with evidence for an earthen long barrow. The Ertebølle layers at Bjørnsholm date from 5050 to 4050 B.C. and the early Neolithic from 3960 to 3530 B.C. The location of residence clearly did not change from the Mesolithic to the Neolithic; in fact, residence continued in virtually the same spot following a 90-year gap in the occupation record. The latest Mesolithic occupations are enormous, extending more than 300 m along the coastline of the fjord and 10–50 m in width. The earliest Neolithic midden, containing funnel beaker pottery, is

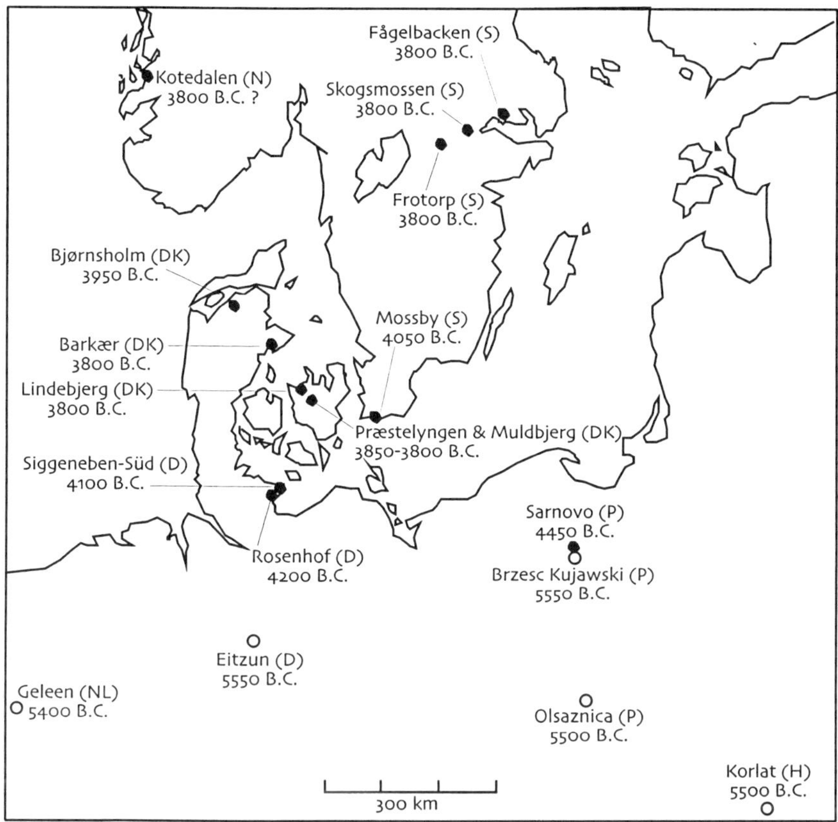

Fig. 14.1. Map of northern Europe and approximate dates (calibrated radiocarbon years B.C.) for selected Linearbandkeramik (LBK–open circle) and Trichterbecher (TRB) early Neolithic sites

smaller and a meter or so lower on the coastline, in response to lowering sea levels. There was a shift from oysters to cockles in the shell midden around the time of the transition, suggesting that changes in the environment may have been taking place. The earliest Neolithic pottery and polished axe fragments, however, were present in the sequence before this shift in the midden occurred (Andersen and Johansen 1992).

The evidence for Early Neolithic settlement at Bjørnsholm comes from two small scatters of artifacts, in close association with an earthen long barrow. Neolithic subsistence remains were very similar to those from the Mesolithic, with the addition of small amounts of wheats, barley, sheep, cattle, and pig (Bratlund 1993). The long barrow held a massive timber chamber with a single burial furnished with a large thin butt polished flint axe and a diabase copy of a central European copper axe (Andersen and Johansen 1992). It very much appears that the major hallmarks of the Neolithic are all found together at the earliest sites of this period.

In terms of the rate of change, the transition to farming in southern Scandinavia was both rapid and gradual. From the perspective of radiocarbon dates and the spread of domesticates, TRB pottery, and polished flint axes, the transition to agriculture in southern Scandinavia was very rapid. Neolithic materials spread quickly across southern Scandinavia to the limits of cultivation (Price 2000b). Radiocarbon dates from southern Jutland, between 4000 and 3800 B.C., are essentially indistinguishable from the dates for earliest TRB in the Stockholm area of Sweden, a distance of more than 800 km (fig. 14.1).

On the other hand, the initial introduction of domesticates into southern Scandinavia was delayed by more than 1,000 years. The first farmers in northern Europe were Linear Pottery groups that spread quickly around 5500 B.C. across the loess lands of central Europe and to the north. The margins of the distribution of the LBK lie inland a few tens of kilometers from the shore of the Baltic and North Seas (fig. 14.1). The spread of this first farming adaptation stopped short of the coasts. Hunter-gatherers living on those coasts resisted the ideas and products of the Neolithic. Certain components were borrowed: Mesolithic pottery, antler axes and combs, bone rings, shoe-last adzes made in central or southeastern Europe, and perhaps copper (e.g., Fischer 1982; Klassen 1997). Clearly there was exchange among foraging and farming groups, but the primary hallmarks of the Neolithic did not enter the region until 4000 B.C. From this perspective, the transition to farming was a slow process, likely because of the success of foraging adaptions in coastal areas.

The transition was also a slow process in terms of the adoption of a Neolithic way of life (fig. 14.2). As noted above, sedentary foragers were established in southern Scandinavia by at least the beginning of the late Mesolithic or Ertebølle period dating to ca. 5400 B.C. During the latter half of this period, certain exports from Neolithic farmers to the south began to appear in Mesolithic contexts. Between 4000 and 3800 B.C. the first domesticated plants and animals, Funnel Beaker pottery, and polished flint axes began to appear as the transition to the early Neolithic TRB took place. During this time, however, there is very limited evidence of land clearance,

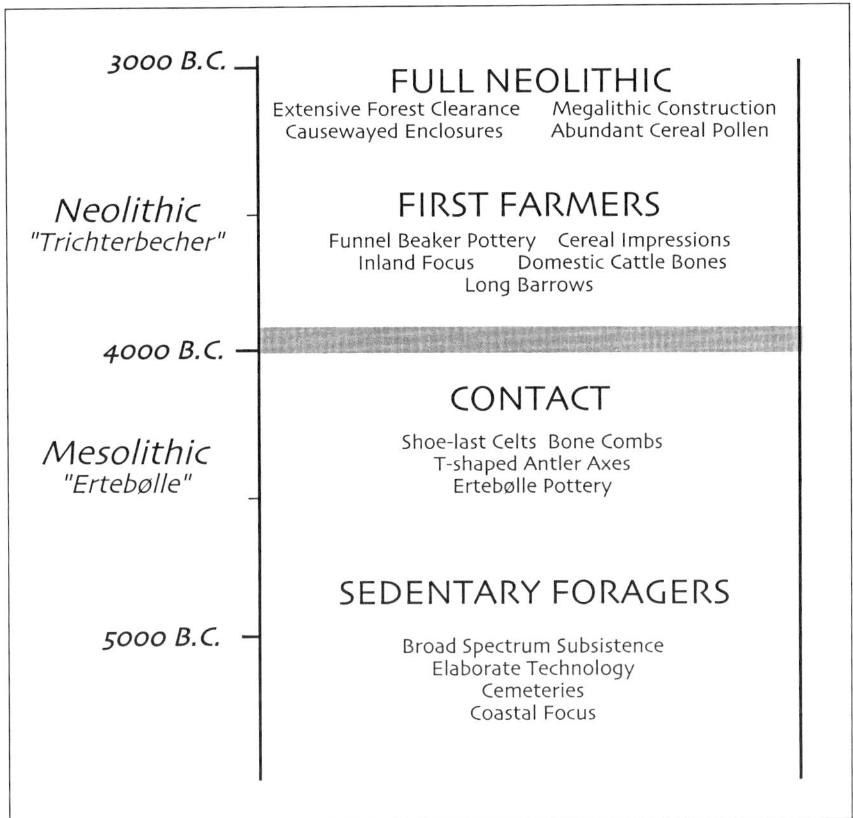

Fig. 14.2. The Mesolithic-Neolithic transition in southern Scandinavia

cultivation, or herding; the settlements of hunter-gatherers continued in use, and wild foods appear to dominate the diet. It is not until the later part of the Early Neolithic that substantial evidence is found for land clearance, cultivation, and larger, inland villages associated with TRB farmers. Thus, it appears that the transition from Mesolithic foraging to Neolithic farming actually took more than 2,000 years, from approximately 5500 B.C. until 3300 B.C. From the perspective of settlement and subsistence then, the transition to agriculture was very slow indeed.

Who Was Involved

The who question concerns the populations that participated in the transition in southern Scandinavia: Were they local or foreign? Was there colonization or the local adoption of farming knowledge, technology, and materials? In many parts of Europe, particularly in the Mediterranean region, it is clear that agriculture was spread primarily through the arrival of new people. In Scandinavia, however, there was a substantial indigenous population of hunter-gatherers and a lengthy period of interaction between these groups, before the introduction of agriculture. The evidence both for and against colonization comes largely from material culture and shows indications of both continuity and change, depending upon perspective. Arguments in favor of colonization cite changes such as the simultaneous introduction of the Neolithic package (e.g., Solberg 1989). Similar lines of evidence—lithic and ceramic technology, settlement location, and burial practice—support an argument for continuity in population and indigenous adoption.

Neolithic flint technology appears to be largely derived from the Ertebølle (Nielsen 1985; Stafford 1999). Throughout much of southern Scandinavia the only way to distinguish between late Mesolithic and early Neolithic surface sites is the presence of fragments of polished flint axes. All of the technology for the production of these axes is present in the Mesolithic (Stafford 1999; Nordqvist 1991). The earliest polished flint axes, the pointed butt forms, morphologically resemble late Mesolithic examples. Indeed, some of the specialized core axes from the last part of the Ertebølle period have polished edges.

The pottery tradition of the Funnel Beaker Culture, and perhaps the technology of food preparation, are relatively new in the Neolithic. The funerary pottery of TRB reflects clear innovation in shape and decoration. Only minor differences exist, however, between the Ertebølle ceramics and the utilitarian wares of the early TRB; the two groups of pottery can be seen as developmental stages within the same tradition of pottery production and there is an intermediate form (Gebauer 1995; Koch 1998). The earliest sites with TRB pottery are often at the same location as the latest Mesolithic sites, emphasizing continuity in settlement (Andersen 1991; Koch 1994). In addition to residential sites, hunting and fishing stations continue in use from the late Mesolithic well into the early Neolithic (e.g., Fischer 1991; Skaarup 1973). Clearly the initial TRB settlement follows the pattern of the latest Mesolithic. At the same time, the first ephemeral, inland settlements began to appear.

Compared to the previous Ertebølle period, the monumental earthen long barrows of the TRB culture are certainly a new phenomenon. These

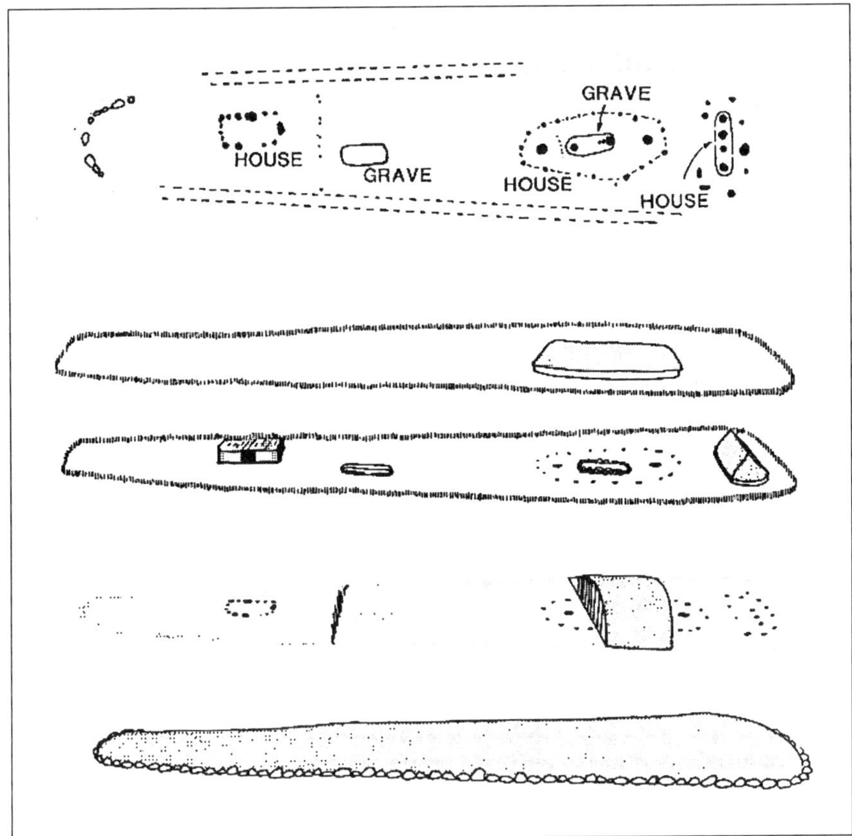

Fig. 14.3. Earthen long barrows were monumental tombs of timber, earth, and stone, constructed in stages. Plan view (top), and four stages of construction. (After Jensen 1982)

graves were framed by a timber palisade and covered by a long, linear mound (fig. 14.3); however, the fully extended position of the body in the central tomb resembled previous Ertebølle practices. The grave goods consisted of similar kinds of artifacts, translated into different materials: amber beads instead of tooth beads, flint axes or a stone battle axe instead of antler axes, pots instead of bark containers (Koch 1998). Part of the burial ritual thus seems to be a continuation of the Ertebølle tradition. The simple, flat graves of the early Neolithic in Eastern Denmark are largely indistinguishable from the late Mesolithic.

The physical anthropology of the Neolithic remains differs only slightly from their Mesolithic predecessors. There are several hundred radiocarbon-dated Mesolithic skeletons and more than 50 early Neolithic individuals (Bennike and Ebbensen 1987). Several minor differences are seen in the early Neolithic skeletons; bones and skulls are less robust and the teeth are smaller (Bennike 1993), but the overall impression is one of continuity rather than replacement.

The bulk of the evidence—from continuity in lithic and ceramic technology, settlement location and distribution, burial patterns, and physical anthropology—suggests that local Mesolithic foragers became the first farmers in southern Scandinavia. While there are substantial changes with the introduction of the Neolithic, and while some new people may have arrived, the predominant theme of the transition in this area is of indigenous adoption rather than colonization.

Why Agriculture Came to Southern Scandinavia

The question of why populations of hunter-gatherers became farmers is one of the most difficult in archaeological research. Most attempts at explanation invoke either external pressures forcing human populations to change or else conscious decisions internal to society. External causes involve forces of nature such as climate change, deterioration in the environment, or inherent population growth. Internal causality involves human choice and social change.

In southern Scandinavia, agriculture arrives at the time of the transition between the Atlantic and Subboreal climatic episodes, a decline in elm trees, and a series of marine transgressions and regressions. One early model for the introduction of agriculture into Scandinavia was closely tied to the decline of the elm forest that Iversen (1941) identified at the end of the Atlantic climatic episode. A shift toward cooler, moister conditions was argued to have caused the marked decrease in elm and created conditions favoring the introduction of agriculture. More recent investigations of the elm decline have convincingly connected this event with the spread of disease, not climatic change (Groenman-Wateringe 1983; Peglar 1993).

Some crisis in resource availability in southern Scandinavia at the time of the transition is implied in a number of articles. Rowley-Conwy (1984), for example, has argued that changes in water levels and salinity reduced the availability of shellfish, causing a food shortage. However, late Mesolithic shell middens are found only in limited areas of southern Scandinavia; shellfish were not important in the diet in many regions and certainly not in much of the Baltic.

Strand Petersen (1992) has documented a decrease in the tidal range and water salinity of the North Sea at the end of the Atlantic period. Others (Larsson and Larsson 1991, Larsson et al. 1992) have argued that stable sea level and marine erosion resulted in the closure of marine estuaries and a decrease in the amount of biomass at the end of the Atlantic climatic episode. A change in water levels in the Kattegat and Baltic would reduce the availability of fish and sea mammals in the long fjords and straits of southern Scandinavia. There is an observed shift from oysters to mussels or cockles from the Mesolithic to the Neolithic layers in shell middens in Jutland (Andersen 1991). As noted earlier, however, Neolithic materials predate the change in marine resources at sites such as Bjornsholm in Denmark.

Clearly, some changes in the environment do take place at the end of the Atlantic episode. Yet there is no substantial evidence to indicate that these had a profound effect on human population. Other evidence exists to suggest that environmental change and food stress were not significant factors in the transition to agriculture. Several climatic indicators, including a species of mistletoe and tortoise today found only in central and southern Europe, were present in Denmark during the Subboreal period, indicating that conditions remained warmer than today (Troels-Smith 1960). Food stress is not seen in skeletal material from the late Mesolithic; individuals were large and robust with few indications of nutritional deficiency (Meiklejohn and Zvelebil 1991). It is also notable that the TRB expanded, one could say exploded, across the southern one-third of Scandinavia into a variety of different environments. Such evidence strongly argues against climatic or environmental factors in the expansion of the Neolithic.

Evidence on human population size in the Mesolithic of Scandinavia is difficult to obtain, but some impressions are available from archaeological surveys and from the distribution of radiocarbon dates. Several survey projects in Northern Europe point to the predominance of coastal settlement. The Ystad project in southernmost Sweden, for example, was carried out in the 1980s as a regional study of the cultural landscape from 4000 B.C. to the present (Berglund 1991; Larsson et al. 1992). Mesolithic settlement was found at lagoons and river mouths along the coast, with little evidence inland. Fewer than 10 settlements from the Mesolithic were reported from the study area; late Mesolithic sites were poorly represented. During the first part of the early Neolithic, settlements were smaller and more widely distributed in sandy soils. These sites are found in two zones: along the coastal strip and in a hummocky, inland landscape. There were more early Neolithic settlements than late Mesolithic, but the total number was no more than 10.

The Saltbæk Vig project was conducted along the coast and immediately adjacent areas of a small inlet near the town of Kalundborg, Denmark (Price and Gebauer 1992). More than 18 km^2 was intensively surveyed and more than 350 localities were mapped and collected. There are more sites from the Early and Middle Ertebølle than there are from the late phase. Specialized core axes, a hallmark of this late phase, occur on no more than seven of the more than 50 Ertebølle sites. There are a number of sites from the early Neolithic, but the vast majority date to the later phase of the Neolithic. There are virtually no Mesolithic settlements away from the seacoast in this area, while early Neolithic sites are found almost exclusively inland. Population size appears to have been small.

In addition, radiocarbon dates can provide an indirect source of information on human population. Figure 14.4 shows the number and distribution of radiocarbon dates from human skeletal material in the period between 6000 and 2000 B.C. calibrated (Persson 1998). The upper curve shows all dates and the lower, darker line is based on averages for individual sites. Several trends can be seen in this information. The overall population of the Neolithic, as represented by dated settlements, appears to be three to four times greater than during the Mesolithic. The major increase in population clearly comes after the beginning of the Neolithic, ca. 4000 B.C. In addition, there appears to be a decline in population in the later Mesolithic, during the fifth millennium B.C., prior to the introduction of agriculture. On the basis of this evidence, it seems to be clear that substantial population growth is a consequence, rather than a cause, of the transition to farming.

In southern Scandinavia there is no clear relationship between changes in population, climate, or environment and the transition to agriculture. Climatic and environmental data for the region do not demonstrate a correlation in timing with the transition to agriculture. Climate changes were gradual and witnessed only slight changes in the seasonal patterns of temperature and precipitation. Changes in the environment such as the elm decline, the disappearance of oysters, or the closing of estuaries cannot be directly related to the introduction of agriculture. Information on population numbers, based on archaeological survey and radiocarbon dates, indicate that population was not substantial compared to the Neolithic and may even have declined toward the end of the Mesolithic.

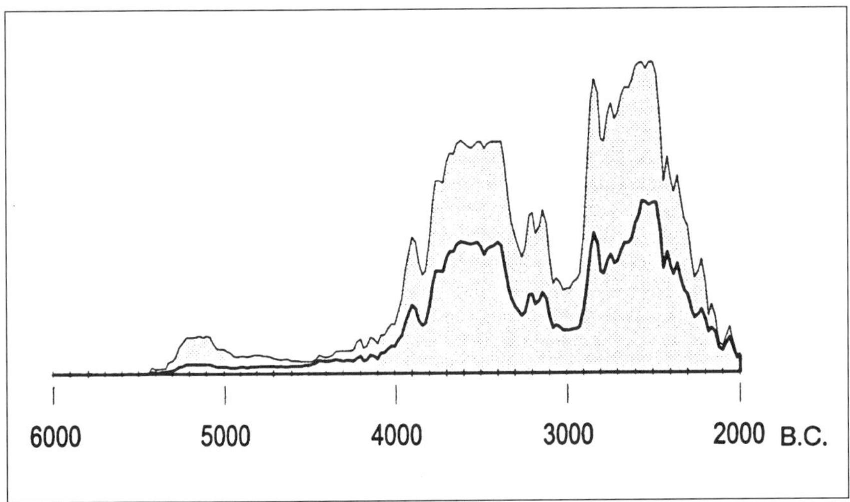

Fig. 14.4. Distribution of calibrated radiocarbon-dated graves from the Mesolithic and Neolithic of Denmark and Sweden. The lighter line indicates all dates; the heavier line indicates dated sites. (From Persson 1998)

In southern Scandinavia the transition is one of adoption, not colonization; changes in subsistence, settlement, and technology are gradual. The major innovations seen are the introduction of monumental tombs, the production of polished axes for trade, and the sacrifice of humans, animals, and other food in the lakes and bogs. These changes suggest that differences between the Mesolithic and the Neolithic lie not in nature, but rather in the social, economic, and religious domains of human society.

The appearance of barrows at the onset of the Neolithic indicates that substantial changes in social organization took place at the time of the transition. Nothing like these elite, mound-covered internments are known from the Mesolithic period. Obviously, a limited segment of the Neolithic population was entitled to monumental burial. This pattern continued during the construction of the megalithic monuments after 3500 B.C. The tombs appear to have been built originally for only a few inhabitants and only later became communal burial places (Skaarup 1988).

Changes in economic organization are also apparent in the early Neolithic (e.g., Price et al. 1995; Perlès 1992). Differences in the amounts of raw materials and exotic goods in Mesolithic and Neolithic contexts likely reflect shifts in the organization of both productivity and exchange. Specialization of production at the settlement level may well be characteristic of the early Neolithic. Evidence for such specialization is seen in several realms, including large-scale fishing, variation in site activities, and access to raw materials (Pedersen 1997). Long distance exchange systems characterize the Early Neolithic; these networks expand dramatically in extent and in the amounts and types of materials moving such as flint, ground and polished axes, amber, and other materials (Larsson 1988). Exotic imports include copper ornaments, jewelry, and axe blades (Klassen 1997; Randsborg 1975, 1979).

Community specialization and long-distance exchange in prestige items were important aspects in the introduction of agriculture in Scandinavia (Simonsen 1975). The simultaneous appearance of status differentiation, sacrifice at special centers, community specialization in production, and trade and exchange of both local and exotic materials reflect a largely new pattern of social and economic organization and belief systems that may be more important hallmarks of the "Neolithic Revolution" than the domestication of plants and animals and new subsistence strategies.

Conclusions

There are a number of lessons to be learned from a study of the Neolithic transition in southern Scandinavia. Perhaps foremost is the recognition that one's experiences strongly determine one's perceptions. I came to the Neolithic from the bottom up, so to speak, with an initial interest in hunter-gatherers, and only later involvement with the transition to agriculture. I suspect there are some advantages in this perspective. For example, instead of looking for origins, I place more emphasis on change and transition. I tend to seek indigenous adoption in the archaeological record rather than the arrival of new folks. In spite of my upbringing among New Archaeologists and human ecologists, I have for some time found explanations that depend upon changes external to human society to be rather abstract and peripheral. To my mind, human population growth is not the cause of anything. Certainly numbers of people provide essential conditions for change but population growth per se is largely a consequence of change in society rather than a cause. Similarly, climate and environmental changes cannot directly bring about fundamental change in human society; there are few mechanisms other than major natural catastrophes that directly influence social organization and structure.

The rather detailed evidence in Scandinavia has led me toward the work of Bender, Hayden, and others of their ilk in the search for causality in the transition to agriculture. I am hard pressed to find evidence for hereditary inequality in human society prior to the origins of agriculture. I envision such structured hierarchical arrangements often spreading with farming (Price 1995). At the same time it is clear that there are dramatic, concurrent changes in religious and economic organization.

My experience in Denmark has led me to rethink questions of colonization versus indigenous adoption as well. A volume entitled *Europe's First Farmers* (Price 2000c) expounds this view. Traditional views of the spread of agricultural as monolithic, either through colonization or indigenous adoption, are far too simplistic. There is little question that the spread took place in different ways in different areas. In the southern half of Europe, much of the evidence indicates that colonization was a principal mechanism. In the north, indigenous adoption may be a more common mode for the transition. Even LinearBandKeramik (LBK), the classic case for agricultural colonization in Europe, is being reconsidered. Recent investigations in central Europe have postulated a greater role for Late Mesolithic inhabitants in the transition and at times challenging the entire notion of an agricultural colonization (e.g., Gronenborn 1994; Lüning 1997; Tillman 1994).

It is also intriguing to note that the remarkable speed of the expansion of LBK, often cited as a characteristic of migration and colonization, is matched in areas such as Scandinavia where indigenous adoption took place. The rapid spread of agriculture is not necessarily associated with colonization. The evidence for the spread of TRB across Southern Scandinavia certainly suggests that rapid spreads can occur in the context of indigenous adoption as well. The radiocarbon dates for the first Funnel Beaker Pottery across southern and central Scandinavia all fall within ± 100 years of 3900 B.C. This is food for further thought.

Certainly the answer to questions about causality in the transition to agriculture is not yet at hand. For the present, it is useful to view the Neolithic as a set of social, economic, and ideological information and material, borrowed or carried, that spread in fits and starts across the European continent. It seems more appropriate that we understand the larger global phenomenon of the origins and spread of agriculture in terms of factors internal to human society. The Scandinavian evidence suggests that we need to accept that climate, environment, and population growth played a minor role, if any, in comparison to the decisions that human groups made with regard to their livelihood and way of life. In this context, the Neolithic transition had less to do with subsistence and technology and more to do with social and economic organization and ideology. I believe that the answer to questions about causality will be found in these arenas in the coming years.

REFERENCES

Ammerman, A.J. 1985. *The Acconia Survey. Neolithic Settlement and the Obsidian Trade.* London.

———. 1995. "The Dynamics of Modern Land Use and the Acconia Survey." *Journal of Mediterranean Archaeology* 8:77–92.

Ammerman, A.J., and G.D. Shaffer. 1981. "Neolithic Settlement Patterns in Calabria." *Current Anthropology* 22:430–2.

Andersen, S.H. 1991. "Bjornsholm: A Stratified Kokkenmodding on the Central Limfjord, North Jutland." *Journal of Danish Archaeology* 10:59–96.

Andersen, S.H., and E. Johansen. 1992. "An Early Neolithic Grave at Bjørnsholm, North Jutland." *Journal of Danish Archaeology* 9:38–59.

Andersen, S.T. 1992. "Early and Middle Neolithic Agriculture in Denmark: Pollen Spectra from Soils in Burial Mounds of the Funnel Beaker Culture." *Journal of European Archaeology* 1:153–80.

Bennike, P. 1993. "The People." In *Digging into the Past: 25 Years of Danish Archaeology,* edited by S. Hvass and B. Storgaard, 34–9. Aarhus.

Bennike, P., and K. Ebbesen. 1987. "The Bog Find from Sigersdal. Human Sacrifice in the Early Neolithic." *Journal of Danish Archaeology* 5:83–115.

Berglund, B.E., ed. 1991. *The Cultural Landscape during 6000 Years in Southern Sweden—The Ystad Project.* Copenhagen.

Biagi, P. 1991. "The Prehistory of the Early Atlantic Period Along the Ligurian and Adriatic Coasts of Northern Italy in a Mediterranean Perspective." *Rivista di Archeologia* 15:46–54.

———. 1999. "Some Aspects of the Late Mesolithic and Early Neolithic in Northern Italy." In *At the Fringes of Three Worlds,* edited by R. Kertesz and J. Makkay. Szolnok, Hungary.

Biagi, P., and R. Maggi. 1994. "Aspects of the Mesolithic Age in Liguria." *Preistoria Alpina* 19:159–68.

Biagi, P., E. Starnini, and B.A. Voytek. 1993. "The Late Mesolithic and Early Neolithic Settlement of Northern Italy: Recent Considerations." *Porocilo o raziskovanju paleolita, neolita in eneolita v Sloveniji* 21:45–67.

Binder, D. 1987. *Le Néolithique Ancien Provençal: Typologie et technologie des outillages lithiques.* Paris.

———. 2000. "Mesolithic and Neolithic Interaction in Southern France and Northern Italy: New Data and Current Hypotheses." In *Europe's First Farmers,* edited by T.D. Price, 117–43. Cambridge.

Bintliff, J.L., and A. Snodgrass. 1985. "The Cambridge/Bradford Boetia Expedition: The First Four Years." *Journal of Field Archaeology* 12:123–61.

Bogucki, P. 1988. *Forest Farmers and Stockholders.* Cambridge.

Bratlund, B. 1993. "The Bone Remains of Mammals and Birds from the Bjørnsholm Shell-Mound: A Preliminary Report." *Journal of Danish Archaeology* 10:97–104.

Cherry, J.F. 1990. "The First Colonization of the Mediterranean Islands: A Review of Recent Research." *Journal of Mediterranean Archaeology* 3:145–221.

Fischer, A. 1982. "Trade in Danubian Shaft-Hole Axes and the Introduction of Neolithic Economy in Denmark." *Journal of Danish Archaeology* 1:7–12.

———. 1991. *Store Åmose—Danmarks hidtil største kulturhistorisk begrundede fredninssag.* Hørsholm.

Gebauer, A.B., 1995. "Pottery Production and the Introduction of Agriculture in Southern Scandinavia." In *The Emergence of Pottery,* edited by W.K. Barnett and J.W. Hoopes, 99–112. Washington, D.C.

Gebauer, A.B., and T.D. Price. 1990. "The End of the Mesolithic in Eastern Denmark: A Preliminary Report on the Saltbæk Vig Project." In *Contributions to the Mesolithic in Europe*, edited by P.M. Vermeersch and P. van Peer, 259–280. Leuven.

———, eds. 1992. *Transitions to Agriculture in Prehistory.* Madison.

Groenman-van Wateringe, W. 1983. "The Early Agricultural Utilization of the Irish Landscape: The Last Word on the Elm Decline." In *Landscape Archaeology in Ireland,* edited by T. Reeves-Smyth and F. Hammond, 217–32. Oxford.

Gronenborn, D., 1994. "Überlegungen zur Ausbreitung der bäuerlichen Wirtschaft in Mitteleuropa—Versuch einer kulturhistorischen Interpretation ältestbandkeramischer Silexinventare." *Praehistorische Zeitschrift* 69 (2):135–51.

Iversen, J., 1941. *Landnam i Danmarks Stenalder: En pollenanalytisk Undersøgelse over det første Lantbrugs Indvirkning paa Vegetationsudviklingen.* København.

Jensen, J. 1982. *The Prehistory of Denmark.* London.

Jochim, M. 1998. *A Hunter-Gatherer Landscape.* New York.

Klassen, L. 1997. "Die Kupferfunde der Nordgruppe der Trichterbecherkultur." *Archäologische Informationen* 20:189–93.

Koch, E., 1994. "Typeneinteilung und Datierung frühneolithischer Trichterbecher aufgrund ostdåanischer Opfergefässe." In *Beiträge zur Frühneolithischen Trichterbecherkultur im westlichen Ostseegebiet*, edited by J. Hoika and J. Meyers Balke, 165–93. Neumünster.

———. 1998. *Neolithic Bog-Pots from Zealand, Møn, and Lolland-Falster, Denmark.* Copenhagen.

Larsson, L. 1988. "The Use of the Landscape during the Mesolithic and Neolithic in Southern Sweden." In *Archeology en Landschap: Bijdragen aan het gelijknamige symposium gehouden op 19 en 20 oktober 1987, ter gelegenheid van het afscheid van H.T. Waterbolk,* 31–48. Groningen.

———. 1990. "The Mesolithic of Southern Scandinavia." *Journal of World Prehistory* 4:257–310.

Larsson, L., and M. Larsson. 1991. "The Introduction and Establishment of Agriculture." In *The Cultural Landscape during 6000 Years in Southern Sweden—The Ystad Project,* edited by B.E. Bjerglund, 293–314. Lund.

Larsson, L., J. Callmer, and B. Stjernquist, eds. 1992. *The Archaeology of the Cultural Landscape.* Lund.

Larsson, M. 1992. "The Early and Middle Neolithic Funnel Beaker Culture in the Ystad Area (Southern Scania): Economic and Social Change 3100–2300 B.C." In *The Archaeology of the Cultural Landscape*, edited by L. Larsson, J. Callmer, and B. Stjernquist, 17–90. Lund.

Lüning, J., ed. 1997. *Ein Siedlungsplatz der Altesten Bandkeramik in Bruchenbrucken, Stadt Freidberg/Hessen.* Bonn.

Martini, F., and C. Tozzi. 1996. "Il Mesolitico in Italia Centro-Meridionale." *XIII UISPP Conference Proceedings.* Section 7, *The Mesolithic,* 47–58. Forli.

Meiklejohn, C., and M. Zvelebil. 1991. "Health Status of European Populations at the Agricultural Transition and the Implications for the Adoption of Farming." In *Health in Past Societies: Biocultural Interpretations of Human Skeletal Remains in Archaeological Contexts*, edited by H. Bush and M. Zvelebil, 129–45. *BAR-IS* 567. Oxford.

Nielsen, P.O. 1985. "De første bønder. Nye fund fra den tidligste Tragtbægerkultur ved Sigersted." *Aarbøger for Nordisk Oldkyndighed og Historie 1984,* 96–126.

Nordqvist, B. 1991. "Reduktionsprocesser av boplatsflinta från Halland. En spatial och kronologisk studie." In *Västsvenska stenålderstudier,* edited by H. Browall, P. Persson, and K.-G. Sjögren, 71–109. Göteborg.

Pedersen, L. 1997. "They Put Fences in the Sea." In *The Danish Storebælt Since the Ice Age,* edited by L. Pedersen, A. Fischer, and B. Aaby, 124–44. Copenhagen.

Peglar, S.M. 1993. "The Mid-Holocene Ulmus Decline at Diss Mere, Norfolk, UK: A Year-by-Year Pollen Stratigraphy from Annual Laminations." *The Holocene* 3:1–13.

Perlès, C. 1992. "Systems of Exchange and Organization of Production in Neolithic Greece." *Journal of Mediterranean Archaeology* 5:115–64.

Persson, P. 1998. "The Beginning of the Neolithic: Investigations of the Introduction of Agriculture into Northern Europe." Ph.D. diss., University of Gothenburg.

Pluciennik, M. 1997. "Radiocarbon Determinations and the Mesolithic–Neolithic Transition in Southern Italy." *Journal of Mediterranean Archaeology* 10:115–50.

Price, T.D. 1991. "The Mesolithic of Northern Europe." *Annual Review of Anthropology* 20:211–33.

———. 1995. "Agricultural Origins and Social Inequality." In *Foundations of Social Inequality*, edited by T.D. Price and G.M. Feinman, 129–51. New York.

———. 2000a. "Human Population in Europe during the Mesolithic." In *Archaeological Papers in Honor of Bernhard Gramsch*, edited by E. Cziesla, T. Kersting, and S. Pratsch, 185–95. Weissbach.

———. 2000b. "The Introduction of Agriculture into Northern Europe." In *Europe's First Farmers*, edited by T.D. Price, 260–300. Cambridge.

———, ed. 2000c. *Europe's First Farmers.* Cambridge.

Price, T.D., and J.A. Brown, eds. 1985. *Prehistoric Hunter-Gatherers: The Emergence of Cultural Complexity.* Orlando.

Price, T.D., and A.B. Gebauer. 1992. "The Final Frontier: First Farmers in Northern Europe." In *Transitions to Agriculture in Prehistory*, edited by A.B. Gebauer and T.D. Price, 97–116. Madison.

———, eds. 1995. *Last Hunters—First Farmers: New Perspectives on the Prehistoric Transition to Agriculture.* Santa Fe.

Price, T.D., A.B. Gebauer, and L.H. Keeley. 1995. "The Spread of Farming into Europe North of the Alps." In *Last Hunters—First Farmers: New Perspectives on the Prehistoric Transition to Agriculture*, edited by T.D. Price and A.B. Gebauer, 95–126. Santa Fe.

Randsborg, K., 1975. "Social Dimensions of Early Neolithic Denmark." *Proceedings of Prehistoric Society* 41:105–18.

———. 1979. "Resource Distribution and the Function of Copper in Early Neolithic Denmark." In *The Origins of Metallurgy in Atlantic Europe,* edited by M. Ryan, 303–18. Dublin.

Rowley-Conwy, P. 1984. "The Laziness of the Short-Distance Hunter: The Origin of Agriculture in Western Denmark." *Journal of Anthropological Archaeology* 3:300–24.

Runnels, C.N. 1996. "The Palaeolithic and Mesolithic Remains from the Berbati-Limnes Survey." In *The Berbati-Limnes Archaeological Survey 1988–1990,* edited by B. Wells and C.N. Runnels, 23–34. Stockholm.

———. 1997. "Soil Erosion and Farming in Prehistoric Greece, Human Effects on the Landscape in Ancient Greece." *Environmental Review* 4 (2):1–8.

Shackleton, J.C., T.H. van Andel, and C.N. Runnels. 1984. "Coastal Paleogeography of the Central and Western Mediterranean during the Last 125,000 Years and Its Archaeological Implications." *Journal of Field Archaeology* 11:307–14.

Simonsen, P. 1975. "When and Why Did Occupational Specialization Begin at the Scandinavian North Coast?" *Papers of the IX International Anthropological Congress: Prehistoric Maritime Adaptations,* 75–85. Chicago.

Skaarup, J. 1973. *Hesselø-Sølager: Jagdstationen der südskandinavischen Trichterbecherkultur.* Copenhagen.

———. 1988. "Burials, Votive Offerings and Social Structure in Early Neolithic Farmer Society of Denmark." *Rivista di Antropologia* 66:435–54.

Solberg, B. 1989. "The Neolithic Transition in Southern Scandinavia: Internal Development or Migration." *Oxford Journal of Archaeology* 8:261–96.

Stafford, M. 1999. *From Forager to Farmer in Flint: Stone Tools and the Transition to Agriculture in Southern Scandinavia.* Aarhus.

Strand Petersen, K. 1992. "Environmental Changes Recorded in the Holocene Molluscan Fauna, Djursland, Denmark." *Unitas Malacologica* 12:129–33.

Tillmann, A. 1994. "Kontinuität oder Diskontinuität? Zur Frage einer bandkeramischen Landnahme im südlichen Mitteleuropa." *Archäeologische Informationen* 17:157–87.

Troels-Smith, J. 1960. "Ivy, Mistletoe, and Elm. Climate Indicators—Fodder Plants." *Danmarks Geologiske Undersøgelse* 4 (4):4–32.

Woodman, P. 2000. "Getting Back to Basics—Transitions to Farming in Ireland and Britain." In *Europe's First Farmers,* edited by T.D. Price, 219–59. Cambridge.

Zilhão, J. 1993. "The Spread Of Agro-Pastoral Economies across Mediterranean Europe: A View from the Far West." *Journal of Mediterranean Archaeology* 6:5–63.

———. 2000. "From the Mesolithic to the Neolithic in the Iberian Peninsula." In *Europe's First Farmers,* edited by T.D. Price, 144–82. Cambridge.

NOTES

[1] The meeting in Venice in October 1998 was one of the more stimulating I have attended. The presentations and discussions continually raised important questions about the transition to agriculture in prehistoric Europe and pointed to topics and areas where evidence was either present or absent, or needed. I am sure that impressions from that meeting and from this volume will influence my thinking about the transition in the coming years. My sincerest thanks go to Albert Ammerman and Paolo Biagi for planning and hosting that delightful conference.

PART V

Human Genes and Languages

— 15 —

Returning to the Neolithic Transition in Europe

Luca L. Cavalli-Sforza

More than 25 years have passed since we first wrote about the wave of advance model (Ammerman and Cavalli-Sforza). The conference in Venice offered a fitting occasion to look back on what we have learned about the transition to agriculture. At the time that Albert Ammerman and I worked on our first article together, it seemed to be of wide enough interest that I had hoped to publish it in the Proceedings of the National Academy of Sciences. A friend, who was a member of the Academy, agreed to send it out for review to two eminent scholars in the field (following the usual procedure for submission). The response was a sample of the intellectual atmosphere of the time: the reviews must have been so negative that their contents were not even shown to me.

Of course, the idea that farming spread by the expansion of farmers was not new. V. Gordon Childe had already proposed it. But the real problem was that he looked at almost every case of diffusion as the result of the spread of people. In the 1960s, a reaction against Childe's reliance upon diffusion had begun to form especially among British and American archaeologists. By 1970, the pendulum of archeological thought had swung to the opposite extreme: no indigenous group was supposed to migrate even over short distances and only traders were allowed to move. This new position soon became dogma—not an infrequent event in the history of science. It has been called "indigenism" by Ammerman, who has made a most interesting analysis of the social and psychological reasons behind it in his opening chapter. Not surprisingly, our suggestion with regard to the spread of agriculture fell into a vacuum, where it remained for many years.

What was novel in our approach was that we had decided to handle the problem quantitatively. We chose to measure the rate of spread of early farming and then to see whether or not the wave of advance model was compatible with it. Agriculture has the potential for supporting sustained population growth and for increasing the level of human population density on the land. The result of repeated cycles of growth together with local migratory behavior is an expansion—a very common biological event in the history of successful species. An expansion is not the same as a "migration" in the sense of a population leaving an area to move to a completely new one. Here the area previously occupied might remain empty, but this is not the case in an expansion. Nor should such an expansion be confused with colonization, a term best reserved for the planned movement to some distant place usually by a group of people of some size. There are good examples of colonization in the classical world (the Greek colonies established in the western Mediterranean) and more recent history.

Our conception of the expansion of farmers was that agriculture brought with it the potential for sustained population growth that had not existed before. In late Paleolithic times, this was not really possible since population levels, as seen on the regional level, probably approached the carrying capacity permitted by a form of subsistence based upon foraging. Like others at the time, I was intrigued by the notion that food production started independently, using local plants and animals, some 10,000 to 6,000 years ago in several different parts of the world (e.g., Harlan 1971). Something interesting must have been happening at that time that led to separate pathways to agriculture. Instead of exploring the origins of agriculture, as many other researchers did in the 1960s and 1970s, we focused our attention on the consequences of food production once it arose. My own thinking turned to the demographic processes that might produce an expansion of early farmers.

Unfortunately, classical demography does not include a theory of demographic expansions. There are historical examples of the phenomenon such as the growth and expansion of French Canadians and of Dutch farmers who established the Cape Colony. Both were started by the initial migration of a relatively small group of a few thousand individuals. In the literature of demography, however, there was no body of theory for cases of this kind. It was my good fortune to have worked in the late 1940s with Ronald A.

Fisher, the Cambridge geneticist and statistician. Thus, I knew his little-known theory of the spread of advantageous genes, which provided an important mathematical instrument (Fisher 1937). It could be easily extended to other related situations. In fact, J. Skellam (1951), an ecologist, had already realized that it could be used to deal with the spread of living organisms. Fisher was able to show that, under simple and reasonably realistic expectations, the rate of radial advance of a population was constant in time and that it could be predicted on the basis of two quantities: the rate of population growth and the rate of local migratory activity. For our purposes, the latter component of his model could be measured approximately as the mean distance between the birthplace of parents and their children. By plotting the ages of the arrival of early farming at archaeological sites in different parts of Europe, we were able to see that there had been an outward spread from the Middle East and to obtain an initial estimate for the rate of radial advance in the model (Ammerman and Cavalli-Sforza 1971). Then we tried out feasible values for local population growth in the expanding "frontier" zone (ones in the range of 1–3% per year; the latter value is essentially the maximum in the case of human populations). The rest is now history (Ammerman and Cavalli-Sforza 1984).

An interesting mathematical extension of Fisher's original formulation has recently been put forward by Fort and Méndez (1999). Fisher's treatment of the diffusionary process assumes that there was a continuous advance of the farmers. But it may be more realistic—if much more demanding in terms of mathematics—to consider that the advance takes place as a series of steps with a time delay between each step. Fort and Méndez did just this in their new formulation based upon Einstein's approach to Fickian diffusion. The time delay can be thought of as the interval between one relocation event and the next for a given household or settlement. While the time delay may vary to some extent from one early Neolithic settlement system to the next, values in the range 15–25 years would seem to be reasonable for this variable. When working values for local population growth, local migratory activity, and the time delay (based on ethnographic parallels) were employed in this new formulation, they turned out to predict a rate of radial advance close to the one observed for the spread of early farming in Europe.

At the start of our work, we had simplified the problem at the conceptual level to a choice between two different modes of diffusion: one based on the movement of the farmers and the other involving the passage of a new technology without the relocation of people. We called the former demic diffusion and the latter cultural diffusion. We insisted from the start that they were not mutually exclusive and that they could both contribute to the spread of agriculture in a given region of Europe. The fit of Fisher's model to the dynamics of the spread of early farming in Europe did not automatically exclude cultural diffusion. Indeed, the dynamics in formal terms (those of a diffusionary process) could be much the same under the two modes of diffusion. But in practical terms, the rate of advance is likely to be quite different in the two cases. Under the wave of advance model, the rate of spread is essentially limited by what is possible in terms of the growth rate for a human population. As mentioned before, this rate cannot be more than about 3% per year and thus sets its limit to the rate of advance (to one of a few kilometers per year). On the other hand, there are fewer limitations when it comes to the cultural transmission of an innovation. Unfortunately, the question of how cultural transmission works is much more difficult to formulate in quantitative terms—even in the case of the rate of acceptance of technological innovations in the modern world (Cavalli-Sforza and Feldman 1981). And few attempts have been made to do this in archaeology so far. One can illustrate the differences in the rates of advance for the two modes of diffusion by comparing the spread of early farming with that of the adoption of pottery in the eastern Mediterranean. In the first case, the rate is measured to be about 1 km per year. By contrast, the rate for the adoption and spread of pottery was much faster. It is well known that pottery first made its appearance in the Middle East at least 1,000 years after the beginning of food production (see contributions in Barnett and Hoopes 1998). Once it appeared, its use spread rapidly to those areas where agriculture was already well established. Thus, in the Near East, agriculture was born aceramic. But the whole agricultural area soon acquired ceramics once the innovation appeared. In short, cultural diffusion can have a much faster tempo than demic diffusion. The slowness of the spread of agriculture in most parts of Europe is one of the things that leads me to prefer the demic hypothesis.

It is heartening to see how the glacial welcome that our hypothesis and model originally received has now changed. We were, of course, aware that the wave of advance model tends to simplify an extremely complex process. For example, it is clear that the expansion of early farming also set in motion a series of other innovations (something not considered in Fisher's model)—

each of which had its own local and general effects. Thus, a likely corollary of a demic expansion is that there will be an increasing number of innovations during and after the expansion. In other words, one can argue, in general terms, that the greater the size of a population, the more frequent will be the appearance of useful innovations or inventions (since they are probably rare events and there will be a greater chance for an innovation to occur when there are more people). In addition, there are likely to be more problems to be solved in a large, developing population. Another corollary to consider is that there will be more encounters with different peoples and different cultures. These can take many forms, as we have said before, and some of them may lead to positive cultural interactions.

It is also clear that the environment is far from isotropic. This is one of the simplifying assumptions both in Fisher's classical model of the wave of advance and in the new formulation proposed by Fort and Mendez (1999). In the real world, a multitude of local factors will enter the picture if one attempts to make a more refined analysis at the micro-geographic level. For example, local environmental conditions can act in a positive way on the migratory component (in the case of rivers and coasts) or in a negative way (barriers such as mountains or forests). It is important to recall that we proposed the wave of advance model as a first-generation model—one which would help, at the general level, to reformulate thinking about the problem of the Neolithic transition in Europe as a whole. There is also a need for the development of second-generation models of demic diffusion that deal more closely with local conditions. Indeed, this is just what van Andel and Runnels (1995) did when they proposed a new model that is specifically tailored to the environmental setting in Greece. This is, by the way, a normal development in model building as a field matures over time. The point here is that the wave of advance should be seen today as a macro-model. It was useful for getting started and drawing the outlines of the larger picture. It also helped in attacking the dogma of anti-diffusionism, which dominated archaeological thought in the 1970s and 1980s, and in introducing a new hypothesis as well as ways of testing it.

I do not feel that my archaeological background is sufficient to entitle me to comment on the archaeological data and conclusions that were presented at the conference, which I have found all extremely interesting. One of the new results that I enjoyed learning about in Venice was the recent work on the rate of recolonization of northern Europe by late Paleolithic hunter-gatherers at the end of the last ice age (Housley et al. 1997). It is remarkable that this rate turns out to be similar to the average rate for the spread of early farming across Europe. We need more quantitative work of this kind to help understand the expansion of behaviorally modern humans out of Africa. Today, at least, well-dated sites of early Homo sapiens sapiens are exceedingly rare. We still have a long way to go before this much earlier expansion in Africa can be measured in the same way. Genetic dates, estimated on the basis of changes in mtDNA and Y chromosomes, suggest rates of expansion in Africa that are similar in their order of magnitude to the two well documented cases in Europe (Quintana-Murci et al. 1999; Underhill et al. 2000). The rate measurement obtained by Housley et al. (1997) may slightly underestimate the potential rate of expansion for hunter-gatherers, since the displacement toward the north was conditioned by the rate at which the glacier sheet melted.

Genetic Approaches to the Neolithic Transition in Europe

If it were easy to compare the genetic types of human remains of late Mesolithic age with those of early Neolithic age, the task of choosing between demic and cultural diffusion would be easy. One could estimate in a direct way the relative importance of the two modes of diffusion. Unfortunately, the conservation of DNA in old bones is usually not good enough. DNA fragmentation is frequent and demands an amount of work that is far from trivial. The analysis of mtDNA has the greatest probability of success since there are many copies of mtDNA in every cell, while there are at most two of any chromosomal (nuclear) gene. There is thus a much better chance of survival in reasonable condition of mtDNA than of nuclear genes. And the chances are still higher if the bones are found in a place with a cold climate that favors the survival of protein.

In this section, I would like to review briefly some of the main steps in the genetic work that we have done on extant human populations in Europe and its relationship to the study of the Neolithic transition. At the time that we put forward the wave of advance model in 1973, there were no synthetic gene maps for Europe. Available at that time were only maps for individual gene systems (such as the ABO group system, which is essential for blood transfusions). And they showed considerable variation from one group to the next. What was really needed was a broader summary of the patterns of variability found among human populations in Europe based on a large number of genes. A major step toward such a synthetic approach was a study that I did in collaboration with Menozzi and Piazza (Menozzi et al. 1978; see also Ammerman and Cavalli-Sforza 1984, 104–80). It involved the analysis of information on 39 genes collected from human populations in different regions of Europe. The analysis of the gene frequency data was done with the statistical method known as principal components, which extracts independent latent patterns present in data with a large number of variables (in our case, the frequencies of many genes from many populations in Europe). Principal components analysis also ranks the latent patterns in order of the percentage of total variation that each of the components explains. We found that the first principal component—the one explaining the most variation—gave a pattern on the map that looked just like the spread of early farming in Europe. This was an encouraging result and led to a further cycle of studies based on more genes and improved methods for generating synthetic gene maps. Our work on the Neolithic transition had begun essentially along theoretical lines: formulating the demic hypothesis and proposing a new model to explore this hypothesis. Now, with the synthetic gene maps, there was the possibility for a more empirical dialogue between human genetics and the archaeological record.

The demic hypothesis for the spread of agriculture carries with it clear implications for the evolution of genes and their patterns of geographic distribution in human populations. If agriculture spread solely by means of cultural diffusion, it should leave the previous gene distribution intact. The populations that were present in Europe before the Neolithic were certainly of small size and accordingly diverse in their genetic character because of the operation of factors such as genetic drift. On the other hand, a complete demic diffusion should lead to the replacement of the human populations that were there before—and consequently their genes—by the new one. A mixture of demic and cultural diffusion (that is, a growing Neolithic population and interaction of the farmers with local Mesolithic populations) should generate a gradient of values or a cline along the pathway of the expansion of early farming. In this case, one would expect the gene frequencies of the original farmers to decrease proportionally as one moves from the Middle East across Europe. This is just what the first principal component of the genetic analysis had shown. In a population simulation study that I did with Sgaramella-Zonta, we had previously developed a method for the detection of such a cline (Sgaramella-Zonta and Cavalli-Sforza 1973). This study also helped to explain why the first principal component has the shape of a fan (that is, a segment of a circle with the pivot in the place of origin). This is just what one should expect to find in the case of demic diffusion.

The result obtained in the 1978 study was confirmed by subsequent analyses that we and others performed. For example, Sokal et al. (1991, 1992) obtained the same result using an entirely different method of statistical analysis. In all of the various studies, the first principal component consistently explained 26–28% of the total variance. Finally, in 1994, we published a synthesis for Europe, which was now based on the analysis of 95 genes (representing a wide range of different genetic systems, which are spread widely across the chromosomes) and a new set of data from the Caucasus (Cavalli-Sforza et al. 1994). Again the first principal component accounted for 28% of the overall genetic variation observed among extant human populations in Europe. In short, it is reasonable to hold that the Neolithic transition left a deep imprint on the genetic structure of those living in Europe. In retrospect, it is perhaps worth adding that the final synthesis of the classical polymorphisms in Europe took a great deal of work by many people over a span of more than 20 years.

It was only in 1981 that one had the possibility to go further and study genetic variation directly on the basis of DNA markers themselves. It would take many years, however, before this work would reach the stage where enough DNA markers and enough human populations had been sampled so that one could attempt a synthesis of the new evidence. One major advantage of the DNA techniques is that they involve genetic material that is transmitted by one parent only: mitochondrial DNA (mtDNA), which is passed only by mothers to their children (of both sexes), and the Y chromosome, which is passed only from fathers to sons. In effect, this permits a substantial simplification of the genetic analysis. At the same time, one must remember that the analysis of these two different genetic systems will not always yield the same result (even for a given population) because of gender differences in patterns of movement at the time of marriage.[1] Genes that occur on chromosomes other than X or Y (also called autosomes) are transmitted equally by both parents; the results obtained from the analysis of such chromosome should comprise an average of those obtained for genes transmitted respectively on the female line and on the male line.

A recent analysis of the D-loop, a small segment (4%) of mtDNA in Europeans, was made in the laboratory of Bryan Sykes at Oxford, which seemed to show that there are no differences among European populations (Richards et al. 1996).[2] The conclusion reached on the basis of this preliminary analysis was that the demic expansion tied with early farming did not exist; however, this conclusion was premature. The study turned out to be based on an insufficient number both of individuals and DNA markers (Cavalli-Sforza and Minch 1997). The extension of the analysis to a larger sample of individuals and to more mtDNA variants produced a different result (Richards et al. 1998). As Sykes notes in his chapter to this volume, the new results show that the demic expansion of farmers is indeed visible in mtDNA and explains about 20% of the variation observed in this genetic system. As mentioned above, this is to be compared with the value of 26–28% obtained for the classical polymorphisms. Thus, the two results are not all that different in the end. And much more work remains to be done on mtDNA. One way to explain the difference between the two values may be in terms of differential patterns of marriage-migration between the sexes. This represents a question that has seldom been studied in any real depth.

Recent work on the Y chromosomes found in European populations gives much the same answer. In 1996, a preliminary study of two specific markers on the Y chromosome revealed a clear pattern across Europe in agreement with the hypothesis of demic diffusion (Semino et al. 1996). Again, as more work was done on other Y chromosome markers and more populations, there would be the opportunity for a more synthetic analysis. The lead here was taken by Semino at the University of Pavia (Italy), who also worked in my laboratory at Stanford in the summer of 1999. By 2000, there was information from 22 binary markers of the nonrecombining Y chromosome and a synthesis could now be made for this genetic system (Semino et al. 2000). Without going into the details of the analysis here, it shows that four haplotypes (EU4, Eu9, Eu10, and Eu11) represent the male contribution of the demic diffusion of farmers from the Middle East to Europe. Together they constitute about 22% of the Y chromosomes in extant European populations. In addition, the analysis of the Y chromosome now makes it possible to go even further. By comparing the cline obtained for populations in southern Europe (Turkey to Portugal) with the one for populations in central and northern Europe, the Y chromosome evidence indicates that demic diffusion played a more active role in the former case than it did in the latter one. This result seems to be in good agreement with what we learned at the Venice meeting—the archaeological evidence for demic diffusion is stronger in the south than it is in the north. It is also worth noting that the value of 22% (for the Y chromosomes connected with the spread of early farmers) is much the same as what we have seen already for the classical polymorphisms and mtDNA. In other words, three independent genetic systems now all point in the same direction. Of course, as more studies are carried out on the Y chromosome and mtDNA over the next decade, the genetic picture can be worked out in greater detail.

Pygmies and the Transition to Agriculture in Sub-Saharan Africa

I cannot resist the temptation of telling my colleagues in archaeology that, if they wish to observe an agricultural transition that is going on today, they can still do so. But they will have to hurry. Such an experience may lead to a better understanding of interactions between foragers and farmers of the kind that were discussed at the Venice meeting. In looking back on my research on the Pygmies in central Africa, there is also a chance to explain how I became interested in the transition in Europe and began working with Albert Ammerman. I first met him at a conference that was held in Romania in 1970.[3] It was there that we first talked about the possibility of doing a collaborative study on the spread of agriculture in Europe. In 1966, I had started a genetic study of Pygmies in Africa. For the next two decades, I traveled extensively in African countries where Pygmies still live. The fieldwork also brought me into regular contact with the local farmers in the same areas. My major aim was to investigate the demography and the genetic structure of those Pygmies who were still living, at least in part, as hunters and gatherers. The contrast with the farmers, almost all Bantus, was striking and of great interest to me.

Agriculture started in sub-Saharan Africa when the Sahara began to dry up some 5,000 years ago. Before it became a desert, it contained an extensive community of farmers and herders, including its more mountainous parts with well-known rock art. As the drying became more pronounced, most of the inhabitants were forced to move toward the south, where they had to develop new crops and face problems with cattle diseases such as sleeping sickness. Harlan indicates a broad band just below the Sahara as the place where sub-Saharan agriculture emerged. This band corresponds approximately to the Sahel and extends from the Atlantic Ocean to the Red Sea (Harlan 1971).[4] Most scholars agree that the expansion of the Bantu farmers was a demic one. It appears to have started some 3,000 years ago in the part of west Africa that today we call Cameroon. The Bantus expanded south both by a western route (arriving in Nigeria around 2,500 years ago) and an eastern one (making it to Lake Victoria around 2,000 years ago) until they reached South Africa. In the process, Bantu farmers came to have a wide geographic distribution in central and southern Africa. Their expansion ended about two centuries ago when the Bantus encountered the descendants of Dutch farmers, who had settled in Capetown around 1650 and then expanded their settlement in a northern direction. As the crow flies, the Bantus spread from north to south over a distance of about 3,000 km in a time of just under 3,000 years. Thus, the average rate of advance again may be on the order of 1 km per year. It is worth adding here that the original crops that they cultivated, yams and sorghum, were replaced by manioc, which was introduced from South America in the 18th century.

The unity of Bantus was first established on the basis of linguistics. Their languages form the latest branching (known as "Narrow Bantu") of the Niger Kordofanian family (Ruhlen 1987, 312–6). The linguistic unity of the Bantus is paralleled by their biological unity. Bantus are similar to but also genetically distinguishable from West Africans, from whom they originated as a peripheral, eastern subset. The Bantus, by expanding to the south and east, grew into a large population. During the course of their expansion, they also mixed to some extent with other indigenous populations. However, they still show greater genetic homogeneity than West Africans. Thus, although the word Bantu has a clear linguistic origin, it can be used to describe these people genetically as well (Cavalli-Sforza et al. 1994, 185).

There were many local groups of hunter-gatherers that once lived along the path of the Bantus expansion. In central Africa, these include groups of Pygmies who survive today in countries such as Cameroon, Congo Brazza, the Central African Republic, the Democratic Republic of the Congo (formerly Zaire), and Rwanda. There they commonly call themselves Aka (Biaka, Beaka), a name also used for them in early Egyptian documents (near images of Pygmy dancers). Between 1966 and 1985, we carried out 10 field seasons of multidisciplinary research on Pygmies and some of their farming neighbors.[5] The study of Pygmies has relevance for our understanding of some of the issues that arise in the study of the Neolithic transition in Europe. I am aware that the forms of agriculture and also the environmental settings are different in Africa and Europe; moreover, in earlier times, the nature of the contacts between Bantu farmers and Pygmy foragers may have been quite different than what we observe today. We have a firsthand opportunity, however, to observe the dynamics of their interactions with one another and the many different forms that such interactions can take from one place to the next.

A major difference between Europe and the situation observed in central Africa is the long survival of hunting and gathering as a way of life in the latter case. When the radiocarbon dates for the latest Mesolithic sites in given region of Europe are compared with those for the first farmers in the same region, there is not much overlap between the two (Ammerman and Cavalli-Sforza 1984, 58–60).[6] Following the arrival of agriculture in most parts of Europe, the Mesolithic way of life did not endure for a long period of time. By contrast, this situation appears to be common in central Africa, where Bantu farmers and Pygmy hunter-gatherers have managed to live more or less side-by-side in some cases for more than 2,000 years. It is possible for Pygmies to maintain their foraging way of life as long as their traditional environment, the forest, still survives. The tropical forest is a much more daunting place than the forests of Europe. Even today there are still large tracts of forest where only Pygmies live. Here it has always been easier to maintain a foraging way of life, where old values, knowledge, and customs remain useful and can continue unperturbed. On the other hand, in smaller tracts of forest (the consequence of farming and more recently the timber industry), conditions are less favorable for the Pygmy way of hunting. When the forest is interrupted by savannas and fragmented to a mosaic, the availability of large game is reduced.[7]

Over the years, my work on Pygmies involved surveying large areas of Cameroon, the Central African Republic, and Rwanda. For example, in Rwanda, there are today around 20,000 Pygmies, but almost all of them are now farmers who also make pottery. Because of the vast scale of deforestation in Rwanda, acculturation was an absolute requirement for the survival of Pygmies there. My most extensive survey was done in the Central African Republic, where I visited practically every area inhibited by Pygmies and had the chance to see all possible degrees of acculturation. In the Central African Republic, we were able to take genetic samples from about one-eighth of the Pygmy population (more than 1,000 individuals in all). I have also observed the continuous withdrawal of the forest toward the south over the course of the last 30 years in this area. There is little prospect for the Pygmies to retreat into the forest and follow their traditional way life once the forest becomes patchy and limited in size.

The relations between Pygmies and farmers vary considerably from place to place. Where the forest is still intact, Pygmies spend the greater part of the year in it and a minor part (usually the winter) close to the farmers' villages. Most farmers go into the forest only a few weeks each year; their women go into the forest on their own to exchange manioc for meat. When living near the farmers, Pygmies normally make their camps at an hour's walk from the village. In some cases, however, a Pygmy camp is located in the area just behind the villagers' huts. In the winter, Pygmies work for the farmers in their plantations, help them to build roofs for their huts, or bring meat to the village from hunting expeditions. A few farmers have many Pygmy servants and consider them to be their property. But most farmers do not "own" Pygmies. Servitude is hereditary: Pygmy children inherit from their parents their relationship to their master. And master's children inherit their father's Pygmies. Farmers usually despise Pygmies; they invariably consider them to be "beasts" and treat them accordingly. But there are limits to ill treatment since Pygmies can leave their master and never return. The relationship is both an economic and a social one. When Pygmies work in the village, they are fed by their masters with the cheapest food (manioc and bananas). Until the early 1980s, they were rarely paid for their work with money. When living near a farmers' village, Pygmies may hunt for themselves or for their masters; they then exchange meat for pottery and iron tools that they do not produce. Other tools and pieces of equipment such as hunting nets, bows, crossbows, arrows, and arrow poisons are made by the Pygmies themselves from forest products.[8]

Pygmy huts are built with twigs and leaves in two or three hours' time. Pygmies can move easily and they may do so every few months in order to change hunting grounds. Their camps consist of 10–50 people, who are usually in a loose patrilineal relationship. Normally, except deep in the forest where there are no resident farmers, the Pygmies are less numerous than the farmers. In areas that one can reach by jeep, I found the population ratio rarely greater than one to four. In the Central African Republic, some Pygmies now work in industries found near the forest (diamond mines and sawmills). When they are still hunter-gatherers, their average exploration range is about 60 km for men and 30 km for women. The range is less when they are partially acculturated (that is, when they farm regularly but still spend substantial time hunting in the forest). For the region as whole, their population density is roughly 30 times lower than that of the farmers. Inside the forest, where the farmers are almost totally absent, Pygmies have a low average population density (only about one or two individuals for every 10 km^2).

Perhaps the most important thing to consider, for our present purpose, is genetic exchange among Pygmies and farmers. Cross marriage is rare and it depends entirely on the customs of the farmers' tribe or village. I found only two small tribes (the Lissongo or Mbati and the Ngbaka) in the Central African Republic that practice mixed marriages with Pygmies. But even in these two cases, mixture is observed only in some villages and only in one direction: farmer males can marry Pygmy women but the opposite is not allowed. The children of a mixed marriage become part of the farmers' tribe. Pygmy women have a reputation for being fertile.[9] In addition, the bride price of a Pygmy woman is lower than that of a Bantu bride. In Rwanda, Tutsi kings are known to have had Pygmy wives, who were recognized as queens. Thus, there is some genetic exchange, but it is limited in many ways. Similar kinds of genetic exchange between foragers and farmers may well have happened during the Neolithic transition in Europe.

In the areas where I have studied Pygmies, there are still major tracts of undisturbed forest. But they are now being rapidly destroyed by the timber industry. This adds a new threat to the ecological niche that sustains the traditional way of life of Pygmies. Until recently, the advance of the Bantu farmers into the forest and the damage that they did with their plantations were limited. However, since the introduction of cash crops such as coffee, the adverse effect of farming has dramatically increased. In the past, penetration of the forest by the Bantu farmers either followed a riverine route or progressed when a new road was opened. In the Ituri forest, there were still only two roads in the 1980s: one from north to south and the other from east to west, crossing in the center at Mambasa. In the forest of Congo Brazza, which is larger than the Ituri, no road crossed the forest until recently. Only rarely do farmers establish a plantation at some distance from a road (that is, in isolation within the forest). Thus, the opening of a new road represents an invitation to farmers to settle in a new area and generate new plantations. Local farmers are not allowed to build huts and plantations along the large roads opened for timber operations (because the roads have to be kept free for large trucks carrying timber at high speed). Farming develops along secondary roads. At the end of a timber concession, the forest is supposed to be allowed to grow back and the major roads should disappear. But over the 20 years in which I have returned again and again to the same places, I have seen a continuous retreat of the forest. Because of it, Pygmies are forced to retreat, acculturate, or die. Toponymy also shows that the Pygmies are disappearing. In the last century alone, Pygmies have had to retreat hundreds of kilometers in response to the disappearance of the forest. In the present century, the process of deforestation, if it continues at the current rate, may well spell the end of their lifeway.

REFERENCES

Ammerman, A.J., and L.L. Cavalli-Sforza. 1971. "Measuring the Rate of Spread of Early Farming in Europe." *Man* 6:674–88.

———. 1973. "A Population Model for the Diffusion of Early Farming in Europe." In *The Explanation of Culture Change*, edited by C. Renfrew, 343–57. London.

———. 1984. *The Neolithic Transition and the Genetics of Populations in Europe*. Princeton.

Barnett, W.K., and J.W. Hoopes, eds. 1998. *The Emergence of Pottery: Technology and Innovation in Ancient Societies*. Washington, D.C.

Cavalli-Sforza, L.L. 1963. "The Distribution of Migration Distances: Models and Applications to Genetics." In *Human Displacement*, 139–58. Entretiens de Monaco en Sciences Humaines, Premiers Session. Monaco.

———, ed. 1986 *African Pygmies*. Orlando.

Cavalli-Sforza, L.L., and M. Feldman. 1981. *Cultural Transmission and Evolution*. Princeton.

Cavalli-Sforza, L.L., P. Menozzi, and A. Piazza. 1994. *The History and Geography of Human Genes*. Princeton.

Cavalli-Sforza, L.L., and E. Minch. 1997. "Paleolithic and Neolithic Lineages in the European Mitochondrial Gene Pool." *American Journal of Human Genetics* 61:247–51.

Fisher, R.A. 1937. "The Wave of Advance of Advantageous Genes." *Annals of Eugenics, London* 7:355–69.

Fort, J., and V. Méndez. 1999. "Time-Delayed Theory of the Neolithic Transition in Europe." *Physical Review Letters* 82:867–70.

Harlan, J.R. 1971. "Agricultural Origins: Centers and Non-Centers." *Science* 174:468–74.

Hodson, F.R., D.G. Kendall, and P. Tautu, eds. 1971. *Mathematics in the Archaeological and Historical Sciences*. Edinburgh.

Housley, R.A., C.S. Gamble, M. Street, and P. Pettitt. 1997. "Radiocarbon Evidence for the Lateglacial Human Recolonisation of Northern Europe." *Proceedings of the Prehistoric Society* 67:25–54.

Menozzi, P.A., A. Piazza, and L.L. Cavalli-Sforza. 1978. "Synthetic Maps of Human Gene Frequencies in Europe." *Science* 201:786–92.

Powledge, T.M., and M. Rose. 1996. "The Great DNA Hunt." *Archaeology* 49 (5):36–42.

Quintana-Murci, L., O. Semino, H.J. Bandelt, G. Passarino, K. McElreavey, and A.S. Santachiara-Benerecetti. 1990. "Genetic Evidence of an Early Exit of *Homo sapiens sapiens* from Africa through Eastern Africa." *Nature Genetics* 23:437–41.

Richards, M., et al. 1996. "Paleolithic and Neolithic Lineages in the European Mitochondrial Gene Pool." *American Journal of Human Genetics* 59:185–203.

Richards, M.B., V.A. Macaulay, H.J. Bandelt, and B.C. Sykes. 1998. "Phylogeography of Mitochondrial DNA in Western Europe." *Annals of Human Genetics* 62:241–60.

Ruhlen, M. 1987. *A Guide to the World's Languages*. Stanford.

Semino, O., G. Passarino, A. Brega, M. Fellous, and A.S. Santachiara-Benerecetti. 1996. "A View of the Neolithic Demic Diffusion in Europe through Two Y Chromosome Specific Markers." *American Journal of Human Genetics* 59:964–8.

Semino, O., et al. 2000. "The Genetic Legacy of Paleolithic *Homo sapiens sapiens* in Extant Europeans: A Y Chromosome Perspective." *Science* 290:1155–9.

Sgaramella-Zonta, L., and L.L. Cavalli-Sforza. 1973. "A Method of Detection of a Demic Cline." In *Populations Genetics Monography III*, edited by N.E. Morton, 128–35. Hawaii.

Skellam, J. 1951. "Random Dispersal in Theoretical Populations." *Biometrika* 38:196–218.

Sokal, R.R., N.L. Oden, and C. Wilson. 1991. "Genetic Evidence for the Spread of Agriculture in Europe by Demic Diffusion." *Nature* 351:143–5.

Sokal, R.R., C. Wilson, and N.L. Oden. 1992. "Patterns of Population Spread." *Nature* 355:214.

Underhill, P.A., et al. 2000. "Y Chromosome Sequence Variation and the History of Human Populations." *Nature Genetics* 26:358–61.

Van Andel, T.H., and C.N. Runnels. 1995. "The Earliest Farmers in Europe." *Antiquity* 69:481–500.

NOTES

[1] For an illustration of such gender differences in the context of a modern farming society in Europe, see Cavalli-Sforza (1962). As discussed in the last section of this chapter, one also observes gender-based differences in patterns of marriage between Pygmies and Bantu farmers in Central Africa. Marriage only occurs between Pygmy women and Bantu men.

[2] Unfortunately, this incorrect result found its way into the popular literature of archaeology; see Powledge and Rose (1996, 42).

[3] The meeting focused on the application of mathematics to archaeology and the historical sciences (Hodson et al. 1971).

4 I suspect that there were at least two major centers of origin on this band: one in Ethiopia and another in the broad area of West Africa shared by Mali, Niger, and Burkina Faso. My persuasion about the importance of the West African area arises from it being the center of a major genetic anomaly that is observed on the maps of several principal components of genetic data (Cavalli-Sforza et al. 1994, 180–94).

[5] Some of the main results of this work are summarized in *African Pygmies* (Cavalli-Sforza 1986).

[6] One exception here is the case of central Portugal, where the Mesolithic may have persisted five centuries after the arrival of agriculture; see the chapter by Zilhão in this volume.

[7] There are still many elephants in the major African tropical forests, just to mention the most desirable animal. Pygmies still hunt them today. Until a century ago, Pygmies were the main suppliers of the ivory consumed by Europeans and Americans.

[8] Pygmies have only recently learned how to make crossbows (entirely of wood and string) from the farmers.

[9] Barry Hewlett has made an important contribution on this topic. He has observed that the total fertility rate for a Pygmy woman is 6.1, while it is 4.5 for a Bantu woman; he also found that 30% of Bantu women have no children.

— 16 —

European Ancestry: The Mitochondrial Landscape

Bryan Sykes

In this chapter I refer to two regions of the world: Europe and Polynesia. Europe, because it is the main focus of our current research efforts and relevant to the meeting, and Polynesia because some of the lessons we learned from that much more straightforward region, genetically speaking, helped our interpretation of events in the more complex European theater.[1]

Our work on mitochondrial variation in Polynesia (Sykes et al. 1995) raised few eyebrows since its main conclusion—that Polynesia had been initially colonized from the West—confirmed the prevailing consensus built up from archaeology, linguistics, and classical genetics. Only Thor Heyerdahl, whose celebrated hypothesis that the major wave of colonization had been from the Americas, would have had cause for disappointment.

The following year, we published the results of an equivalent study of Europe whose main conclusion was that the bulk of the extant mtDNA variation had its origins in the Paleolithic, and the reaction to these results was very different. This time we appeared to be in conflict with the widely held view that the most important influence on the modern European gene pool had been a large influx of farmers during the Neolithic transition. Thus it was with some trepidation that I presented my paper to the meeting that marked the 25th anniversary of the demic diffusion model from which this view traces its origins. At the same time, I congratulated the architects of that model, and the organizers of this meeting, Albert Ammerman and Luca Cavalli-Sforza, for their intellectual generosity in allowing me to present my case.

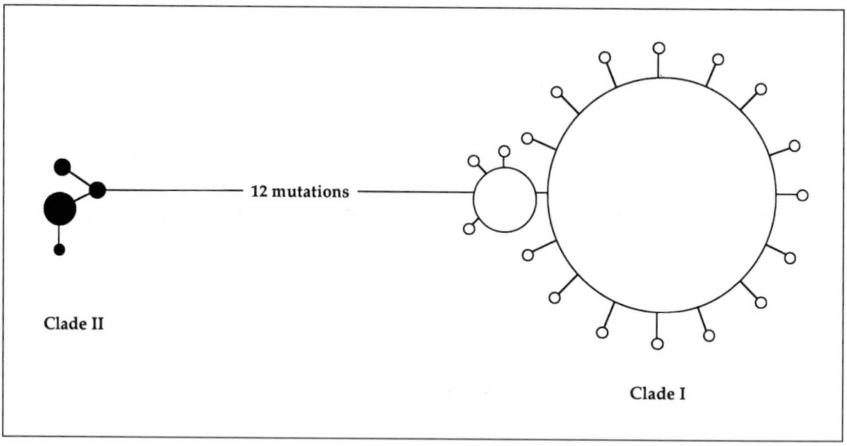

Fig. 16.1. mtDNA phylogeny for Polynesia. Circles represent different haplotypes determined by control region sequence with areas proportional to their frequency in the sample. Distances between haplotypes reflect the number of mutations. (Data from Sykes et al. 1995)

My argument is built around the use of a genetic system whose properties allow us to attempt to reconstruct an evolutionary relationships between *individuals*. In that property it differs from what I have described elsewhere as the "classical approach," where the comparisons are based on differences in gene frequency between *populations* (Sykes 2000). This may not seem like a particularly profound distinction, but it does fundamentally change the interpretation.

The new approach can be traced to the late Allan Wilson's article "Mitochondrial DNA and Human Evolution" (Cann et al. 1987). The initial furor which greeted its appearance centered around the claims that all modern humans had a comparatively recent African origin, and the skirmishes that ensued were principally fought around that issue. It also introduced the new way of treating human populations—not as effectively separate subspecies—but as collections of individuals, whose genes have their own genealogical histories. I illustrate this concept with data from Polynesia.

The mitochondrial variation of Polynesia, assessed by control region sequence, was divided into two clades each with a few closely related haplotypes (Sykes et al. 1995) (fig. 16.1). There was such a big mutational distance between them that it is inconceivable that they all shared a common origin within Polynesia, which has only been inhabited for the past 2,000–3,000 years. Comparisons with mtDNA from potential source pop-

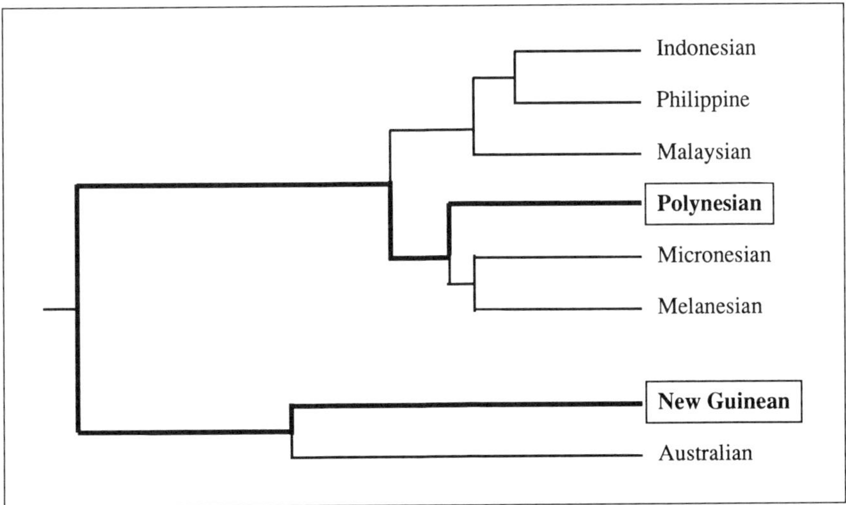

Fig. 16.2. The positions of Polynesian and New Guinean populations on a genetic population tree. (Redrawn from Cavalli-Sforza et al. 1994)

ulations showed that the most frequent clade, accounting for 95% of sequences, could be traced to East Asia, possibly Taiwan, whereas the other, with a frequency of 4%, came from the highlands of New Guinea. The other 1% was a mixture of haplotypes from a variety of sources. Figure 16.2 highlights the position of Polynesia and New Guinea on a diagram that compares genetic difference between populations. They are a very long way apart. The mitochondrial result indicates that Polynesians actually have a mixture of mtDNA from two different sources, an important result for the colonization of remote Oceania. Although the population diagram misses this connection, it does not make it wrong *unless* it is literally interpreted as representing an evolutionary history of populations.

In our study on Europe we constructed a mitochondrial phylogeny using the sequence variation contained in about 350 b.p. of the first hypervariable (HVS I) segment of the control region (Richards et al. 1996). We imagined we saw six clusters in the diagram. They were indistinct and often only separated by a single mutation. We then calculated the divergence time for each cluster and saw that most of them dated back well into the Paleolithic. One, which we called 2A, stood out as having very clear Middle Eastern ancestry and lower diversity, and we thought this cluster might reasonably be attributed to the Neolithic farmers. But the frequency of this group was only 15%,

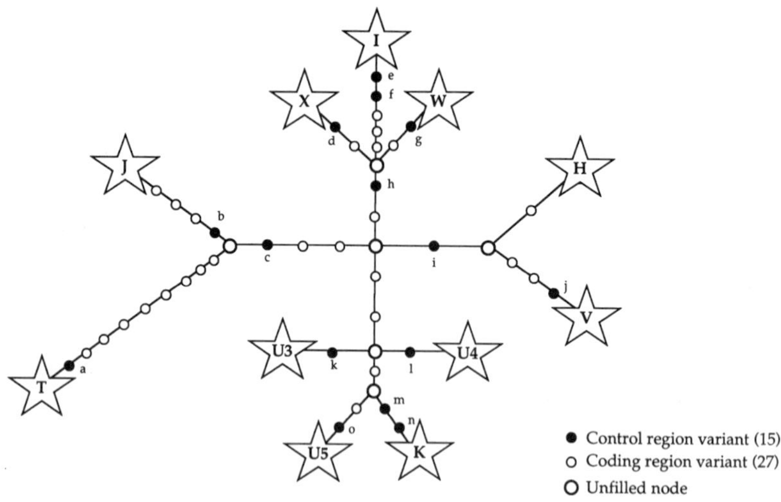

Fig. 16.3. European haplotype clusters (shown as stars) and the characters that separate them. Mitochondrial control region positions are as follows: a = 16294, b = 16069, c = 16126, d = 16278, e = 16129, f = 16391. g = 16292, h = 16223. I = 00073. j = 16298, k = 16343, l = 16356, m = 16224, n = 16311, o = 16270.

hence our conclusion that the data are best explained by a relatively small-scale Neolithic contribution and a much larger surviving Paleolithic component. The controversy that ensued, to which I have already referred, has been widely covered (e.g., Lewin 1997). And it has led to a lively correspondence (Cavalli-Sforza and Minch 1997; Richards et al. 1997; Barbujani et al. 1998; Richards and Sykes 1998). Four main objections were raised.

First, the male and female contributions of Neolithic immigrants were different. It is inescapable that mtDNA only has things to say about females whereas the demic diffusion model, built on nuclear-encoded classical marker frequencies, considers an average of maternal and paternal contributions. It remains to be seen what the results of ongoing Y-chromosome surveys will reveal in Europe, and they may indeed show considerable differences in the sexes. As an example of this, we find in Polynesia that although at least 99% of mtDNA predates European arrival, at least a third of Y-chromosomes come from Europeans (Hurles et al. 1998).

Second, the phylogeny was incorrect. As I have already mentioned, the clusters were defined by the control region sequence. They were not separated by long branches; it follows, therefore, that they had low statistical support. We have since added more characters from outside the control region in order to resolve the issue (Macaulay et al. 1999). Some of these are RFLP variants, while others are sequence dimorphisms in coding genes that cannot be revealed by restriction enzymes. The result is a much more robust phylogeny with all the reticulations resolved and clusters much better separated (fig. 16.3). There is an excellent agreement between this phylogeny and our earlier effort defined by the control region alone. All the earlier clusters survive intact and can now be broken down into further subdivisions. We have abandoned our earlier Eurocentric nomenclature in favor of the worldwide alphabetical notation (Torroni et al. 1996). The correspondence between them is shown in table 16.1. The principal differences between the original and revised phylogenies are the following: (1) group 4 (now K) joins others from group 5 (now U), (2) group 1 is broken down into H and V, and (3) group 3 is broken down into I, W, and X.

Table 16.1. Correspondence of Numerical (1996) and Alphabetical (1998) Classification of European Haplotype Clusters

Classification (1996)	Classification (1998)
1	H and V
2A	J
2B	T
3	I, W, X
4	K
5	U

All the groups can still be recognized by their control region sequences alone and only one site (bp 00073) from the second hypervariable segment of the control region (HV II) is required to distinguish haplogroup H from the rare ancestral U haplotype.

Third, the mutation rate estimate was wrong. There has been speculation recently that the mutation rate used for estimating mtDNA divergence is too slow by almost an order of magnitude (Howell et al. 1996). The faster rate was arrived at by extrapolation from a few pedigrees segregating for the mitochondrial disease phenotype LHON. Some individuals within the pedigrees had more than one mitochondrial allele—a state known as heteroplasmy. Heteroplasmy is the inevitable transition state between the time a new allele arises, presumably by mutation of a single DNA molecule, and when it becomes fixed in the maternal line. Heteroplasmy can persist for several generations as the new and old alleles battle it out—not literally though, since the process is entirely random—until one is triumphant and the other eliminated. The mutation rate, estimated from the LHON pedigrees, was about eight times higher than that used for divergence estimates. The main reason for this discrepancy was that, aside from the small sample size and the reporting bias (pedigrees without new alleles tending not to be reported), the new mutations were in HVS II whereas divergence date estimates use data from HVS I. Studies on HVS I heteroplasmy in our laboratory and elsewhere found mutation rates compatible with the rates we and others used in estimating divergence times (Bendell et al. 1996; Jazin et al. 1998). In addition, field data from Polynesia supported the usual rate where new alleles arising from the common central haplotype (see fig. 16.2) did so at a rate that aged the cluster at about 3,000 years, a date compatible with the archaeological dates for first colonization (Macaulay et al. 1997). So it seems that the rate is about right despite the flurry of anxiety.

Fourth, dating the clusters was incorrect. In our 1996 article we used the pairwise differences (π) to estimate diversity within a cluster. The improved phylogeny, which identifies the cluster founder, enables us to employ the simpler and more reliable statistic (ρ) to estimate diversity. (ρ) is the average number of mutations that have accumulated from the cluster ancestor, and it can be converted directly to a divergence date by using a mutation rate estimate which, for the HVS I region we sequence, is 1 per 20,000 years. By itself, this makes very little difference to the previous cluster dates and still leaves them, except for 2A (now J), firmly embedded in the Paleolithic.

Divergence dates, however, are not the same as arrival times. If there were already some diversity within the clusters before they arrived in Europe, then the arrival dates estimated without taking this into account would be too old. One possible way around this is to identify haplotypes within the clusters that were already in the source population. The only Middle Eastern data available for our 1996 study was from a small Bedouin sample with a very high frequency of cluster J, including founders of several sub-clusters. It was this correspondence that gave J a young divergence date in Europe. It was clearly important to substantially increase the sample size of the source population, which for Europe meant Anatolia and the Middle East (or very roughly the region east of the Bosphorus). We now have data from many more sites in the region as well as more in Europe and have found haplotypes in most of the clusters that are found in Europe. It is straightforward to correct (ρ) for this preexisting diversity by subtracting the shared founder hap-

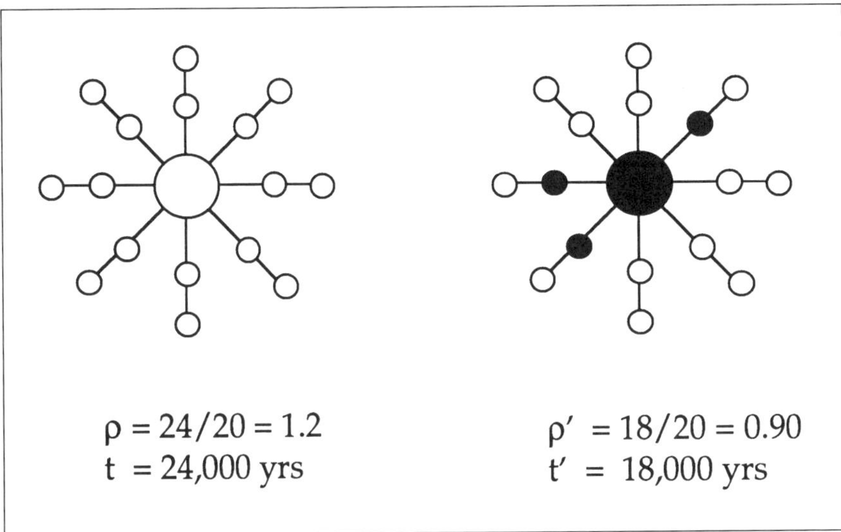

Fig. 16. 4. Calculation of (ρ) uncorrected and corrected for multiple founders

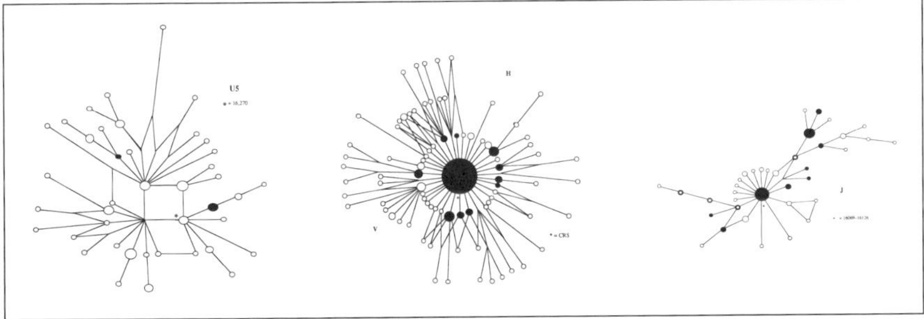

Fig. 16.5. Three European clusters showing haplotypes shared with the potential source populations of Anatolia and the Near and Middle East. Haplotypes found only in Europe are in white; shared haplotypes are in black. Nodes on the phylogeny that are present in the source populations but not Europe are shown as white with bold borders.

lotypes. In the theoretical example (fig. 16.4), the presence of three shared haplotypes reduces the divergence date from 24,000 to 18,000 years B.P.

Figure 16.5 shows three examples that illustrate the effect on actual clusters. In U5 there are virtually no shared haplotypes between Europe and the Middle East so that date remains at about 50,000 years B.P. Cluster H has several shared nodes, all toward the center, as one would expect from a relatively distant common ancestry. This brings down the date in Europe from 21,500 to 12,500 years B.P. Lastly, J has multiple shared haplotypes, already taken into account in the 1996 analysis, which reduces the overall divergence date of 28,000 to a corrected European date of 8,000 years B.P. Table 16.2 and figure 16.6 show the corrected and uncorrected dates for the clusters. A striking feature is the way that, apart from U5 and X, all the uncorrected Paleolithic dates now move forward to the period 11,000–13,000 yrs B.P. Once again, only J is Neolithic. X, a curious and rare group also found in Native Americans, remains at 20,000 years (but with a wide confidence interval because of the small sample size).

In summary, the phylogeny and mutation rate are largely confirmed. However, correcting for preexisting diversity does have a significant effect on the cluster dates for Europe, which brings most of them into the Late Upper Paleolithic—but not quite into the Neolithic.

Can we now offer any context for these revised results? Only U5 remains stubbornly Early Upper Paleolithic, with the extant diversity developing over 50,000 years in Europe. This is a good match to the first appearance in the European archaeological record of anatomically modern humans, including Cro-Magnon, who brought with them the Aurignacean lithic culture. They shared the continent with the Neanderthals until about 28,000 years B.P. when the last Neanderthal disappeared from southern Spain. We have now examined over 3,000 European mtDNA sequences without finding a single one that is sufficiently distinct to be credibly Neanderthal. It is now probably safe to assume that there was no interbreeding with female Neanderthals.

By 18,000–20,000 years B.P., Europe was firmly in the grip of the last Ice Age. There is a distinct lack of authenticated archaeological sites in northern Europe between 22,000 and 14,000 years B.P., and it is thought that the population, which we predict as having been in group U5, moved into core areas either in southwest France and Cantabria or in the Ukraine to the east in order to escape the worst of the conditions. As the climate warmed and the ice retreated, there was a re-expansion across northern Europe which, by this time, supported large herds of big game. Good radiocarbon dates show the first archaeological sites at 13,000–14,000 years B.P. in northern Europe (Housley et al. 1997). The majority of the Paleolithic clusters have their European divergence dates at about this time so our interpretation is that this late glacial expansion from the core areas distributed the mitochondrial ancestors of most modern Europeans. Quantitatively, we believe it was this event, and not the Neolithic, that was the most significant in shaping the modern mitochondrial gene pool.

Finally, group J is still the only convincingly Neolithic cluster. The striking distribution of two important sub-clusters, J1a and J1b, roughly shadow the two major farming routes into Europe: one along the Mediterranean and Atlantic coasts and the other through the river valleys of central Europe. This is the most marked geographical distribution that we have yet detected for any cluster, and we have no reason yet to revise our earlier suggestion that cluster J is a signal of the Neolithic farmers (table 16.2). Group J, which is, in our opinion, entirely Neolithic in Europe, makes up only 16% of the modern mtDNA lineages. This does not necessarily mean that the Neolithic farming pioneers were composed exclusively of group J—indeed, it would be very surprising if they were. There are also small sub-clusters of H, T, and K that also have young dates in Europe, and we are currently examining whether these too might be Neolithic in origin. In other words the overall Neolithic contribution to the mtDNA gene pool might well exceed 20%.

Table 16.2. A Summary of Three Main Waves of European Colonization

Component	Dates (years B.P.)	Main Associated Clusters	Contribution to Modern Gene Pool
Neanderthal	300,000	Unclassified	0%
Early Upper Paleolithic	50,000	U5	10%
Late Upper Paleolithic	11–14,000	H, V, I, W, T, K	70%
Neolithic	8500	J (+ more of H, T, K ?)	20%

Does this substantially conflict with the demic diffusion model, which was the subject of the anniversary conference here in Venice? It was refreshing to be reminded in Albert Ammerman's presentation at the meeting that the hypothesis they had proposed 25 years ago was exactly that—a model to act as a framework for further work (this volume, Introduction). In that capacity, it has been enormously useful and influential stressing, as it did, the importance of the Neolithic in Europe when the prevailing intellectual climate strongly favored indigenism. Only in subsequent transformations in the literature did the Neolithic transition come to be seen as an overwhelming tide that effectively eliminated the indigenous Mesolithic populations. I suspect that this has arisen from a simple misunderstanding. R.A. Fisher's general mathematical treatment of expansion outward from a growing center, which covered everything from animals to ideas and on which the demic diffusion model is based, is called the "wave of advance." All too easy, then, for enthusiasts to inadvertently transfer this stirring title, with its much more dramatic metaphorical connotations, to events in Neolithic Europe. It is refreshing to be reminded by Albert Ammerman that nowhere in the original hypothesis was there an insistence on the demic diffusion being overwhelming. Certainly, if we equate the Neolithic diffusion with the famous cline of classical gene frequencies constructed from the first principal component, we account for around 26–28% of the variance. Luca Cavalli-Sforza recently proposed that this value could be considered to be a reasonable estimate of the Neolithic contribution (Cavalli-Sforza and Minch 1997). This approaches the value for the Neolithic contribution derived from mtDNA. Evidence emerging from the Y-chromosome may close the gap even further.

In general, Y-chromsome variation is more geographically specific than mtDNA. This geographical limitation can be explained by a combination of a restricted migration of males (females tending to transfer between groups on marriage more often than males) and the variance in the number of offspring (whereby some men can have very many offspring and others have few or none at all, while women are biologically restricted). The overall effect of this differential is that genetic drift operates more severely for paternally transmitted than maternally transmitted alleles. The two phenomena will work in the same direction to increase local specificity of YDNA variation while decreasing it for mtDNA. YDNA clines are therefore expected to be steeper than for mtDNA with the nuclear gene gradients being an average of the two. If this mechanism were at work in Europe, the 26–28% first PC nuclear gene clines would be an average of shallower mtDNA and steeper YDNA components. Although this could reflect a greater male contribution to the Neolithic demic diffusion, it might also be a product of the behavioral mechanisms discussed above.

The demic diffusion model has been immensely valuable in providing a framework for thinking and research both in genetics and archaeology for the past quarter of a century. Despite the initial disagreement, our mtDNA work, which does demonstrate a powerful Neolithic component, now sits comfortably with the original model, if not with its more extreme metaphorical derivatives. The dialogue over the last few years has been useful in encouraging us to refine our thinking on the phylogeny, the mutation rate, and the effect of common founding haplotypes.

For the future, the comparative pattern of mtDNA and YDNA variation looks to be an extremely fruitful avenue for further research in Europe and elsewhere. Ancient DNA still has a lot more to offer in Europe since the acid test of any hypothesis based on modern variation is the testing of predictions using well-preserved assemblages from relevant time depths in the right places. The prospects for the next 25 years are bright indeed.

REFERENCES

Barbujani, G., G. Bertorelle, and L. Chikhi. 1998. "Evidence for Paleolithic and Neolithic Gene Flow in Europe." *American Journal of Human Genetics* 62:488–91.

Bendall, K.E., V.A. Macaulay, J.R. Baker, and B.C. Sykes. 1996. "Heteroplasmic Point Mutations in the Human mtDNA control region." *American Journal of Human Genetics* 59:1276–87.

Cann, R.L., M. Stoneking, and A.C. Wilson. 1987. "Mitochondrial DNA and Human Evolution." *Nature* 325:31–6.

Cavalli-Sforza, L.L., P. Menozzi, and A. Piazza. 1994. *The History and Geography of Human Genes*. Princeton.

Cavalli-Sforza, L.L., and E. Minch. 1997. "Paleolithic and Neolithic Lineages in the European Mitochondrial Gene Pool." *American Journal of Human Genetics* 61:247–51.

Housley, R.A., C.S. Gamble, M. Street, and P. Pettitt. 1997. "Radiocarbon Evidence for the Lateglacial Human Recolonisation of Northern Europe." *Proceedings of the Prehistoric Society* 63:25–54.

Howell, N., I. Kubacka, and D.A. Mackey. 1996. "How Rapidly Does the Human Mitochondrial Genome Evolve?" *American Journal of Human Genetics* 59:1–9.

Hurles, M.E., C. Irven, J. Nicholson, P.G. Taylor, F.R. Santos, J. Loughlin, M.A. Jobling, and B.C. Sykes. 1998. "European Y-Chromosomal Lineages in Polynesia: A Contrast to the Population Structure Revealed by mtDNA." *American Journal of Human Genetics* 63:1793–806.

Jazin, E., H. Soodyall, P. Jalonen, E. Lindholm, M. Stoneking, and U. Gyllensten. 1998. "Mitochondrial Mutation Rate Revisited: Hot Spots and Polymorphism." *Natural Genetics* 18:109–10.

Lewin, R. 1997. "Ancestral Echoes." *New Scientist* 2089:32–7.

Macaulay, V.A., M.B. Richards, P. Forster, K.E. Bendall, E. Watson, B. Sykes, and H.-J. Bandelt. 1997. "mtDNA Mutation Rate—No Need to Panic." *American Journal of Human Genetics* 61:983–6.

Macaulay, V., M. Richards, E. Hickey, E. Vega, F. Cruciani, V. Guida, R. Scozzari, B. Bonne-Tamir, B.C. Sykes, and A. Torroni. 1999. "The Emerging Tree of West Eurasian mtDNAs: A Synthesis of Control Region and RFLPs." *American Journal of Human Genetics* 64:232–49.

Richards, M.R., H. Côrte-Real, P. Forster, V. Macaulay, H. Wilkinson-Herbots, A. Demaine, S. Papiha, R. Hedges, H.-J. Bandelt, and B.C. Sykes. 1996. "Paleolithic and Neolithic Lineages in the European Mitochondrial Gene Pool." *American Journal of Human Genetics*

59:185–203.

Richards, M.R, V. Macaulay, B. Sykes, P. Pettitt, R. Hedges, P. Forster, and H.-J. Bandelt. 1997. "Reply to Cavalli-Sforza and Minch." *American Journal of Human Genetics* 61:251–4.

Richards, M., and B. Sykes. 1998. "Reply to Barbujani et al." *American Journal of Human Genetics* 62:491–2.

Sykes, B. 2000. "Using Genes to Map Population Structure and Origins." In *The Human Inheritance*, edited by B. Sykes, 93–117. Oxford.

Sykes, B.C., A. Leiboff, J. Low-Beer, S. Tetzner, and M. Richards. 1995. "The Origins of the Polynesians: An Interpretation from Mitochondrial Lineage Analysis." *American Journal of Human Genetics* 57:1463–75.

Torroni, A., K. Huoponen, P. Francalacci, M. Petrozzi, L. Morelli, R. Scozzari, D. Obinu, M.L. Savontaus, and D.C. Wallace. 1996. "Classification of European mtDNAs from an Analysis of Three European Populations." *Genetics* 144:835–50.

NOTE

[1] I am grateful to Martin Richards and Vincent Macaulay for advice during the preparation of this presentation. This work has been supported by grants from the Wellcome Trust, the European Union, and the Royal Society.

— 17 —

The Neolithic Transition in Europe: Linguistic Aspects

Colin Renfrew

More than 25 years have passed since Ammerman and Cavalli-Sforza (1973) introduced the wave of advance model to Europe and more than 25 years also since the publication of the article (Renfrew 1973) in which I applied an early form of the farming dispersal/ language dispersal model to prehistoric Europe and used it to argue that the Proto-Indo-European language accompanied the spread of farming to Europe from Anatolia, and that this was the first and most important determinant of the ultimate distribution of the languages of the Indo-European family. That paper was delivered at a conference held in Sheffield in March 1970, well before the Sheffield conference of December 1971, where "demic diffusion" and the "wave of advance" model were presented. It was not this that I had in mind in my own article, but the map by Grahame Clark (1965), which very clearly showed the spread of the farming economy, although it was prepared before the calibration of radiocarbon dating. The initial formulation was as follows (Renfrew 1973, 270):

> The IE languages or dialects ancestral to those of later Europe were brought to Europe by the first farmers, arriving from Anatolia with their new subsistence pattern, and spread by their descendants along the north coast of the Mediterranean, up the Danube of central and northern Europe and eastwards, along the lands north of the Black Sea to central Russia. Enabled by their new subsistence technique to support a much higher population density than their hunter-fisher

> predecessors in the same area, they effectively assimilated or replaced that population, or at least replaced its languages with their own.

It was not intended, however, to give an inflexible view, and following discussions at the 1971 meeting I added the following comment (Renfrew 1973, 273):

> It would be easy to modify the theory proposed above, of a Neolithic spread of IE languages in Europe. . . . This original spread of IE languages through the Aegean and Balkans, up the Danube (by the agency of the Danubian I culture) and in the western Mediterranean (by the "Impressed Ware" Neolithic) need not have embraced western and northern Europe, areas where there is good evidence for a flourishing Mesolithic population. We could postulate that the inhabitants of the areas which soon adopted what has been termed a "Western Neolithic" culture, did not at this time adopt the new languages. They kept, perhaps, in both language and culture, more of their own original character than was general elsewhere. Is it indeed mere coincidence that collective burial under long mounds or in built chamber-graves emerged in Europe precisely (and exclusively) in these areas?

The model proposed by Ammerman and Cavalli-Sforza seemed to me a very compelling one (and indeed still does) for its simplicity and its explicit nature. Of course it is a first order model—Europe is not an anisotropic plane—and does not take into account factors beyond its rather simple assumptions. But its beauty is that it contains a mechanism. I regard it as one of the most successful models yet applied in the field of prehistoric archaeology. And I later introduced it into my more developed treatment of the Indo-European question (Renfrew 1987). But at the same time, I added a qualification (Renfrew 1973, 273): "To insist that the main plants and animals were imported to Europe is not, however, the same as to demonstrate that the early farmers were themselves immigrants. It is perfectly possible to argue that exchange systems between the local, Mesolithic population and their farming neighbours could have provided the former both with the necessary domesticates and with the stimulus to use them."

This was the position adopted and developed by Zvelebil and Zvelebil

(1988) in their thoughtful and interesting critique, and I believe that there is much of value in their treatment. If the simple "wave of advance" picture is a good first approximation, their map (1998, fig. 1) seems a better fit. In effect, it allows for a "wave of advance" as far as the extent of the spread of the Linearbandkeramik culture, but not further. Beyond that point there were processes of acculturation at work, and the model of demic diffusion may be less appropriate. In the Mediterranean likewise the situation is more complicated, as Lewthwaite (1986) indicated. In Iberia, the picture now seems different with local enclaves of farmers who are likely to have traveled by sea (Zilhão 1993).

These matters have been much debated. Cavalli-Sforza and his colleagues were the first to use genetic evidence to clarify the problem. They recognized the first principal component in the synthetic maps which they prepared using classical genetic markers (Menozzi et al. 1978) as a strong cline from southeast to northwest and identified this with the spread of farming (although they did not at that point associate it with the spread of Proto-Indo-European). More recently mitochondrial DNA studies (Richards et al. 1996; Torroni et al. 1998) have called this into question, arguing that the greater part of the variation is of Paleolithic date. In their view the clines seen are indeed the result of colonization from Anatolia, resulting from a much earlier episode than the coming of farming, namely the original peopling of the continent by Homo sapiens. The matter has already produced a lively debate. It would seem that Y-chromosome studies (Malaspina et al. 1998) give results that harmonize broadly with the mtDNA data (Semino et al. 1996). But some recent studies (Chikhi et al. 1998) have utilized nuclear DNA markers to give results very close, it would seem, to those originally proposed by Cavalli-Sforza. These matters currently are the subject of controversy in the field of genetics and remain to be resolved before interpretations for the archaeologist can be securely offered. But already Sykes (1999) and his colleagues in Oxford (including Richards) have proposed what seems a compromise position. They argue that the greater part of the variation is indeed to be dated back to the Paleolithic, but a significant element, perhaps one quarter, is indeed to be dated to the early Neolithic. This may give a figure close to the 28% of the total variability that was proposed for the first principal component in the first place. It seems likely, then, that the genetic evidence may prove to be broadly in harmony with the original wave of advance model as modified by Zvelebil and Zvelebil.

Linguistic Aspects

In examining how the linguistic evidence fits with other models of the Neolithic transition, Zvelebil has produced a model of contact-induced language change that harmonizes with the earlier work, and offers a mechanism for the spread of the new language with farming, even when there is very little gene flow (Zvelebil and Zvelebil 1988). The original language/farming dispersal model is made by him to operate without demic diffusion. In view of the genetic evidence, however, a mixture of the two models, as proposed by Zvelebil and Zvelebil (1988), may be a viable compromise.

The linguistic objections prove weightier. One objection, made by a number of commentators, was that the Greek language contains a high proportion of non-Greek words, and which are sometimes claimed to be non-Indo-European. This has generally been taken as an indication of a substantial "pre-Greek" population at the time of the coming of the Greeks, which advent it would therefore be inconvenient to set as early as 6500 B.C. (the approximate date of the inception of farming in Greece). Many of these vocabulary words, however, relate to sophisticated concepts that would not be at home in the Neolithic or early Bronze Age (as they would have to be if they were "pre-Greek"). Rather they pertain to a more advanced civilization; many or most of these words are likely to be derived from the Minoan language of Crete (Renfrew 1998a). It would therefore be a linguistic *adstratum* acquired through long-lasting interactions between Crete and the Mainland during the Bronze Age. The case for a "pre-Greek" language is very much weakened.

The date of ca. 6500 B.C. for the arrival of farming in Europe has seemed to some commentators too early for the existence of Proto-Indo-European. But the circularity of most arguments for the dating of Proto-Indo-European are increasingly recognized. As the linguist R.M.W. Dixon (1997) has recently argued:

> Why couldn't proto-Indo-European have been spoken about 10,500 years ago? This would correlate with a major socio-economic development, the introduction of agriculture, which archaeologists date about 10,500 BP for this part of the world. . . . The received opinion of a date around 6000 B.P. for proto-Indo-European—with dates for other proto-languages being calibrated on this scale—is an ingrained

> one. I have found this a difficult matter to get specialists to even discuss. . . . This is a question that demands careful re-examination with a full range of possibilities being discussed and compared.

This is in many ways a crucial point, and it is one which linguists have not yet fully addressed. It is notable also that the recent discovery of remarkably well-preserved desiccated corpses in Chinese Turkestan, dating from shortly after 2000 B.C. and with features that seem to many observers to be Caucasoid, has reopened a debate on the origins of the Tocharian language (Mair 1998). The Tocharian language is not attested until very much later, the eighth century A.D., but it may be necessary to consider the possibility of a proto-Tocharian presence in this region much earlier than had previously been considered likely.

The linguistic objections to the origin within Anatolia of the "Anatolian" languages (including Hittite) have also come under scrutiny recently. The "Indo-Hittite" hypothesis of Sturtevant (1962), whereby the Anatolian languages would be the first to break off from the ancestral Proto-Indo-European, has found strong support from the numerical analysis by Warnow (1997). The most obvious location for such a branching in the Indo-European family tree would be in Anatolia, as Gamkrelidze and Ivanov (1995) have argued on linguistic grounds. Moreover, the Hattic objection seems to have been removed. The Hattic language, which is non-Indo-European, is found among the records at the Hittite capital of Bogazköy (Hattusas), and it has often been taken as evidence that the Hatti were indigenous and the Hittites intrusive to Anatolia. But Ivanov (1985) has shown that Hattic is related to the North Caucasian language family and may thus itself well be intrusive to the Anatolian plateau.

Finally, a number of linguists (Adrados 1992) have argued that Indo-European should not be seen as a single "flat" entity, but rather that it is stratified, with a series of phases. In this way, some of the features like the *centum/satem* distinction should not so much be explained geographically as chronologically. I have myself suggested (Renfrew 1999) that the Balkans may have acted as a linguistic area in just that time which Marija Gimbutas (1973) designated "Old Europe" and that convergence processes may have continued there until the fragmentation of this prehistoric Balkan *Sprachbund*. It is not suggested, however, that the demise of this linguistic area was followed by movements of peoples. In the main, the constituent languages continued to be spoken in situ, but were no longer subjected to the same convergence effects, with a high degree of borrowing, which they had experienced up to about 3000 B.C.

At the same time, it is increasingly realized that exaggerated claims have been made in relation to the horse. Although horses may have been intensively exploited at such Ukrainian sites as Dereivka as early as the early fourth millennium B.C., the claims that they were being ridden at this time have proved unfounded (Mallory and Adams 1997). Evidence shows that the first horse-drawn vehicles were chariots with spoked wheels, first seen around 2000 B.C., and that horses were not ridden for military purposes until later (Renfrew 1998). The warrior horseman can no longer be considered as the driving force for some hypothetical "Kurgan" expansion.

Farming Dispersals

When reviewing the geographical distribution of the world's language families, some patterns are evident. In particular, some families show a very wide geographical distribution, with indications of relatively shallow time depth, while others are more compact, with indications of greater time depth. The former are sometimes said to have "spread" distributions and the latter "mosaic" distributions (Renfrew 2000). The Indo-European language family falls within the spread category. Languages having spread distributions frequently owe their distribution to the process of farming dispersal, while those with the compact distribution have in many cases been in their current regions for very much longer, indeed since the Pleistocene period. Peter Bellwood (1996) has argued this position in a number of cases, dealing in considerable detail with the Austronesian languages, and it has been argued (Renfrew 1992) that the farming dispersal process has been most influential at a world level. It may be suggested, therefore, that the Indo-European case falls within a larger category or class. But the dynamics of language change are not yet well understood. Linguists are now reexamining them using a wider range of historical models than the traditional one of simple migration (Dixon 1997). The study of molecular genetics as an indicator of population history is likely to help clarify linguistic change and spread.[1]

REFERENCES

Adrados, F.R. 1992. "The New Image of Indoeuropean, the History of a Revolution." *Indogermanische Forschungen* 97:5–28.

Ammerman, A.J., and L.L. Cavalli-Sforza. 1973. "A Population Model for the Diffusion of Early Farming in Europe." In *The Explanation of Culture Change: Models in Prehistory*, edited by C. Renfrew, 343–57. London.

Bellwood, P. 1996. "The Origins and Spread of Agriculture in the Indo-Pacific Region: Gradualism and Diffusion or Revolution and Colonization." In *The Origins and Spread of Agriculture and Pastoralism in Eurasia,* edited by D.R. Harris, 465–98. London.

Chikhi, L., G. Destro-Bisol, G. Bertorelle, V. Pascalli, and G. Barbujani. 1998. "Clines of Nuclear DNA Markers Suggest a Largely Neolithic Ancestry of the European Gene Pool." *Proceedings of the National Academy of Sciences of the USA* 95:9053–8.

Clark, J.G.D. 1965. "Radiocarbon Dating and the Spread of Farming Economy." *Antiquity* 31:57–73.

Dixon, R.M.W. 1997. *The Rise and Fall of Languages.* Cambridge.

Gamkrelidze, T.V., and V.V. Ivanov. 1995. *Indo-European and the Indo-Europeans.* Trends in Language Studies Monograph 80. The Hague.

Gimbutas, M. 1973. "Old Europe c. 7,000–3,500 BC: The Earliest European Civilisation before the Infiltration of the Indo-European Peoples." *Journal of Indo-European Studies* 1:1–21.

Ivanov, V.V. 1985. "Ob otnosenii xattskogo jazyka k severozapanokavkazskim." *Drevnjaja Anatolija* (Moscow) 2:26–59.

Lewthwaite, J. 1986. "The Transition to Food Production: A Mediterranean Perspective." In *Hunters in Transition*, edited by M. Zvelebil, 53–66. Cambridge.

Mair, V.H., ed. 1998. *The Bronze Age and Early Iron Age Peoples of Eastern Central Asia.* New York and Philadelphia.

Malaspina, P., F. Cruciani, B.M. Ciminelli, L. Terrento, P. Santolamazza, A. Alonso, J. Banyko, R. Brdicka, P. Mondich, P. Moral, R. Qamar, S.Q. Mehdi, A. Rogusa, G. Stefanescu, M. Caraghin, C. Tyler-Smith, R. Scozzari, and A. Noveletto. 1998. "Network Analysis of Y-Chromosomal Types in Europe, North Africa and West Asia Reveal Specific Patterns of Geographical Distribution." *American Journal of Human Genetics* 63:847–60.

Mallory, J.P., and D.Q. Adams, eds. 1997. *Encyclopedia of Indo-European Culture.* London.

Menozzi, P., A. Piazza, and L.L. Cavilli-Sforza. 1978. "Synthetic Maps of Human Gene Frequencies in Europe." *Science* 201:786–92.

Renfrew, C. 1973. "Problems in the General Correlation of Archaeological and Linguistic Strata in Prehistoric Greece: The Model of Autochthomous Origin." In *Bronze Age Migrations in the Aegean*, edited by R.A. Crossland and A. Birchall, 263–76. London.

———. 1987. *Archaeology and Language: The Puzzle of Indo-European Origins.* London.

———. 1992. "World Languages and Human Dispersals: A Minimalist View." In *Transition to Modernity, Essays on Power, Wealth, and Belief,* edited by J.A. Hall and I.C. Jarvie, 11–68.

———. 1998a. "Word of Minos: The Minoan Contribution to Mycenaean Greek and the Linguistic Geography of the Bronze Age Aegean." *Cambridge Archaeological Journal* 8:239–64.

———. 1998b. "All the King's Horses: Assessing Cognitive Maps in Later Prehistoric Europe." In *Creativity in Human Evolution and Prehistory*, 260–84. London.

———. 1999. "Time, Depth, Convergence Theory and Innovation in Proto-Indo-European: 'Old Europe' as a PIE Linguistic Area." *Journal of Indo-European Studies* 27:257–93.

———. 2000. "At the Edge of Knowability: Towards a Prehistory of Languages." *Cambridge Archaeological Journal* 10:7–34.

Richards, M., H. Côrte-Real, P. Forster, V. Macaulay, H. Wilkinson-Herbots, A. Demaine, S. Paphia, R. Hedges, H.-J. Bandelt, and B.C. Sykes. 1996. "Palaeolithic and Neolithic Lineages in the European Mitochondrial Gene Pool." *American Journal of Human Genetics* 59:185–203.

Semino, O., G. Passarino, A. Brega, M. Fellous, and S. Santachirara-Benerecetti. 1996. "A View of the Neolithic Diffusion in Europe through Two Y-Chromosome-Specific Markers." *American Journal of Human Genetics* 59:964–8.

Sturtevant, F.H. 1962. "The Indo-Hittite Hypothesis." *Language* 38:376–82.

Sykes, B. 1999. "The Molecular Genetics of European Ancestry." In *Molecular Information and Prehistory*, edited by M. K. Jones, D.E.G. Briggs, G. Eglington, and E. Hagelberg, 131–9. Philosophical Transactions of the Royal Society, Series B, 354. London.

Torroni, A., H.-J. Bandelt, L. D'Urbano, P. Lahermo, P. Moral, Dr. Sellitto, C. Regno, P. Forster, M.L. Savontaus, B. Bonné-Tomir, and R. Scozzari. 1998. "mtDNA Analysis Reveals a Major Late Paleolithic Population Expansion from Southwestern to Northeastern Europe." *American Journal of Human Genetics* 62:1137–52.

Warnow, T. 1997. "Mathematical Approaches to Comparative Linguistics." *Proceedings of the National Academy of Sciences of the USA* 94:6585–90.

Zilhão, J. 1993. "The Spread of Agro-Pastoral Economies across Mediterranean Europe: A View from the Far West." *Journal of Mediterranean Archaeology* 6:5–63.

Zvebil, M., and K.V. Zvelebil. 1988. "Agricultural Transition and Indo-European Dispersals." *Antiquity* 62:574–83.

NOTE

[1] This chapter was written in 1999; except for a few references, it has not been updated by the author since that time.

PART VI

Looking Forward

— 18 —

Summary of the Discussion

In taking a longer view of how research is done on the Neolithic transition in Europe, it is useful to look forward as well as back. The last session of the conference was devoted to exploring directions in the field over the next 20 years. Participants were asked to consider what still remains to be done on the question. What basic gaps in knowledge still need to be closed? What are the main issues that stand before us and what lines of investigation offer the most promise in resolving them? And what new developments, in a broader sense, might make a significance contribution to the advancement of the field in the near future? The session was divided into two panels for this purpose.[1] After the panelists introduced briefly those topics of special interest to them, each panel turned to a wider discussion. This section of the book concisely summarizes the main topics and themes discussed in the closing session. Clearly, the future is not easy to read. We can only try to do so with the limited means that are available to us. This exercise in looking forward serves to place our work in the stream of time, and it may prove to be of historical interest to those in the next generation, who will have the chance one day to look back and compare the topics considered here with what actually transpired in the field.

AMS Dating

Everyone at the meeting recognized the importance of this new method of radiocarbon dating. Since its advent in the early 1980s, it has changed the approach to ^{14}C dating at Neolithic sites—an approach now oriented toward obtaining dates directly from seeds and bones. What is called for here, as Housley noted, is for the best practice today to become the norm rather than the exception. By this, he means the dating of samples from well-excavated contexts, the analysis of an ample series of samples from a given site, and the prompt and proper publication of the dates. If a sustained commitment is made to AMS dating over the next 20 years, questions about the pattern and the rate of spread of early farming in Europe can be answered with a degree of resolution that is not possible with the evidence currently available. In addition, for any given region of Europe, the systematic dating of the remains of domesticated plants and animals should make it possible to resolve the issue of whether the domesticates were introduced there singly or as a package. Unfortunately, there are still few laboratories that can run such dates, and the cost is often high. This puts the researcher in a country without such a laboratory at a distinct disadvantage. There is thus the need for an international program of assistance to foster the broad implementation of AMS dating. A resolution in support of such an initiative was drafted at the meeting; it was unanimously approved by the participants (see below).

The Recovery and Analysis of Floral and Faunal Remains

The need for further work along these lines—with an emphasis on the study of a large number of samples from a given site—was underscored by Moore. Again, it is a matter of translating the best case studies done into common practice. All of this will call for a substantial investment of time and effort. While none of these recommendations are new or surprising, such work is essential for advancing the study of the Neolithic transition.

Detailed Micro-Regional Studies

Kozłowski drew attention to the importance of studies of this kind for addressing the question of local continuity over the transition, and a number of the other participants agreed. In most basic terms, the comparative study of the late Mesolithic and the early Neolithic in a given area requires first the establishment of the settlements patterns for both times (usually requiring the intensive coverage of the land by means of surveys as well as the excavation of several sites) and then the fine-grained analysis of a number of lithic assemblages (from several different sites in the area) that date to each time. When this is done, there is a robust framework for arguing either for or against continuity over the Neolithic transition in the area. Work of this kind takes time and a large team; it is best done in the context of a major, multi-year project.

The Generic Mesolithic

One of the points to emerge from the meeting was the need to focus more closely on the late or final Mesolithic when researching the Neolithic transition. There is a need to break the habit of speaking about "the Mesolithic" in generic terms. In most parts of Europe, the Mesolithic period lasted for several thousand years and experienced its own development over time. Accordingly, it may not be all that meaningful to talk or write about the Mesolithic in general terms. What really counts for the question of the Neolithic transition in a given place is the nature of the latest Mesolithic in that place.

The Final Mesolithic in Southern Europe

As revealed by the chapters of this volume that examine the situation in Greece, Italy, and Portugal, there are whole areas of southern Europe where the late Mesolithic is either still missing or poorly documented. This is notwithstanding considerable fieldwork directed toward the recovery of such sites over the last 30 years. As pointed out by Renfrew, we still need to learn a good deal more about the late foragers in Turkey as well. Even in the case of southern France, as Guilaine noted, one may begin to wonder about the shortage of final Mesolithic sites. In other words, there is a fundamental gap for future fieldwork to close. If the absences persist after subsequent fieldwork is done, and little evidence comes to light for complex, late foragers (who should leave a clear signature in the archaeological record) over much of southern Europe, then we may have to change some of our ideas about the late Mesolithic there. Indeed, there would be a sharp contrast with the situation in northern Europe. Until now, our expectations for the south have often drawn upon analogies with the north.

The Coastal Question

One of the factors that may limit the visibility of late Mesolithic sites along the coasts of the Mediterranean is the rise in sea level that occurred during the first half of the Holocene. Biagi drew attention to this question; Rowley-Conwy observed that it is not a problem in Scandinavia because of the phenomenon of isostatic recovery there. In particular, if late Mesolithic communities in the Mediterranean world specialized in the exploitation of marine resources and chose to locate their sites on the coast for this reason, this may help to account for why their sites are so hard to find. One implication here would be the need to pay greater attention to coastal contexts and the underwater archaeology of submerged coasts. Van Andel, however, voiced a word of caution about holding too high expectations for the latter (since this is a demanding line of investigation). Instead, it may be more practical and productive to conduct land-based studies in carefully selected places where the coast happens to drop off more steeply (that is, where the coast some 9,000 to 7,000 years ago once stood in a position much the same as the coastline today). It is worth adding that active sea-level rise (which only began to level off some 6,000 years ago) also places limits on our knowledge of the full distribution of early Neolithic settlements in some coastal areas of the Mediterranean.

Bone Chemistry

Another way to look at the question of change in subsistence strategies over time, as Price observed, is by means of the chemical analysis of human skeletal remains. For example, values of ^{13}C (obtained at the same time that human bone is dated by the AMS method) can indicate whether or not an individual's diet was based to a large extent on the consumption of aquatic or marine resources. The values would be different for those whose diets were based on agriculture in Europe. To date, this approach has been used mainly in northern Europe. In central and southern Europe, the opportunity for making comparisons of this kind over the transition is limited by the small number of burials of late Mesolithic age recovered so far. We hope that this situation will improve in the near future, and such studies in bone chemistry can be extended to the rest of Europe.

Mesolithic-Neolithic Interactions

How populations of late foragers and early farmers interacted with one another represents another question of importance. Keeley looks forward to the time when it will be possible to examine this question in greater depth and detail. The interactions can take many different forms, and they may vary from one part of Europe to the next (and even within the same region). So far, there has been a tendency to place emphasis on the positive side of such contacts (mutualism, cooperation, and acculturation). Perhaps more attention will be paid to the other side (disease, competition, and warfare) in coming years. In the history of medicine, new diseases are often closely linked with the keeping of domesticated animals. As Ammerman proposed, the large early settlements in Turkey and elsewhere in the Middle East (with a package of domesticated animals) may well have served as a breeding ground for new animal-related diseases in Neolithic times. In turn, exchange systems reaching out toward the western Mediterranean may have contributed to the circulation of such diseases and put the late foragers there (with lower immunity to them) at far greater risk than we have previously thought. In short, diseases such as influenza reaching new areas in the west, in connection with the periodic exchange of prestige items between populations, may have caused the decline of local Mesolithic populations at the time of contact. A further suggestion was made by Cavalli-Sforza: the archaeologist may want to take greater advantage of the opportunity to learn directly about the nature of interactions between hunters-gatherers and farmers through the study of living populations in Africa.

Human Population Genetics

The participants, in a second resolution (see below), recognized the contribution of human genetics to the field and expressed their support for the collaboration between archaeology and genetics. In the field of human population genetics, one can expect major gains to be made across a broad front over the next decade—a point that Cavalli-Sforza and Sykes both fully agreed upon. Indeed, the tempo of research in genetics and the biomedical field as a whole is such that it is now transforming basic notions about the human condition as held by the academic community at large. As a consequence, the kinds of questions that the archaeologist will ask about the Neolithic transition are likely to change over time as well. This may even mean moving beyond the current preoccupations of prehistoric archaeology with social organization and cultural matters and toward the dynamics of populations and their long-term histories.

The Genetics of Prehistoric Populations

In drawing up a wish list for the future, one of the items on it, as Runnels commented, would surely have to be the possibility of doing DNA analyses directly on Mesolithic and Neolithic skeletal remains. This would provide a clear way for testing the different hypotheses about the Neolithic transition in Europe. From the initial trials that have been made so far on human remains from the remote past, the results are mixed at best. On the other hand, some new positive findings on burials of Neolithic age (from the

Linear Pottery culture) were presented by Sykes at the meeting. The archaeologist can only wait and see what will happen next.

Explanation

In looking forward, the panelists expressed a strong interest in why the Neolithic transition took place. Explanation thus promises to be a subject that will continue to receive significant attention. Only a few of the topics that came up in the discussion will be mentioned here. In the case of demic diffusion, the wave of advance model, as mentioned earlier, is now seen as a first-generation model; it offers a starting point for building second- and third-generation models. Zilhão would foresee the explanation of the Neolithic transition as becoming less abstract and taking a more historical turn in the years to come. Guilaine would look forward to a map—one showing the earliest Neolithic cultures in different parts of Europe—that will be much richer in its complexity. In turn, there is, as he noted, the challenge of explaining the sources of diversity that act to generate such a map. Harris made the case for an attempt at new theory-building, which would concern itself specifically with the ecology of the spread of early agriculture. The need to identify the leading social and economic imperatives operating behind the Neolithic transition in Europe was emphasized by several of the speakers. Others were somewhat less optimistic in this regard and stressed the role of contingency in human behavior and history. In the final analysis, what the spread of early agriculture in Europe derives from is the movement of people on the landscape: either the relocation of households or settlements (under demic diffusion) or else the contact and exchange between people living in different places (under cultural diffusion). At the same time, even in the case of a given society, the factors that may induce the movement of individuals on the Neolithic landscape are likely to be many and diverse.

Publication and Communication

Finally, major changes are afoot in how we publish archaeological data and how we communicate with one another. Bogucki drew attention to the rapid developments that are taking place today because of computers and other new information technologies. In the next 20 years, one can expect further innovations to occur in the recording of archaeological data in the field and the laboratory as well as in their graphic display and dissemination. For the international community of scholars, distance is no longer a serious obstacle to the open circulation of information and ideas. In this sense, we can all look forward to a doubly widening harvest in the years ahead.

Appendix: Two Resolutions

1. The Venice meeting on the Neolithic Transition in Europe (October 1998) recommends the inception of a dating program to undertake the study of the Early Neolithic in Europe and its Western Asian antecedents through the medium of AMS (accelerator mass spectrometry) radiocarbon determinations to be conducted upon well-provenienced and well-identified samples of (a) plant remains of wheat, barley, and rye, and (b) bones of sheep and goat. Due attention is to be given to those regions of Europe in which access to AMS dating facilities is not yet readily available.

2. The archaeologists convened in Venice for the meeting on the Neolithic Transition in Europe consider that genetic data can contribute to the reconstruction of prehistoric events in ways that can be highly useful for understanding the processes that were involved. They therefore urge the collection of genetic data on living human populations and archaeological specimens necessary to improve on present knowledge of the subject.

NOTE

[1] Those in the first panel were Bogucki, Guilaine, Kozłowski, Moore, Price, Rowley-Conwy, and Runnels; it was chaired by Watson. The second panel included Biagi, Cavalli-Sforza, Harris, Keeley, Renfrew, Sykes, Van Andel, and Zilhão; it was chaired by Ammerman.